Ecological Studies, Vol. 130

Analysis and Synthesis

Ecological Studies

Volumes published since 1992 are listed at the end of this book.

Springer

Berlin
Heidelberg
New York
Barcelona
Budapest
Hong Kong
London
Milan
Paris
Santa Clara
Singapore
Tokyo

K. Dettner G. Bauer W. Völkl (Eds.)

Vertical Food Web Interactions

Evolutionary Patterns and Driving Forces

With 82 Figures and 37 Tables

Springer

Professor Dr. KONRAD DETTNER
Lehrstuhl für Tierökologie II
Universität Bayreuth
Universitätsstraße 30
D-95440 Bayreuth, Germany

Professor Dr. GERHARD BAUER
Institut für Biologie I
Universität Freiburg
Hauptstraße 1
D-79104 Freiburg

Dr. WOLFGANG VÖLKL
Lehrstuhl für Tierökologie I
Universität Bayreuth
Universitätsstraße 30
D-95440 Bayreuth, Germany

ISSN 0070-8356
ISBN-13: 978-3-642-64528-0 e-ISBN-13: 978-3-642-60725-7
DOI: 10.1007/978-3-642-60725-7

Cataloging-in-Publication Data applied for
Die Deutsche Bibliothek - CIP-Einheitsaufnahme

Vertical food web interactions: evolutionary patterns and driving forces; with 37 tables / K. Dettner ... (ed.). - Berlin; Heidelberg; New York; Barcelona; Budapest; Hong Kong; London; Milan; Paris; Santa Clara; Singapore; Tokyo: Springer, 1997
(Ecological studies; Vol. 130)

NE: Dettner, Konrad [Hrsg.]; GT

Cover design: Design & Production, Heidelberg
Typesetting: Best-set Typesetter Ltd., Hong Kong
SPIN 10529187 31/3137 5 4 3 2 1 0 - Printed on acid-free paper

Preface

In the past years, much work has been carried out on either life-history evolution or structure and function of food webs. However, most studies dealt with only one of these areas and often touched upon the other only marginally. In this volume, we try to synthesize aspects of both disciplines and will concentrate on how the interactions between organisms depend on their life-history strategies. Since this is a very comprehensive topic, this volume will focus on vertical interactions to remain within a clearly arranged field. We present some scenaria based on life-history variation of resource and consumer, and show how particular patterns of life-history combinations will lead to particular patterns in trophic relationships. We want to deal with the selective forces underlying these patterns: the degree of specificity of the consumers determines the dependence on its resource, and its adaptation to the spatial and temporal availability of the resource. In this respect, the spatial structure of the resource and its "quality" may play an important role. The impact of natural enemies is another important selective force which may influence the evolution of interactions between species and the structure of communities. Here, the acquirement of an enemy-free space may provide selective adavantages. The importance of the impact of enemies is also expressed by the development of numerous and sometimes very subtle defense strategies. This will be demonstrated especially for various aspects of chemical ecology. Whether these strategies may be anticipated by counterstrategies, eventually leading to an arms race, depends on the symmetry of the interactions, but also on the different generation times. In some cases, the consumer may have a drastic impact on its resource, rendering it suitable as a biocontrol agent. However, the opposite case is more likely, i.e. a strong influence from the resource on the consumer or temporally varying interactions. The ratio of the generation times is especially important if the interspecific relationships are very close, since generation time is often negatively correlated with the rate of evolution. In host-parasite interactions, the parasite's generation time is usually shorter or equal to that of its host, but, exceptionally, the parasite may have longer generation times.

This volume is dedicated to Prof. Helmut Zwölfer, who was among the first ecologists to point out that food webs are not random assemblages of species but are structured by various rules, so that they may be evolutionarily highly stable (Zwölfer 1987, 1988; Zwölfer and Arnold-Rinehart 1993). His conclusions were based on his extensive studies on Cardueae-insect food webs. These insect-plant complexes constitute ecological microsystems, which represent "evolutionary replicates of a particular type of food web" (Zwölfer 1994). They offer the opportunity to compare ecological variants along different geographical transects, under different ecological and climatological conditions and of different phylogenetic age. As one major result of his studies, Helmut Zwölfer could show that the organization patterns of the insect guilds feeding in Cardueae flower heads follow specific rules and are highly predictable. He also presented striking examples of subsequent phases in the evolution of insect-plant systems: the ready accumulation of unspecialized phytophages after the introduction of a plant into a new biogeographic region, various micro- and macro-evolutionary processes which lead to specialization, biotype formation and speciation in phytophagous insects, and finally the formation of predictable, highly organized insect guilds. Recently, the analysis of various insect-plant systems has shown that the conclusions drawn from the analysis of the Cardueae-insect food web are not an exception, but can be transferred to various other plant-insect systems.

With this book, we would like to acknowledge the comprehensive scientific work of our colleague and teacher, Helmut Zwölfer, who retired in 1994, and express our gratitude also in the name of all his friends, colleagues and students who did not participate in the current issue. We hope that we can contribute to the understanding of evolutionary processes and thus to the ideas of Helmut Zwölfer.

K. Dettner, G. Bauer, and W. Völkl
Bayreuth and Freiburg, April 1997

References

Zwölfer H (1987) Species richness, species packing, and evolution in insect-plant systems. Ecological studies, vol 61. Springer, Berlin Heidelberg New York, pp 301–319

Zwölfer H (1988) Evolutionary and ecological relationships among the insect fauna of thistles. Annu Rev Ent 33:103–122

Zwölfer H (1994) Structure and biomass transfer in food webs: stability, fluctuations, and network control. In: Schulze ED (ed) Flux control in ecological systems. Academic Press, San Diego, pp 365–419

Zwölfer H, Arnold-Rinehart J (1993) The evolution of interactions and diversity in plant-insect systems: the Urophora-Eurytoma food web in galls on Palearctic Cardueae. Ecological studies, vol 99. Springer, Berlin Heidelberg New York, pp 211–233

Contents

Part C Aspects of Chemical Ecology in Different Food Chains

Contributors

Achtziger, R.

Lehrstuhl für Tierökologie I, University of Bayreuth, P.O. Box 10 12 51, D-95440 Bayreuth, Germany.
Present address: TU Bergakademie Freiberg, Interdisziplinäres Ökologisches Zentrum (IÖZ),
Agricolastraße 22, D-09596 Freiberg

Bauer, G.

Institut für Biologie I, Universität Freiburg, Hauptstraße 1, D-79104 Freiburg/Breisgau, Germany

Boller, E.F.

Eidgenössische Forschungsanstalt für Obst, Wein und Gartenbau, CH-8820 Wädenswil, Switzerland

Bush, G.L.

Department of Zoology, Michigan State University, East Lansing, Michigan 48824, USA

Carr, T.

Section of Ecology and Systematics, Corson Laboratories, Cornell University, Ithaca, New York 14853-2701, USA

Derby, J.L.

Research Centre, Agriculture and Agri-Food Canada, 107 Science Place, Saskatoon, Saskatchewan, S7N 5E1, Canada

Dettner, K.

Lehrstuhl für Tierökologie II, Universität Bayreuth, P.O. Box 10 12 51, D-95440 Bayreuth, Germany

Dixon, A.F.G.

School of Biological Sciences, University of East Anglia, Norwich, NR4 7TJ, UK

Ebert, D.

Zoologisches Institut, Universität Basel, Rheinsprung 9, CH-4051 Basel, Switzerland

Goeden, R.D.

University of California, Department of Entomology, College of Natural and Agricultural Sciences, Riverside, California 92521, USA

Gut, D.

Eidgenöss. Forschungsanstalt für Obst, Wein und Gartenbau, CH-8820 Wädenswil, Switzerland

Horak, I.G.

Department of Veterinary Tropical Diseases, Faculty of Veterinary Scince, PB X04, Onderstepoort 0110, South Africa

Mackauer, M.

Center for Pest Management, Department of Biological Sciences, Simon Fraser University, Burnaby, B.C., V5A 1S6, Canada

Otto, M.

Center for Pest Management, Department of Biological Sciences, Simon Fraser University, Burnaby, B.C., V5A 1S6, Canada

Payne, R.J.H.

Department of Zoology, University of Oxford, South Parks Road, Oxford OX1 3PS, UK

Peschken, D.P.

Research Station, Agriculture and Agri-Food Canada, Box 440, Regina, Saskatchewan, S4S 1R5, Canada

Petney, T.N.

Lehrstuhl für Parasitologie, Hygiene-Institut, Universität Heidelberg, D-69120 Heidelberg, Germany

Price, P.W.

Department of Biological Sciences, Northern Arizona University, P.O. Box 5640, Flagstaff, Arizona 86011-5640, USA

Remund, U.

Eidgenöss. Forschungsanstalt für Obst, Wein und Gartenbau, CH-8820 Wädenswil, Switzerland

Roininen, H.

Department of Biology, University of Joensuu, P.O. Box 111, SF-80101 Joensuu, Finland

Romstöck-Völkl, M.

Lehrstuhl für Tierökologie I, Universität Bayreuth, P.O. Box 10 12 51, D-95440 Bayreuth, Germany

Sequeira, R.

Center for Pest Management, Department of Biological Sciences, Simon Fraser University, Burnaby, B.C., V5A 1S6, Canada

Smith, J.J.

Department of Zoology, Michigan State University, East Lansing, Michigan 48824, USA

Spurr, D.T.

Research Station, Agriculture and Agri-Food Canada, 107 Science Place, Saskatoon, Saskatchewan, S7N 5E1, Canada

Stadler, B.

Bayreuther Institut für Terrestrische Ökosystemforschung (BITÖK), P.O. Box 10 12 51, D-95440 Bayreuth, Germany

Stechmann, D.H.

Museum am Schölerberg, Am Schölerberg 8, D-49082 Osnabrück, Germany

Tomaschko, K.H.

Universität Ulm, Abteilung Allgemeine Zoologie, Am Eselsberg, D-89069 Ulm, Germany

Topp, W.

Zoologisches Institut der Universität Köln, Physiologische Ökologie, Weyertal 119, D-50923 Köln, Germany

Völkl, W.

Lehrstuhl für Tierökologie I, University of Bayreuth, P.O. Box 10 12 51, D-95440 Bayreuth, Germany

Weisser, W.W.

Imperial College at Silwood Park, Department of Biology, Silwood Park, Ascot, Berks, SL5 7PY, UK

Part A

Plant-Insect Relationships: Life-History and Evolution of Tephritid Flies

1 The Sympatric Origin of Phytophagous Insects

GUY L. BUSH and JAMES J. SMITH

1.1 Introduction

Global estimates of the number of insect species now range from 10 to 30 million and the tally keeps growing. This means that roughly 75–95% of all living eukaryotic organisms are insects. No matter which figure you care to choose, the numbers are impressively large. What is it about insects that accounts for this inordinately large number of species? An assessment of their biological attributes provides at least three important clues. The most important concerns their relatively high degree of resource specialization. Approximately 70% of British insects, which are probably representative of the world's insect fauna, are parasitoids or parasites on animals and plants (Price 1980). Of these about half feed on plants, with the majority infesting one or a few closely related hosts (Strong et al. 1984). A second important clue is that when sister species of these host specialists are found coexisting sympatrically or parapatrically they are almost always feeding on different host plant species. Finally, a third important characteristic shared by many of these host specialists is that they use their host plant or their host plant's habitat as a rendezvous site for locating a mate (Bush 1975b; Zwölfer 1975).

The strong linkage between host and mate choice coupled to the fact that sister species of host specialists coexist in close proximity on different host plants led several biologists to suggest that the colonization of new hosts can result in the non-allopatric origin of new species (Walsh 1864; Brues 1924; Thorpe 1945; Dethier 1954; Bush 1975a; Zwölfer and Bush 1984; Tauber and Tauber 1989). Acceptance of this view has met with strong resistance from those adhering to the touchstone of the Evolutionary Synthesis, i.e., that the origin of new species requires the accumulation of many changes, each with small effect (micromutations) *in complete geographic isolation* (Mayr 1942; Carson 1975; Futuyma and Mayer 1980).

1.2 The Debate over Allopatric and Nonallopatric Speciation

1.2.1 Contrasting Views on the Importance of Geographic Isolation

The origin and evolution of sympatric sister species of animals and plants have been the subject of major controversy among evolutionary biologists

Ecological Studies, Vol. 130
Dettner et al. (eds.) Vertical Food Web Interactions

ever since Charles Darwin published *On the Origin of Species* in 1859. Darwin concluded that species could evolve either with or without geographic isolation, although he was vague about the process and circumstances for either mode of speciation. Such details were soon forthcoming. To account for the fact that closely related species of phytophagous insects occurring in close proximity tend to feed on different host plants, Benjamin Dann Walsh (1864) proposed that new species evolve in the absence of geographic barriers as they colonize and adapt to new host plants. This view contrasted sharply with that of Moritz Wagner who as early as 1841 (Mayr 1988) had observed that closely related species of predaceous, flightless carabid ground beetles are often distributed in a disjunct fashion, and concluded that species can only evolve during periods of geographic isolation (Wagner 1868).

These two views on the origin of species, one emphasizing that populations can diverge in sympatry as a direct outcome of selection during the course of adaptation and specialization on different host plants or in different habitats, and the other maintaining that divergence can occur only during periods of geographic isolation, set the stage for subsequent debate. Unfortunately, the majority of naturalists focused their research on either large, usually conspicuous vertebrates or showy, easily collected or reared invertebrate groups such as butterflies, predaceous or polyphagous insects like carabid beetles and saprophagous *Drosophila*. Because habitat or host choice in such animals is not usually involved directly in locating mates, those working with these groups seldom appreciated the fact that habitat or host preferences in phytophagous parasites play a crucial role in mate choice. Also, there was little appreciation for the effects such ecological barriers have on limiting gene flow between sister populations of phytophagous insects and parasitoids specializing on different hosts, or the speed with which new species of these host specialists can evolve.

1.2.2 Gene Flow and Sympatric Speciation

Another recurring theme among those who believe that sympatric speciation is unlikely to occur is that even a very small amount of gene flow between populations adapting to different host plants prohibits divergence (Futuyma and Mayer 1980). This is exemplified by Mayr's (1963) assertion that "Since there are normally very numerous ecological niches within the dispersal area of a deme, niche specialization is impossible without continued new pollution in every generation by immigrants". It is not surprising, given this perception, that the consensus which emerged among neo-Darwinians favored allopatric speciation as the universal mode of speciation in animals. Although it is generally true that gene flow may have an homogenizing effect on neutral alleles, it is certainly inappropriate to attribute the same outcome for alleles at loci subject to strong selection such as those responsible for fitness, mate recognition, and habitat or host preference. These are the very kinds of genes that promote genetic divergence, population subdivision, and speciation as a result of adaptation to different hosts or habitats.

In fact, we have very little data on the actual level of gene flow and its outcome in response to selection between natural populations in the process of adapting to new hosts or habitats. Even so, there are a growing number of examples suggesting that genetically distinct populations can evolve sympatrically in a short period of time even in the face of considerable gene flow (Bush 1993b; Feder et al. 1994; Guldemond and Mackenzie 1994; Emelianov et al. 1995). Laboratory experiments (Rice and Hostert 1993) as well as theoretical (Liou and Price 1994) and computer simulation studies (Rice 1984; Kondrashov and Mina 1986; de Meeûs et al. 1993; Johnson et al. 1996) also support the idea that sympatric genetic divergence can occur under conditions encountered in nature. Finally, recent biochemical and molecular genetic studies reveal patterns of genetic divergence between populations of host and habitat specialists adapting to different resources that are compatible with their nonallopatric origin (Feder et al. 1988, 1993; Menken et al. 1991; Mitter et al. 1991). Interestingly, these include some vertebrate examples (Schliewen et al. 1994; Schluter 1996).

1.3 The Plausibility of Sympatric Speciation

Unfortunately, the allopatric paradigm is so deeply imbedded in our current concept of the speciation process that even strong evidence of nonallopatric host race formation and speciation is either rejected out of hand on its assumed implausibility or because some allopatric scenario, no matter how speculative and circuitous, can be invoked to explain the origin of sympatric sister species (Bush and Howard 1986). However, evidence that new species arise sympatrically in a relative short period of time is now so compelling that the simplistic rationalization that sympatric sister species are always the product of secondary contact after allopatric speciation can no longer be accepted without supporting evidence. In the future those who espouse such a view must substantiate their claims with the same exacting and unequivocal evidence of allopatric divergence demanded in the past of those proposing a nonallopatric interpretation. In cases of sympatric sister species it is more parsimonious to reject the hypothesis that they evolved by a nonallopatric process using convincing evidence of how and when allopatric isolation and speciation occurred than to assume allopatric divergence on the basis of unsupported conjecture as has often been the case. Likewise, it is essential that the evidence gathered in cases where allopatric sister species are restricted to different hosts be sufficiently strong to reject the hypothesis that colonization of one or the other host and speciation occurred sympatrically when their host plant ranges overlapped.

1.3.1 Allopatric or Sympatric Speciation in the Beetle Genus *Ophraella*?

Such evidence is rarely provided. For example, in their analysis of speciation in the chrysomelid leaf beetle genus, *Ophraella*, Futuyma (Futuyma et al. 1995) and his associates (Funk et al. 1995) conclude that although speciation in the genus has invariably been accompanied by a host shift to related plants mediated by genetic constraints, biogeographic and phylogenetic evidence is consistent with a peripatric origin of species. However, the pattern of host shifting, present-day host plant distributions, phylogenetic relationships of the species and the genetic evidence are also consistent with sympatric or parapatric modes of speciation. The pertinent question is, or should be, "which model best explains the data and observations?" Futuyma et al. (1995) reason that in peripatric speciation "new species will often be more closely related to certain populations of the 'parent species' than some populations of the latter are to each other". Because this is the pattern of divergence revealed in their analysis of mtDNA of the *O. communa* subclade, they infer that *Ophraella* speciated peripatrically.

We would argue that the data are equally, if not more, consistent with the pattern of divergence expected as an outcome of sympatric or parapatric speciation. The current distributions of sister species in the genus *Ophraella* for which sufficient collection data exist (LeSage 1986; Futuyma 1990, 1991), suggest that these species are sympatric or broadly parapatric with no convincing evidence of when, for how long or even if their host plants were isolated in the past. Nor is genetic evidence of founder events presented in support of their inference that speciation in *Ophraella* occurred only in small, peripherally isolated populations. On the contrary, allozyme data (Futuyma and McCafferty 1990) indicate that speciation involved no genetic bottlenecks or a founder-flush process. Finally, the fact that both larvae and adults are restricted exclusively to feeding and mating on their respective host plants is consistent with recent sympatric host race formation and speciation in other phytophagous insects (Bush 1993a, 1994; Guldemond and Mackenzie 1994; Emelianov et al. 1995; Feder et al. 1996b; Mackenzie 1996). In the light of failures to demonstrate the likelihood of peripatric speciation experimentally (Galiana et al. 1993; Moya et al. 1995), the controversy surrounding its plausibility (Carson and Templeton 1984; Barton 1989; Slatkin 1996), and the lack of tangible and tenable hard genetic evidence of its occurrence in natural populations (Coyne 1994), it seems illogical to reject the possibility that nonallopatric speciation has contributed to the origin of *Ophraella* species.

1.3.2 Allopatric or Sympatric Speciation in the Tephritid Fruit Fly Genus *Rhagoletis*?

1.3.2.1 Biology and Distribution

Another system in which nonallopatric modes of speciation should not be rejected merely because the data are (or can be seen as) consistent with allopatric models is the tephritid fruit fly genus *Rhagoletis*. There are over 64 described and several undescribed species of these flies whose larvae are frugivores on a wide range of fruiting plants throughout the Holarctic and Neotropical regions. Several species are major fruit pests, infesting either apples, cherries, blueberries, walnuts, or tomatoes. Details on the biology of *Rhagoletis* are summarized by Boller and Prokopy (1976). In North America, *Rhagoletis* consists mainly of members of five major species groups of which the *R. pomonella* group is the most speciose with at least nine morphologically almost indistinguishable species (Table 1.1). Eight of these species are broadly sympatric in eastern North America, while the ninth, *R. zephyria*, a primarily western species, is broadly parapatric with *R. pomonella* in Minnesota and southeastern Manitoba (Bush 1993a).

1.3.2.2 Sympatric Host Race Formation

The recent development of *Rhagoletis* populations adapted to introduced hosts provides a good example of how host races evolve sympatrically in phytophagous insects (Bush 1969, 1975b, 1993a; Bush and Diehl 1982; Feder et

Table 1.1. Host plant use and geographic distributions of species in the *Rhagoletis pomonella* species group

Species	Host plant(s)	Distribution
pomonella	*Crataegus* spp.	E NA
	Malus spp.	E NA
Mayhaw fly[a]	*Crataegus opaca*	S Cent NA
nr. *pomonella*	*Crategus mexicana*	Cent Mexico
zephyria	*Symphoricarpos* spp.	NW NA - E NA
mendax	*Vaccinium angustifolium*	E NA
	Vaccinium corymbosum	E NA
	Vaccinium stamineum	SE NA
Sparkleberry fly	*Vaccinium arboreum*	SE NA
cornivora	*Cornus amomum*	NE NA
	Cornus obliqua	NE NA
Flowering dogwood fly	*Cornus florida*	E NA
Flatwoods plum fly[a]	*Prunus umbelata*	SE NA

NA, North America; Cent, Central.
[a] Not included in Fig. 1.1.

al. 1988, 1989; Diehl and Bush 1989). One species in particular, *R. pomonella*, illustrates the biological attributes and processes involved in sympatric host race formation (Bush 1993a). In the early 1860s, *R. pomonella*, whose native host is hawthorn (*Crataegus* spp.) was found infesting apples (*Malus pumila*) in a Hudson River Valley apple orchard (Bush et al. 1989). Infestation of apple quickly spread and within about 50 years covered an area reaching from Nova Scotia west to southeastern Manitoba and south to a line running roughly from southern Iowa to Virginia where the apple population dips south along the Appalachian Mountains to northern Georgia. The flies on *Crataegus* cover this same area, but their range extends south to eastern Texas and along the gulf coast to central Florida.

The apple and haw populations now have established two genetically and biologically distinct races that meet the criteria for the recognition of host races (Jaenike 1981; Bush 1993a). Although there is limited gene flow between the races (6% or less; Feder et al. 1993), the apple- and hawthorn-infesting populations of *R. pomonella* represent distinct genotypic clusters (Feder et al. 1988, 1990) maintained by selection (Feder et al. 1996a,b) and positive assortative mating (Feder et al. 1994).

1.3.2.3 Sympatric Speciation: a Molecular Perspective

Since sympatric host race formation and sympatric speciation may be viewed by some biologists as separate phenomena, we will also examine the divergence of two species in the *R. pomonella* species group, *R. pomonella* and *R. mendax*. Do the data and observations better fit an allopatric or a nonallopatric speciation model?

R. pomonella infests hawthorns (*Crataegus* spp.) and apples (*Malus pumila*) in eastern North America while *R. mendax* infests blueberries (*Vaccinium* spp.) in the same geographic area. *R. pomonella* and *R. mendax* are closely related, and their status as distinct species has been contentious and questioned as recently as 1986 (Diehl and Prokopy 1986). However, distinct species status for *R. pomonella* and *R. mendax*, as judged even by the stringent criteria of the biological species concept, has now been firmly established (Feder and Bush 1989; Feder et al. 1989). These studies showed that although *R. pomonella* and *R. mendax* were not fixed for unique alternate alleles at any allozyme locus, both *R. pomonella* and *R. mendax* were paraphyletic at several loci, i.e., they had "private" alleles that were unique to one species but with an alternate allele that was shared with the other species. This pattern of genetic divergence is expected subsequent to speciation in recently diverged species as alleles undergo stochastic lineage sorting and gene trees make the transition from polyphyly through paraphyly to reciprocal monophyly (see Avise 1994, p. 131).

Further information pertinent to the speciation of *R. pomonella* and *R. mendax* comes from our recent phylogenetic analysis of *Rhagoletis* species using nucleotide sequences of the mitochondrial cytochrome oxidase II (COII) gene (Smith and Bush 1996). Like the allozyme data, the COII data also suggest that the divergence of *R. pomonella* and *R. mendax* was quite recent, as these species are not reciprocally monophyletic with respect to mitochondrial DNA haplotype (Fig. 1.1). In addition, the COII phylogeny shows that *R. pomonella* and *R. mendax* shared a common ancestor much more recently than either species did with the *Crataegus*-infesting *R.* nr. *pomonella* from Mexico. (The species status of *R.* nr. *pomonella* from Mexico, whose present day host is *Crataegus mexicana*, will be considered below.) The placement of *R.* nr.

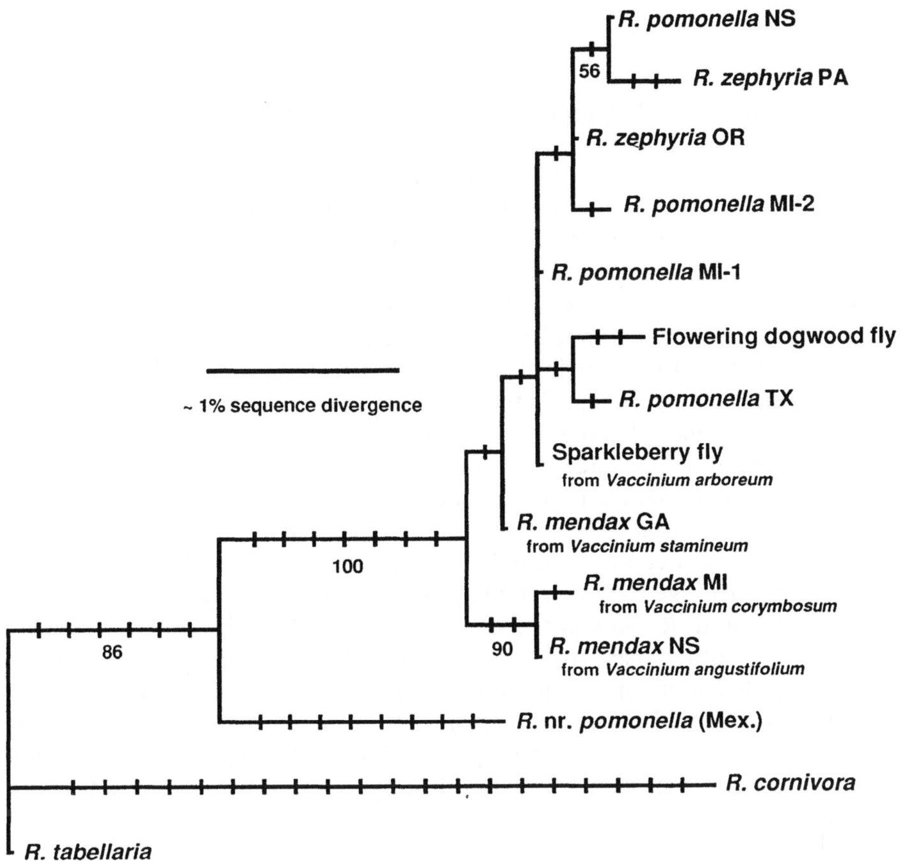

Fig. 1.1. Phylogeny of *Rhagoletis pomonella* group species inferred from nucleotide sequences of the mitochondrial cytochrome oxidase II gene. The phylogram shown is a randomly chosen representative of 28 most parsimonious trees of length 35 (CI = 0.71; RI = 0.78). *Tick marks* on branches represent nucleotide changes along that branch, while *numbers* below branches are bootstrap values based on 100 replicates. Phylogenetic placement of taxa in clades without bootstrap support is arbitrary and should not be considered to be resolved. Additional data will be required to resolve the mitochondrial DNA relationships of these more closely related species

pomonella from Mexico as a sister taxon to *R. pomonella* and *R. mendax* is well-supported (Fig. 1.1), and pairwise nucleotide divergences between the COII sequences can be used to estimate divergence times. For example, the COII sequence from *R.* nr. *pomonella* from Mexico differs from the COII sequence of an *R. pomonella* individual from Nova Scotia (NS) by 22 transition substitutions (3.2% divergence). Using a rate estimate for mtDNA of 2.1% sequence divergence per million years (Brower 1994) as a rough approximation, we calculate that *R.* nr. *pomonella* from Mexico shared a common ancestor with *R. pomonella* and *R. mendax* approximately 1 million years ago. One the other hand, pairwise *R. pomonella-R. mendax* COII divergences ranged from 0.2–1.2%, suggesting that these two species have diverged much more recently.

1.3.2.4 No History of Geographic Isolation Between Sister Species

When we examine the geographic ranges of the plants infested by *R. pomonella* and *R. mendax*, we find that the present day ranges of the *Vaccinium* spp. (Vander Kloet 1988) infested by *R. mendax* are encompassed completely by the ranges of the *Crataegus* spp. (Phipps 1983) infested by *R. pomonella*. The same is true of the flies themselves, with the range of *R. mendax* being contained within the range of *R. pomonella*. In the absence of evidence that these host plants have had disjunct distributions within the time frame of species divergence, our working hypothesis is that the host plants (i.e., *Crataegus* spp. and *Vaccinium* spp.) have been sympatric, the flies have been sympatric, and therefore species divergence was sympatric. The recent occurence of a sympatric host shift and host race formation in the *R. pomonella* species group supports these hypotheses.

Granted, the data can (and should) also be viewed in an allopatric framework, with alternative hypotheses proposed and tested. It may turn out that *R. pomonella* and *R. mendax* evolved allopatrically, via some peripatric mechanism or otherwise. However, to invoke such a mechanism as the working hypothesis requires assumptions for which we have no evidence, i.e., that host plant distributions were disjunct, or that peripheral populations were isolated. Our current working hypothesis that speciation occurred by host shifting in sympatry is the simplest hypothesis that is consistent with the data.

1.4 Peripatric Speciation Unsupported by Convincing Evidence

We do not intend to imply that all phytophagous insect sister species arise sympatrically. Some sister species adapted to different host plants have certainly evolved in geographic isolation, possibly as a result of peripatric founder events. Unfortunately, it is difficult to identify cases of species divergence in phytophagous insect species where the data at hand unequivocally favor an allopatric over nonallopatric interpretation. Thus, we challenge researchers in the field to examine their data from both an allopatric and a nonallopatric perspective, and see which model is better supported.

It has been postulated that peripatric speciation in phytophagous insects occurs in situations where the original preferred host becomes rare or dispersing individuals find it absent (Futuyma et al. 1995). Adaptation to the new host will, in this view, occur when selection for adopting the new host is strong and individuals are sufficiently "preadapted" for positive population growth on the new host. After a period of genetic divergence in isolation the new species may become symaptric with the parent species. Just how the evolution of host specificity and divergence in the mate recognition system during the founder-flush or genetic transilience process plays out during peripatric speciation has not been made clear. In fact, we have no estimates of how frequently peripatric speciation has occurred.

One argument in support of the allopatric origin of sister species specializing on different hosts is the fact that such geographically isolated species can be found in nature. However, it is difficult, if not impossible, to reject the possibility that such allopatric sister species colonized a new host and completed speciation sympatrically or parapatrically before contact between their hosts was broken. It is more biologically plausible for a successful shift and adaptation to a new host to come about when colonization occurs in a zone of contact with the old host. In such sympatric or parapatric situations various allelic combinations can be tested repeatedly over extended periods of time until a genotype capable of sustaining a population on the new host is established. Once this occurs in a host specialist that mates on its host plant, the stage is set for continued adaptive divergence and speciation (Kondrashov 1986; Kondrashov and Mina 1986; Diehl and Bush 1989; Bush 1994; Feder et al. 1995; Johnson et al. 1996).

1.5 Sympatric Speciation and Negative Trade-Offs

One of the arguments that is used to reject sympatric divergence in the genus *Ophraella* is the lack of evidence for negative trade-offs in feeding trials (Futuyma et al. 1995). However such trade-offs have been demonstrated in other insects such as aphids (Mackenzie 1996) and tephritid flies (Feder et al. 1996b) which in early experiments also appeared to show no evidence of host related trade-offs. Not only do trade-offs in host plant usage play a critical role in host race formation, but other factors such as environmental constancy, mating rendezvous limitations or enemy-free space also are contributory factors (Mackenzie 1996). Therefore, a lack of evidence of negative trade-offs does not appear to be sufficient reason for rejecting nonallopatric divergence and speciation.

1.6 Phytophagous Species and Host Races: Is There a Difference?

1.6.1 Species as Genotypic Clusters

A criticism of sympatric speciation via a host shift and host race formation is that the existence of host races has not been demonstrated (Futuyma and Mayer 1980). However, establishing the biological status of *sympatric* sister populations utilizing *different* hosts is fairly straightforward and objective. When such populations maintain distinctly different phenotypic and genetic identities from one generation to the next they satisfy the test of sympatry and are clearly distinct species. We agree with Mallet (1995) who has argued convincingly that sister species should be recognized only when they maintain sympatric genotypic clusters. Although sympatric sister species may exchange genes (Carson and Kaneshiro 1989; Feder et al. 1994), they maintain distinct genotypes and phenotypes and are recognized as distinct species. Defining species objectively on the basis of sympatric genotypic clusters is also assumption-free, and avoids the circularity, inconstancies and untestability of the biological species concept and other assumption-laden definitions (Mallet 1995).

Applying the genetic cluster approach to species delineation also helps recognize the presence of host induced polyphenisms or host associated genetic polymorphisms. A single rather than bimodal genetic cluster characteristic of a species would be expected in cases of host induced polyphenisms which occur when an identical genotype produces distinct phenotypes on different hosts. Host associated genetic polymorphisms are discontinuous phenotypes among individuals within a freely interbreeding population that are the result of allelic variation at a frequency higher than can be maintained by recurrent mutation. Such polymorphisms may produce phenotypically distinct populations on different host plants which are maintained by strong genetic disequilibrium between a few loci responsible for host preference and fitness even when mating between individuals is random. Distinct populations might be maintained when host plant and mate choice are independent. Although larval populations on different host species may differ in alleles at a few loci, they would belong to a single species because adults, mating at random, would form a single genetic cluster. For example, a butterfly species with a hill-topping strategy of mate location and random mating, may have females genetically predisposed to place their eggs on different hosts where strong selection favors slightly different larval genotypes and phenotypes.

1.6.2 Genotypic Clusters, Allopatric Populations, and the "Test of Sympatry"

Resolving the biological status of *allopatric* populations utilizing the *same* host is a far more abstruse problem. In the absence of postmating reproductive incompatibility, an attribute resolvable only by laboratory hybridization trials, the degree of genetic independence between geographically isolated populations using the same host can be established only under condition of natural parapatry or sympatry. In such natural "tests of sympatry" which might result from range expansion or accidental or experimental transplantation, four outcomes are possible in zones of contact between previously allopatric sister taxa sharing the same hosts: (1) the taxa may fuse into one freely interbreeding population resulting in a single genetic cluster at least in the zone of overlap; (2) if mate recognition differences, postmating incompatibility or both are sufficiently well developed so that little hybridization occurs, but competition for host resources is strong, then one taxon will go extinct or (3) the two taxa may establish a more or less stable hybrid zone; (4) if, during the period of geographic isolation, both ecological and mate recognition systems of the taxa have diverged sufficiently, they may be able to maintain their phenotypic and genetic identity in sympatry by utilizing the same host in slightly different ways, thus reducing competition. In outcomes 2–4 the sister taxa have the attributes of species.

If such a test of sympatry cannot be made, and experimental tests reveal little or no hybrid incompatibility between closely related allopatric sister populations adapted to the same hosts, then it is simply a subjective call as to whether to treat them as geographic races or species. Strict application of the genetic cluster species concept would require recognizing such populations as distinct species if they are represented by distinct genotypic clusters although, because the future is unpredictable, there is no guarantee of evolutionary permanence (Mallet 1995). The populations may fuse if contact between the allopatric populations is established in the future. The same may actually be true of some allopatric populations utilizing different hosts. We currently lack either suitable genetic, phylogenetic or biological criteria for dealing with this "allopatric" problem among habitat specialists. *R.* nr. *pomonella* from Mexico and *R. pomonella* from eastern North America, which both infest *Crataegus* spp., provide an example of this problem. These populations (species?) are genetically distinct and resolution of their species status will require tests of cross-hybridization and genetic analyses.

1.6.3 What Is a Host Race?

As we have already stressed, when the host plant serves as the site of courtship and mating, colonization of a new host may rapidly lead to the evolution of a genetically distinct population sympatric with the original species. There are now several well documented examples of such host shifts by insects and other invertebrates that have occurred within the past 100–150 years (Vouidibio et al. 1989; Yoshihisa 1991; Bush 1993b; Stanhope et al. 1993; Guldemond and Mackenzie 1994; Mackenzie 1996). Following past conventions these genetically distinct populations are called host or habitat races (Bush 1993a). However, as we learn more about the origin and evolution of host races, conventional views on how to distinguish a host race from a species may be somewhat arbitrary. It is becoming increasingly clear that the genetic differences between a host race and sister species exploiting different hosts is one of degree.

Mayr (1963) views host races as "non-interbreeding sympatric populations that differ in biological characteristics but not, or scarcely, in morphology [which are] ... prevented from interbreeding by a preference for different food plants or other hosts." This definition uses the same criterion, reproductive isolation, which Mayr uses to define species in his biological or isolation species concept. A host race is at the same time a species and therefore indistinguishable. Genetic and biological studies of populations recently colonized on new hosts or habitats (Bush 1993b, 1994; Feder et al. 1994) indicate that a host race is more appropriately regarded as a population of a species that is genetically distinct from other *sympatric* conspecific sister populations between which gene flow is greatly reduced, but not eliminated, as a direct consequence of adaptation to a host or group of closely related hosts (Diehl and Bush 1984). Host races may share the same alleles at most loci, but such races will differ primarily in allele frequencies at loci specifically involved with adaptation to the different hosts including those involved in mate recognition. Unlike a host related polymorphism, both adults and larvae on the same host will belong to the same genotypic clusters.

Initially, there may be relatively few key loci showing allelic differences that are maintained by selection in the face of gene flow as a population adapts to new host conditions. Recent genetic simulation of host race formation (Johnson et al. 1996) indicates that gene flow, which may at first be substantial between populations, rapidly diminishes as the new host race becomes better adapted and strengthens mate recognition in response to a combination of disruptive selection against hybrids as well as from divergent selection resulting from intrademic competition and assortative mating.

1.6.4 When Does a Host Race Become a Species?

Unfortunately, we know very little about the genetics of this key transition phase in the speciation process of phytophagous insects or for that matter any other animal species. Since both host races and species form genotypic clusters, determining if sufficient genetic differences have accumulated in development, breeding and ecological specialization to insure an independent evolutionary future for each host race irrespective of changes in environmental conditions is usually elusive. Obviously, it is impossible to foretell future ecological changes or to identify those key biological traits that will insure the continued maintenance of separate gene pools. For this reason the designation of biological status remains somewhat arbitrary for many closely related taxa. As Darwin (1859) correctly recognized, races grade imperceptibly into species. The difference is one of degree rather than the sharp boundary between species and non-species based on complete reproductive isolation of Mayr (1994), Templeton's multidimensional criterion of genetic cohesion (Templeton 1987) or the recognition concept of Paterson (1985).

Those who regard geographic isolation as an essential prerequisite for speciation do not believe host races can exist in nature (Mayr 1963; Futuyma and Mayer 1980). Proponents of the allopatric paradigm argue that gene flow between populations on different hosts will prevent any tendency for host race formation in response to selection for adaptation to each host. To acknowledge that divergence can be initiated and proceed in the face of gene flow is to acknowledge that host races and species can arise sympatrically through the process of adaptation and reinforcement, a process which now seems prone to occur between populations exploiting different hosts or habitats (Schluter and McPhail 1992; Butlin 1995; Johnson et al. 1996; Taylor et al. 1996). It also requires the rejection of the biological species concept as a means of recognizing species by undermining its major premise of requiring complete reproductive isolation. Although this isolation concept offers a simple, seemingly clear-cut way of classifying populations on the basis of the presence or absence of gene flow between sympatric populations, it is not biologically meaningful or realistic. The same can be said for definitions which may have utilitarian advantages, such as the cladistic phylogenetic species concept (Cracraft 1987), but which lack biological content and perspective. Defining species on the basis of gene clusters makes no judgment as to the evolutionary future of the population, only that they are currently maintaining genetically distinct populations. To determine the biological status of genotypic clusters requires that each pair of sister populations be treated as a unique problem of divergence and differentiation which only can be determined after adequate biological knowledge is at hand.

1.7 Conclusions

That sympatric speciation occurs in some, possibly all, groups of phytophagous insects now seems clear, but this was not always so. Only in the last few years, with the application of biochemical and molecular tools, have we been able to reconstruct patterns of divergence, obtain an accurate picture of phylogenetic relationships and establish the actual distribution of sister species. Preliminary results of these molecular investigations, coupled with what we know about the recent origin of host races and their biology, are compatible with the sympatric origin of many phytophagous insects and suggest that speciation in such organisms is often rapid. Sympatric speciation is probably common, particularly in insects that specialize on one or a few hosts and which mate on their host plant. The ability to rapidly form new host races and species sympatrically in the course of colonizing and adapting to a new host is probably a major factor in the generation of insect biodiversity.

References

Avise JC (1994) Molecular markers, natural history and evolution. Chapman and Hall, New York

Barton NH (1989) Founder effect speciation. In: Otte D, Endler JA (eds) Speciation and its consequences. Sinauer, Sunderland, pp 229–256

Boller E, Prokopy RJ (1976) The biology and management of *Rhagoletis*. Annu Rev Entomol 112:289–303

Brower AVZ (1994) Rapid morphological radiation and convergence among races of the butterfly *Helliconius erato* inferred from patterns of mitochondrial DNA evolution. Proc Natl Acad Sci USA 91:6491–6495

Brues CT (1924) The specificity of food-plants in the evolution of phytophagous insects. Am Nat 58:127–144

Bush GL (1969) Sympatric host race formation and speciation in frugivorous flies of the genus *Rhagoletis* (Diptera, Tephritidae). Evolution 23:237–251

Bush GL (1975a) Modes of animal speciation. Annu Rev Ecol Syst 6:339–364

Bush GL (1975b) Sympatric speciation in phytophagous parasitic insects. In: Price PW (ed) Evolutionary strategies of parasitic insects and mites. Plenum, New York, pp 187–206

Bush GL, Diehl SR (1982) Host shifts, genetic models of sympatric speciation and the origin of parasitic insect species. In: Visser JH, Minks AK (eds) Proc 5th Int Symp on Insect-plant relationships. Pudoc, Wageningen, The Netherlands, pp 297–305

Bush GL (1993a) Host race formation and sympatric speciation in *Rhagoletis* fruit flies (Diptera: Tephritidae). Psyche 99:335–357

Bush GL (1993b) A reaffirmation of Santa Rosalia, or why are there so many kinds of small animals? In: Lees DR, Edwards D (eds) Evolutionary patterns and processes. Academic Press, London, pp 229–249

Bush GL (1994) Sympatric speciation in animals: new wine in old bottles. Trends Ecol Evol 9:285–288

Bush GL, Howard DJ (1986) Allopatric and non-allopatric speciation: assumptions and evidence. In: Karlin S, Nevo E (eds) Evolutionary processes and theory. Academic Press, New York, pp 411–438

Bush GL, Feder JL, Berlocher SH, McPheron BA, Smith DC, Chilcote CA (1989) Sympatric origins of *R. pomonella*. Nature 339:346

Butlin RK (1995) Reinforcement: an idea evolving. Trends Ecol Evol 10:432–434

Carson HL (1975) The genetics of speciation at the diploid level. Am Nat 109:83–92

Carson HL, Kaneshiro KY (1989) Natural hybridization between the sympatric Hawaiian species *Drosophila silvestris* and *Drosophila heteroneura*. Evolution 43:190–203

Carson HL, Templeton AR (1984) Genetic revolutions in relation to speciation phenomena: the founding of new populations. Annu Rev Ecol Syst 15:97–131

Coyne JA (1994) Ernst Mayr and the origin of species. Evolution 48:19–30

Cracraft J (1987) Species concepts and the ontology of evolution. Biol Philos 2:63–80

Darwin C (1859) On the origin of species. A 1964 Facsimile. Harvard University Press. Cambridge de Meeûs T, Michalakis Y, Renaud F, Olivier I (1993) Polymorphism in heterogeneous environments, evolution of habitat selection and sympatric speciation: hard and soft selection models. Evol Ecol 7:175–198

Dethier VG (1954) Evolution of feeding preferences in phytophagous insects. Evolution 8:33–54

Diehl SR, Bush GL (1984) An evolutionary and applied perspective of insect biotypes. Annu Rev Entomol 29:471–504

Diehl SR, Bush GL (1989) The role of habitat preference in adaptation and speciation. In: Otte D, Endler J (eds) Speciation and its consequences. Sinauer, Sunderland, pp 345–365

Diehl SR, Prokopy RJ (1986) Host-selection behavior differences between the fruit fly sibling species *Rhagoletis pomonella* and *R. mendax* (Diptera: Tephritidae). Ann Entomol Soc Am 79:266–271

Emelianov I, Mallet J, Baltensweiler W (1995) Genetic differentiation in *Zeiraphera diniana* (Lepidoptera: Tortricidae), the larch budmoth: polymorphism, host races or sibling species. Heredity 75:416–424

Feder JL, Bush GL (1989) A field test of differential host plant usage between two sibling species of *Rhagoletis pomonella* fruit flies (Diptera: Tephritidae) and its consequences for sympatric speciation. Evolution 43:1813–1819

Feder JL, Chilcote CA, Bush GL (1988) Genetic differentiation between sympatric host races of *Rhagoletis pomonella*. Nature 336:61–64

Feder JL, Chilcote CA, Bush GL (1989) Are the apple maggot, *Rhagoletis pomonella* and the blueberry maggot, *R. mendax* (Diptera: Tephritidae) distinct species? Implications for sympatric speciation. Entomol Exp Appl 51:113–123

Feder JL, Chilcote CA, Bush GL (1990) Regional, local and microgeographic allele frequency variation between apple and hawthorn populations of *Rhagoletis pomonella* in western Michigan. Evolution 44:595–608

Feder JL, Hunt TA, Bush GL (1993) The effects of climate, host plant phenology and host fidelity on the genetics of apple and hawthorn infesting populations of *Rhagoletis pomonella* (Diptera: Tephritidae). Entomol Exp Appl 69:117–135

Feder JL, Opp SB, Walzlo B, Reynolds K, Go W, Spisak S (1994) Host fidelity is an effective premating barrier between sympatric races of the apple maggot fly, *Rhagoletis pomonella*. Proc Natl Acad Sci USA 91:7990–7994

Feder JL, Reynolds K, Go W, Wang EC (1995) Intra-and interspecific competition and host race formation in the apple maggot fly, *Rhagoletis pomonella* (Diptera: Tephritidae). Oecologia 101:416–425

Feder JL, Roethele JB, Walzlo B, Berlocher SH (1996a) Selective maintenance of allozyme differences between sympatric host races of the apple maggot fly. Proc Natl Acad Sci USA (submitted)

Feder JL, Stolz U, Lewis KM, Perry WM, Roethele JB, Rogers A (1996b) Host plant-associated fitness trade-offs in the apple maggot fly: The interaction of host phenology and winter on the genetics of apple and hawthorn races of *Rhagoletis pomonella* (Diptera: Tephritidae). Evolution (submitted)

Funk DJ, Futuyma DJ, Ortí G, Meyer A (1995) A history of host associations and evolutionary diversification for *Ophraella* (Coleoptera: Chrysomelidae): new evidence from mitochondrial DNA. Evolution 49:1008–1017

Futuyma DJ (1990) Observations on the taxonomy and natural history of *Ophraella* Wilcox (Coleoptera: Chrysomelidae), with a description of a new species. J NY Entomol Soc 98:163–186

Futuyma DJ (1991) A new species of *Ophraella* Wilcox (Coleoptera: Chrysomelidae) from the southwestern United States. J NY Entomol Soc 99:643–655

Futuyma DJ, Mayer GC (1980) Non-allopatric speciation in animals. Syst Zool 29:254–271

Futuyma DJ, McCafferty SS (1990) Phylogeny and the evolution of host plant associations in the leaf beetle genus *Ophraella* (Coleoptera, Chrysomelidae). Evolution 44:1885–1913

Futuyma DJ, Keese MC, Funk DJ (1995) Genetic constraints on macroevolution: the evolution of host affiliation in the leaf beetle genus *Ophraella*. Evolution 49:797–809

Galiana A, Moya A, Ayala FJ (1993) Founder-flush speciation in *Drosophila pseudoobscura*; a large-scale experiment. Evolution 47:432–444
Guldemond JA, Mackenzie A (1994) Sympatric speciation in aphids. I. Host race formation by escape from gene flow. In: Leather SR, Walters KFA, Mills NJ, Watt AD (eds) Individuals, populations and patterns in ecology. Intercept, Andover, Hampshire, pp 367–378
Jaenike J (1981) Criteria for ascertaining the existence of host races. Am Nat 117:830–834
Johnson P, Hoppensteadt F, Smith J, Bush GL (1996) Conditions for sympatric speciation: a diploid model incorporating habitat fidelity and non-habitat assortative mating. Evol Ecol 10:187–205
Kondrashov AS (1986) Multilocus model of sympatric speciation. III. Computer simulations. Theor Popul Biol 29:1–15
Kondrashov AS, Mina MV (1986) Sympatric speciation: when is it possible? Biol J Linn Soc 27:201–223
LeSage L (1986) A taxonomic monograph of the Nearctic galerucine genus *Ophraella* Wilcox (Coleoptera: Chrysomelidae). Memoirs of the Entomological Society of Canada No 133. Entomol Soc Canada, Ottawa
Liou LW, Price TD (1994) Speciation by reinforcement of premating isolation. Evolution 48:1451–1459
Mackenzie A (1996) A trade-off for host plant utilization in the black bean aphid, *Aphis fabae*. Evolution 50:155–162
Mallet J (1995) A species definition for the new synthesis. Trends Ecol Evol 10:294–299
Mayr E (1942) Systematics and the origin of species. Columbia University Press, New York
Mayr E (1963) Animal species and evolution. Harvard University Press, Cambridge
Mayr E (1988) Toward a new philosophy of biology: observations of an evolutionist. Harvard University Press, Cambridge
Mayr E (1994) Reasons for the failure of theories. Philos Sci 61:529–533
Menken SBJ, Herrebout WM, Wiebes JT (1991) Small ermine moths (Yponomuta): their host relations and evolution. Annu Rev Entomol 37:41–66
Mitter C, Farrell B, Futuyma DJ (1991) Phylogenetic studies of insect-plant interactions: insights into the genesis of diversity. Trends Ecol Evol 6:290–293
Moya A, Galiana A, Ayala F (1995) Founder-effect speciation theory: failure of experimental evidence. Proc Natl Acad Sci USA 92:3983–3986
Paterson HEH (1985) The recognition concept of species. Species and speciation. In: Vrba ES (ed) Transvaal Museum Monograph, Pretoria, pp 21–29
Phipps JB (1983) Biogeographic, taxonomic and cladistic relationships between Asiatic and North American *Crataegus*. Ann Mo Bot Gard 70:667–700
Price PW (1980) Evolutionary biology of parasites. Princeton University Press, Princeton
Rice WR (1984) Disruptive selection on habitat preference and the evolution of reproductive isolation: simulation studies. Evolution 38:1251–1260
Rice WR, Hostert EE (1993) Laboratory experiments on speciation: what have we learned in 40 years? Evolution 47:1637–1653
Schliewen U, Tautz D, Pääbo S (1994) Sympatric speciation suggested by monophyly of crater lake cichlids. Nature 368:629–632
Schluter D (1996) Ecological causes of adaptive radiation. Am Nat 148:S40–S64
Schluter D, McPhail JD (1992) Ecological character displacement and speciation in sticklebacks. Am Nat 140:85–108
Slatkin M (1996) In defense of founder-flush theories of speciation. Am Nat 147:493–505
Smith JJ, Bush GL (1997) Phylogeny of the genus *Rhagoletis* (Diptera: Tephritidae) inferred from DNA sequences of mitochondrial cytochrome oxidase II. Mol Phylo Evol 7:33–43
Stanhope MJ, Hartwick B, Baillie D (1993) Molecular phylogenetic evidence for multiple shifts in habitat preference in the diversification of an amphipod species. Mol Ecol 2:99–112
Strong DR, Lawton JH, Southwood TRE (1984) Insects on plants: community patterns and mechanisms. Blackwell, Oxford

Tauber CA, Tauber MJ (1989) Sympatric speciation in insects. In: Otte D, Endler JA (eds) Speciation and its consequences. Sinauer, Sunderland, pp 307–344
Taylor EB, McPhail JD, Schluter D (1996) History of ecological selection in sticklebacks: uniting experimental and phylogenetic approaches. In: Givnish TJ, Sytsma KJ (eds) Molecular evolution and adaptive radiation. Cambridge University Press, Cambridge (in press)
Templeton AR (1987) Species and speciation Evolution 41:233–235
Thorpe WH (1945) The evolutionary significance of habitat selection. J Anim Ecol 14:67–70
Vander Kloet SP (1988) The genus *Vaccinium* in North America. In: Research Branch, Agriculture Canada No 1828, Ottawa
Vouidibio J, Capy P, Defaye D, Sandrin E, Csink A, David JR (1989) Short-range genetic structure of *Drosophila melanogaster* populations in an Afrotropical urban area and its significance. Proc Natl Acad Sci USA 86:8442–8446
Wagner M (1868) Die Darwinsche Theorie und das Migrationsgesetz der Organismen. Dunker und Humbolt, Leipzig
Walsh BD (1864) On phytophagic varieties and phytophagic species. Proc Entomol Soc Philadelphia 3:403–430
Yoshihisa A (1991) Host race formation in the gall wasp *Andricus mukaigawae*. Entomol Exp Appl 58:15–20
Zwölfer H (1975) Artbildung und Ökologische Differenzierung bei phytophagen Insekten. Verh Dtsch Zool Ges 1975:394–401
Zwölfer H, Bush GL (1984) Sympatrische und parapatrische Artbildung. Z Zool Syst Evolutionsforsch 22:211–233

[illegible]

2 Host Race Formation in *Tephritis conura*: Determinants from Three Trophic Levels

M. Romstöck-Völkl

2.1 Introduction

Geographic differentiation in host-plant choice is a common feature in phytophagous insects (e.g. Fox and Morrow 1981; Hsiao 1982; Scriber 1983; Zwölfer and Romstöck-Völkl 1991). Observed differences between the actual and the potential host plant range may simply reflect differences in host-plant abundance or quality. However, they may be also the result of evolutionary processes which may lead to speciation.

Phytophagous insects have long been the subject of speciation models. Bush (1966, 1969, 1992) proposed that speciation in phytophagous insects might occur sympatrically via host shifts. Although it is difficult to prove sympatric speciation since there is a formidable amount of information required, a number of recent studies provide strong evidence that divergent host-plant preferences in adults may lead to reproductive isolation (e.g. Dodson and George 1986; Prokopy et al. 1988; Feder et al. 1989; Katakura et al. 1989; Messina 1989; Bush 1992; Berlocher et al. 1993; Craig et al. 1993; Payne and Berlocher 1995). "Host races", or biotypes, are partially reproductively isolated, conspecific populations specialized on alternative hosts (Diehl and Bush 1984; Bush and Smith, this Vol.). Host-race formation is proposed as a critical intermediate step in the sympatric speciation of phytophagous insects. Its minimal requirements are changes in the preferences for oviposition sites, suitable physiological adaptations to the new host and habitat or host related assortative mating to allow maintenance of differences (Tauber and Tauber 1989). One prerequisite for the formation of host races is a successful host shift. The colonization of a new host may be mediated by non-heritable environmental factors like the escape from natural enemies (Hairston et al. 1960; Zwölfer 1982; Bernays and Graham 1988; Brown et al. 1995; Feder 1995), by the avoidance of inter- and intraspecific competition (Feder et al. 1995) and by local differences in host abundancy and availability.

In this chapter, I report on differences in host-plant exploitation and host-race formation in the tephritid fly *Tephritis conura* Loew (Diptera: Tephritidae). *T. conura* is a univoltine species whose larvae develop endophytically in the flower heads of several species in the host-plant genus *Cirsium* (Zwölfer 1965; Zwölfer and Romstöck-Völkl 1991). Adult flies emerge in mid-summer and hibernate in the adult stage. Host plants are recolonized every spring when suitable oviposition sites – buds in a certain developmental

Ecological Studies, Vol. 130
Dettner et al. (eds.) Vertical Food Web Interactions

stage – are available (Romstöck-Völkl 1990a). Eggs are laid in batches; larvae feed gregariously and induce callus growth of the receptacle. The oviposition period lasts about 4 weeks, and adults die after having finished egg laying. First instar larvae consume the developing florets, second and third instar larvae the receptacle area of the flower head. The ecology and microevolution of *T. conura* have been studied in detail in a number of field and laboratory experiments (Romstöck 1987; Romstöck and Arnold 1987; Romstöck-Völkl 1990a,b; Romstöck-Völkl and Wissel 1989; Zwölfer and Romstöck-Völkl 1991), and here I present an analysis of essential factors contributing to regional differences in host-plant use and evidence for the formation of distinct host races.

2.2 Resource Utilization in Different Host Plants

The host-plant range of *T. conura* covers at least nine species of the genus *Cirsium*, all belonging to the section *Cirsium*: *C. heterophyllum*, *C. oleraceum*, *C. acaule*, *C. erisithales*, *C. palustre*, *C. spinosissimum*, *C. canum*, *C. filipendulum*, *C. tuberosum* (Zwölfer 1965, 1988; Zwölfer and Romstöck-Völkl 1991). All host-plant species but *C. palustre* (a biennial) are perennial and usually form dense stands of rhizomal clones. Not all host plants are attacked throughout their complete European range. For example, *C. palustre* is not attacked in the lowlands of central Europe although it is very abundant in this area and grows in mixed stands with *C. oleraceum* or *C. heterophyllum*, which

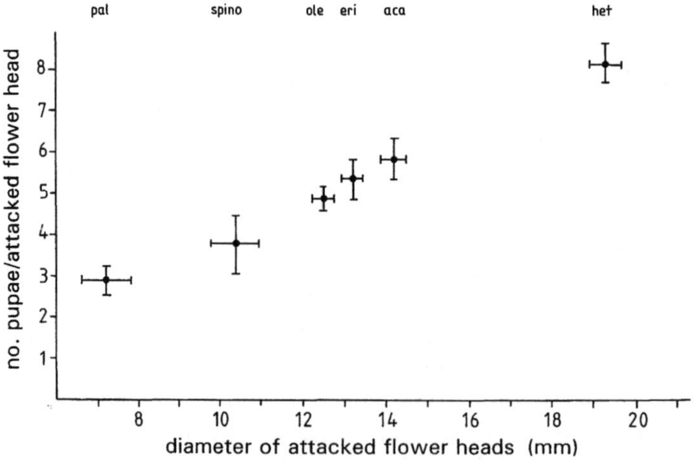

Fig. 2.1. Mean density of *T. conura* pupae per attacked flower head in relation to flower head diameter in six host plants. (*pal*, *C. palustre*; *ole*, *C. oleraceum*; *eri*, *C. erisithales*; *spino*, *C. spinosissimum*; *aca*, *C. acaule*; *het*, *C. heterophyllum*)

are both commonly used by *T. conura*. Also, *C. heterophyllum* in the Pyrenees or *C. acaule* in southern England are not attacked by *T. conura*.

Resource-related population densities varied between host plants and geographic regions (Table 2.1), and consist of two components: (1) the percentage of attacked flower heads per sampling site (the attack rate) and (2) the mean number of pupae per attacked flower head. The percentage of attacked flower heads also serves as an estimate for the impact of the fly on its host plant (reduced seed production and attacked flower heads as metabolic sinks). The attack rates varied considerably both within and between host plants (Table 2.1). There were, however, two obvious tendencies. First, attack rates decreased in host-plant species with a long period of bud production (e.g. *C. palustre*, *C. oleraceum*). Second, attack rates show a relationship to the geographic location of the sampling site and decreased with increasing latitude (Romstöck 1984).

The number of pupae per attacked flower head differed significantly between host plants (Fig. 2.1: oneway ANOVA: $F = 34.99$, $df = 5$, $p < 0.001$). I

Table 2.1. Percent attack rates (mean and range of percent attacked flower heads) and population density (= mean number of fly pupae per 100 flower heads) of *T. conura* in six host plants and in different geographic areas. n_{tot} = total number of samples, n_{at} = number of samples yielding *T. conura*. Attack rates are based on n_{at}, larval densities on n_{tot}

Plant	Area	n_{tot}	n_{at}	Attack rate	Population density
C. heterophyllum	Scotland	24	24	34 (4–83)	238
	S/C Scandinavia	19	19	52 (11–89)	690
	N Scandinavia	11	8	18 (<1–32)	120
	N Bavaria	107	107	45 (<1–92)	400
	C Alps	14	14	48 (27–68)	416
C. palustre	Scotland	43	43	15 (<1–36)	43
	N Bavaria	18	–	–	–
	C Alps	10	1	12	32
C. erisithales	Massif Central	4	4	28 (15–46)	154
	Swiss Jura	2	2	48 (45–51)	278
	C Alps	4	4	12 (13–18)	13
	E Alps	3	3	8 (<1–24)	31
C. oleraceum	N Bavaria	36	36	26 (1–49)	122
	E France	5	4	4 (<1–13)	41
	French Jura	2	2	20 (18–22)	108
	C Alps	6	6	27 (19–43)	113
	E Alps	3	3	37 (29–49)	190
C. acaule	N Bavaria	34	31	13 (<1–43)	63
	N Eifel	15	15	37 (12–73)	247
	French Jura	2	2	41 (37–45)	232
	C Switzerland	2	2	39 (25–53)	198
	C Alps	1	1	45	160
C. spinosissimum	C Alps	12	9	17 (5–32)	51

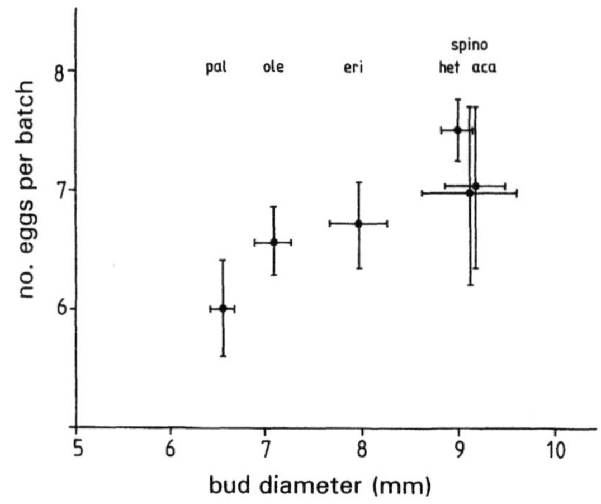

Fig. 2.2. Relationship between average egg batch size and bud diameter of *T. conura* in six host plants (*abbreviations*, see Fig. 2.1)

found a characteristic distribution for each *Cirsium* species which was independent of the location. Only *C. heterophyllum* showed a relationship between pupae/flower head and location (Kruskal-Wallis-ANOVA: $Chi^2 = 7.24$, $df = 6$, $p = 0.027$): Samples from Alpine stands and from Scandinavia had higher densities than samples from northern Bavaria, and Scottish samples yielded the lowest number of pupae per flower head. The mean number of larvae per flower head was correlated with flower head size (Fig. 2.1; $r_s = 1.0$, $n = 6$, $p < 0.001$). It was highest for *C. heterophyllum*, the host plant with the largest buds and largest flower heads, and lowest for *C. palustre*, the thistle with the smallest buds and heads. Similarly, the mean size of an egg batch was characteristic for each host plant and correlated with average bud diameter at the time of oviposition (Fig. 2.2; $r_s = 0.828$, $n = 6$, $p = 0.042$). Batch size differed between host plants even if one fly population exploited two host plants, like *C. heterophyllum* and *C. palustre* in Scotland (Romstöck and Arnold 1987).

2.3 Mortality Risks

2.3.1 Bud Abortion

In all plant populations, *C. heterophyllum* and *C. erisithales* abort some buds, and larvae dwelling in them perish. The probability of bud abortion is mainly determined by bud position and by the total number of buds produced by a plant: the more buds formed, and the later a bud is formed, the higher is the abortion risk regardless of whether or not it is attacked (Romstöck-Völkl 1990a). Bud abortion accounted for a mortality of approx. 20–30% of egg batches in *C. heterophyllum* (Romstöck-Völkl 1990a) and 10–20% of egg batches in *C. erisithales*. The importance of bud abortion as a mortality factor depended on the synchronization of the oviposition period with bud development. Its impact increased with an increasing proportion of ovipositions in phenologically "late" buds.

Bud abortion seems to play no important role as mortality factor in *C. acaule*, *C. spinosissimum*, *C. oleraceum* and *C. palustre*, i.e. less than 1% of the egg batches were affected.

2.3.2 Intraspecific Competition

Intraspecific competition as a mortality factor depended on flower head size. In *C. heterophyllum*, intraspecific competition for space and food was negligible and accounted for less than 5% larval mortality (Romstöck-Völkl 1990a). In *Cirsium* species with smaller flower heads (see Figs. 2.1, 2.2), interspecific competition was much more important. Estimated values varied between 20–30% in *C. acaule*, 25–40% in *C. erisithales* and 30–35% in *C. spinosissimum*. *C. oleraceum* shows an extremely high variablity in flower head size, and mortality ranged between 10–60% (Romstöck-Völkl, unpubl.). The highest values (35–60%) were estimated for *C. palustre* (Romstöck and Arnold 1987), the host plant species with the smallest flower heads.

2.3.3 Parasitization Risks in Different Host Plants

The larvae of *T. conura* are regularly attacked by two endoparasitic wasps, *Pteromalus caudiger* Graham (Hymenoptera, Pteromalidae) and *Eurytoma* sp. near *tibialis* Boheman (Hymenoptera, Eurytomidae). *P. caudiger* was found in all host-plant species and all geographic areas, while *Eurytoma* was lacking in all populations of *C. oleraceum*, *C. spinosissimum* and *C. palustre*, and in Scottish populations of *C. heterophyllum* and French populations of *C. erisithales*.

A detailed analysis of parasitization rates of *T. conura* by *P. caudiger* and *Eurytoma* sp. in *C. heterophyllum* showed that parasitization risk depends crucially on flower head diameter and larval position within flower heads (Romstöck-Völkl 1990b). Large flower heads provide some kind of structural refuge, since the ovipositor length of both parasitoid species is only 3 mm. If this relationship is also inferred for a comparison between host-plant species, average parasitization rates should increase with decreasing flower head diameter. Figure 2.3 shows this relationship for *P. caudiger*, the parasitoid species that occurred in all host plants. In this figure, *C. spinosissimum* is presented twice. Individual flower heads have a small size (mean diameter: 10–11 mm) but the densely aggregated heads (mean aggregation diameter: >30 mm) should be considered as a functional unit for the parasitoid, which is not able to forage between the heads. Additionally, *T. conura* larvae in *C. spinosissimum* feed mainly at the flower head's base near the stem where they are out of the reach of the parasitoids' ovipositor. If *C. spinosissimum* is assigned

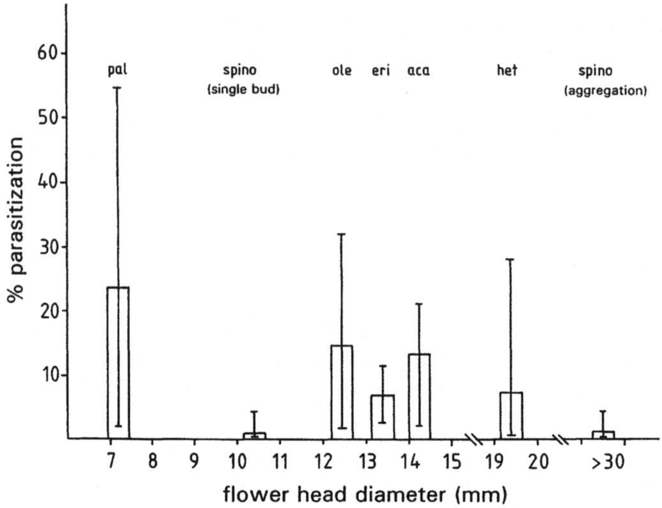

Fig. 2.3. Relationship between parasitization of *T. conura* by *P. caudiger* and mean flower head size. Values give mean, maxima and minima of parasitism of all studied populations of the respective host-plant species (*abbreviation*, see Fig. 2.1). Total parasitization rates in *C. heterophyllum*, *C. acaule* and *C. erisithales* are considerably higher due to parasitism by *Eurytoma* sp.

therefore to the highest "flower head" diameter, there is a significant decrease in parasitism with increasing flower head diameter of the host plant (mean parasitization rates: $r_s = -0.771$, $p = 0.05$; maximal parasitization rates: $r_s = -0.871$, $p < 0.025$). *T. conura* suffers from the highest parasitization risk by *P. caudiger* in small flower heads of *C. palustre* (54.3% in an Alpine population), While *C. spinosissimum* and Scottish *C. heterophyllum* provide an almost "enemy-free space" with maxima of 4.5% parasitism and 5.0% parasitism, respectively.

The importance of the host plant's flower head size for parasitization risk could also be shown experimentally. Experimental populations of *T. conura* in *C. palustre* at two sites in northern Bavaria – *C. palustre* is virtually not attacked by *T. conura* in this region although the plant is very abundant – suffered from 43 and 57% parasitism by *P. caudiger*, while sympatric populations in *C. heterophyllum* and *C. oleraceum*, showed only 5 and 11% parasitism, respectively.

2.4 Differentiation Among *T. conura* Populations

2.4.1 Fly and Host-Plant Phenology

T. conura females only accept buds in a very early developmental stage for oviposition (Romstöck 1987; Romstöck-Völkl and Wissel 1989). Thus, the length of the oviposition period depends on the availability of suitable buds. Time and length of bud production differs considerably for the different host plants (Fig. 2.4). Bud availability and oviposition period were almost identical in *C. heterophyllum* but bud availability exceeded the oviposition period by far in most other host plant species. In the latter case, the percentage of attacked buds decreased as the season progressed. The production of late buds provides some kind of temporal refuge for *C. palustre*, *C. oleraceum* and *C. acaule* (Zwölfer and Romstöck-Völkl 1991). Furthermore, a short period of bud availability requires an exact temporal synchronization of fly phenology with bud phenology. This synchronization should be most exact with *C. heterophyllum* (Fig. 2.4).

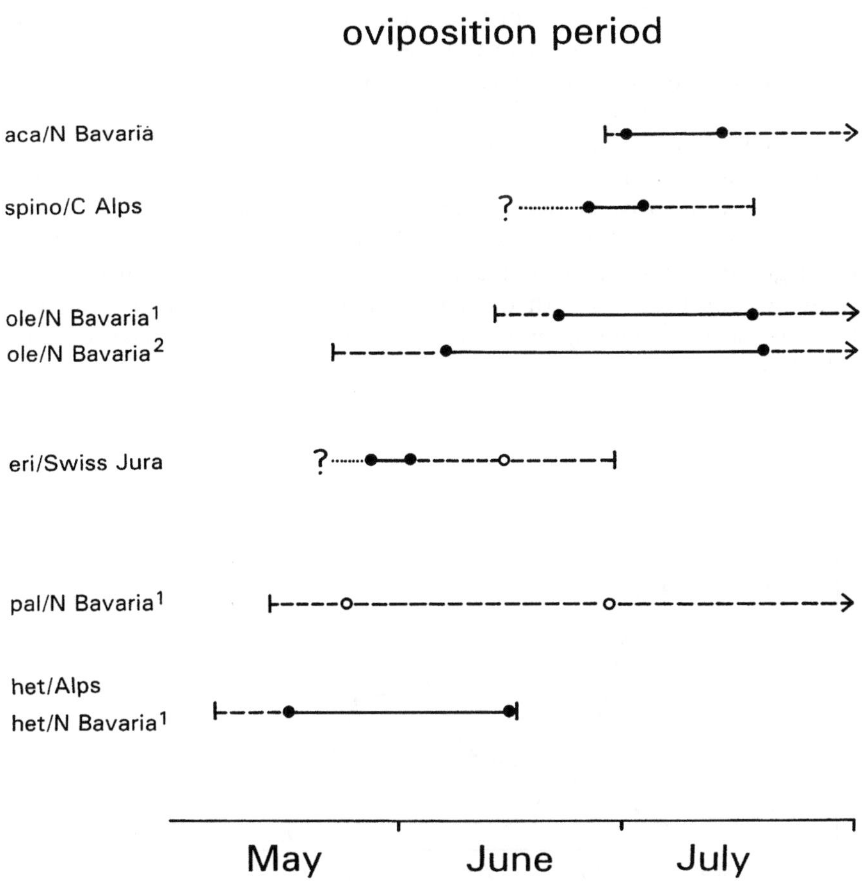

Fig. 2.4. Bud availability and oviposition period of *T. conura* in different host plants and in different geographic areas. *Filled circles* and *solid lines* indicate the observed oviposition period in the field, *open circles* indicate ovipositions in the laboratory. *Dotted* and *broken lines* show observed periods of bud availability in the field before and after the oviposition period of *T. conura*. *1* Northern Bavaria/Fichtelgebirge; *2* northern Bavaria/Upper Main valley

Also, differences in bud availability between sympatric host plants would require a high adult longevity to allow an attack of both host plants by the same fly population. This aspect was studied at a site in the Fichtelgebirge (northern Bavaria, Germany) where *C. heterophyllum* and *C. oleraceum* were growing in a wet meadow less than 50 m apart. The abundancies of adult *T. conura* caught on *C. heterophyllum* and *C. oleraceum* showed two distinct peaks which occurred over a roughly 3-week period (Fig. 2.5). However, there was an overlap of about 10 days during which flies were caught on both host plants. During this interval, none of 85 marked specimens caught on *C. heterophyllum* was recaptured later on *C. oleraceum*.

2.4.2 Mating Preferences

In the field, mating of *T. conura* occurs exclusively on the host plant. Since most males prefer to stay on or near the buds, most matings take place while females examine buds for their suitability for oviposition.

If 20 males and 20 females caught from *C. heterophyllum* and *C. oleraceum* (40 individuals from each species), respectively, were kept together without host plants in a small cage (size 0.5 × 0.5 × 1 m), males did not discriminate significantly in mate choice between females from different host plants: *C. heterophyllum* males mated in 62.5% of all observations with females originating from this host plant, while *C. oleraceum* males mated in 60% of the observations with *C. oleraceum* females. If host plants were added to the cage, *C. heterophyllum* males mated in 82% of cases with *C. heterophyllum* females.

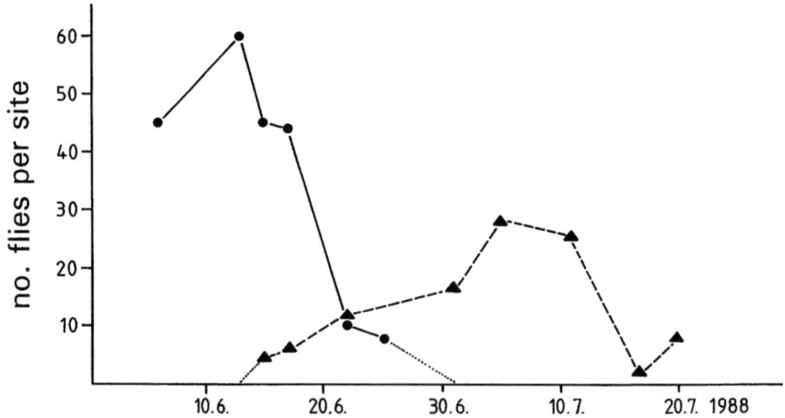

Fig. 2.5. Incidence of *T. conura* on *C. heterophyllum* (*circles*) and *Cirsium oleraceum* (*triangles*) at a stand with mixed host-plant populations at the edge of the Fichtelgebirge, northern Bavaria. The study site consisted of approx. 200 shoots of *C. heterophyllum* and 50 shoots of *C. oleraceum*

2.4.3 Larval Development in Different Host Plants

In the laboratory, adult flies caught from *C. heterophyllum* and *C. oleraceum* accepted a wide range of host plants for oviposition (Table 2.2). For such females with a high oviposition pressure, the phenological state of the bud was obviously more important than the host-plant species (as long as the offered plant belonged into the section *Cirsium*). There were, however, big differences in the developmental success of larvae (Table 2.2). Flies originating from *C. heterophyllum* developed normally only in *C. palustre*, which was not accepted by *C. oleraceum* flies. By contrast, *C. oleraceum* flies developed successfully in *C. acaule*, a host plant which was rejected by *C. heterophyllum* flies. *C. heterophyllum* flies also produced a few offspring in *C. oleraceum*.

If two host plants were offered simultaneously, I observed distinct preferences. *C. heterophyllum* flies oviposited more often in *C. heterophyllum* (93%) than into *C. palustre* (7%). *C. oleraceum* and *C. filipendulum* (a host plant of Spanish populations) were generally rejected if *C. heterophyllum* was offered simultaneously. If *C. heterophyllum* females had a choice between *C. palustre* and *C. oleraceum*, they generally oviposited into *C. palustre*. By contrast, *C. oleraceum* flies generally rejected *C. heterophyllum* and *C. palustre* but chose regularly *C. acaule* (12.5%).

Similarly, overwintered flies from *C. heterophyllum* and *C. oleraceum* without a previous contact with suitable buds chose those host plants for resting and searching in which they had developed ($Chi^2 = 35.52$, $df = 1$, $p < 0.001$). However, this result should be interpreted cautiously since neither males nor females showed normal gonad development after hibernation in the laboratory.

2.4.4 Morphological Adaptations in Different Populations

Zwölfer (1986) and Möller-Joop (1988) showed for *Urophora* species (Diptera: Tephritidae) in different host plants a relationship between the length of the

Table 2.2. The response of *T. conura* females to and larval development in alternative host plants

Origin of females	New host for larvae	Accepted for oviposition	Successful development
C. heterophyllum	*C. palustre*	Yes	Normal
	C. oleraceum	Yes	Weak
	C. erisithales	Yes	No
	C. acaule	No	No
	C. filipendulum	Yes	No
C. oleraceum	*C. palustre*	No	No
	C. acaule	Yes	Medium
	C. canum	Yes	No
	C. heterophyllum	Yes	?

oviscapt and the host plants' bud diameter when suitable for oviposition. Similar relationships were found for *T. conura* populations in different host plants (Fig. 2.6). The average oviscapt length increased with average bud diameter. The only exception were Scottish populations developing in *C. heterophyllum*. They had a considerably shorter oviscapt than central European or Alpine populations from *C. heterophyllum*, indicating that the major host of this population is *C. palustre* (Romstöck and Arnold 1987). Furthermore, it should be noted that buds of *C. spinosissum* are clustered to dense aggregations when suitable for oviposition. Thus, a long ovipositor of *T. conura* populations originating from this host plant may indicate the ability to reach interior buds of an aggregation for oviposition.

2.5 Morphological and Genetic Differentiation Between Populations

An analysis of 30 characters of wing venation, wing colour patterns and oviscapt size from 533 flies of 39 populations from 6 host plants revealed 4 major groups (Fig. 2.7). The first group was comprised mainly of flies reared from Alpine stands. The second group consisted of the Scottish populations, which are characterized by short oviscapts (Fig 2.6), while the third and fourth groups included mainly extra-Alpine populations from *C. heterophyllum* and *C. oleraceum*/*C. erisithales*, respectively.

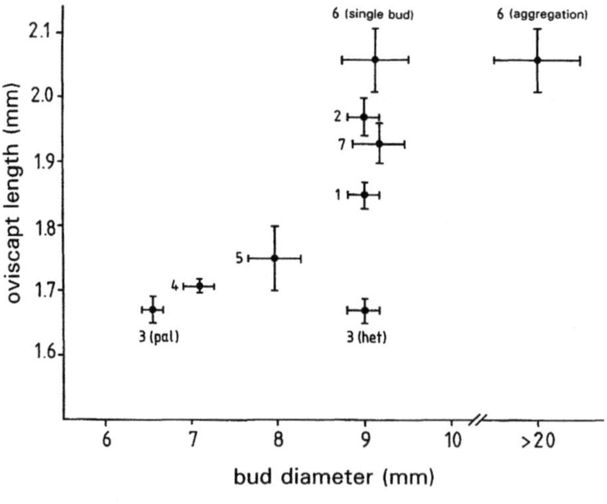

Fig. 2.6. Relationship between oviscapt length of *T. conura* and bud size when available for oviposition; *1*, *C. heterophyllum*/central Europe: *2*, *C. heterophyllum*/Alps; *3*, *C. palustre* (pal) + *C. heterophyllum* (het)/Scotland; *4*, *C. oleraceum*/central Europe; *5*, *C. erisithales*/Swiss Jura + Alps; *6*, *C. spinosissium*/Alps; *7*, *C. acaule*/central Europe

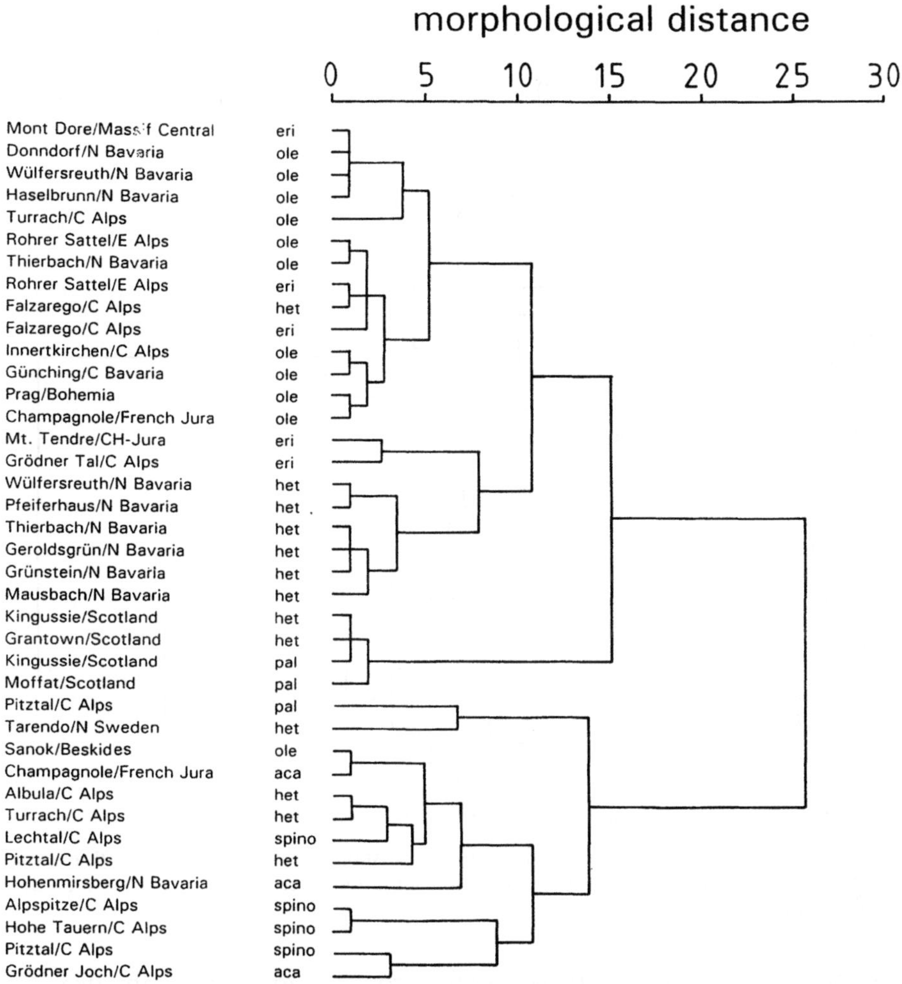

Fig. 2.7. Morphometric distances between 39 *T. conura* populations from different host-plant species and from different geographic areas based on the analysis of 30 selected morphological characters

The analysis of 11 enzymes with polymorphic loci (including 29 populations) by starch gel electrophoresis produced two major groups with genetic distances (Nei 1972) of 0.07 (Fig. 2.8). Flies reared from *C. oleraceum* showed – independent of their geographic origin – only small genetic distances and formed one distinct group, together with two populations from *C. acaule*. Within the second group, the populations from Tarendo/northern Sweden on *C. heterophyllum* and from the Grödner Joch/central Alps on *C. acaule* were somewhat separated. Except for this geographically very distant Swedish population, there were only slight differences between populations from *C. heterophyllum* originating from northern Bavaria and from the Alps. The four

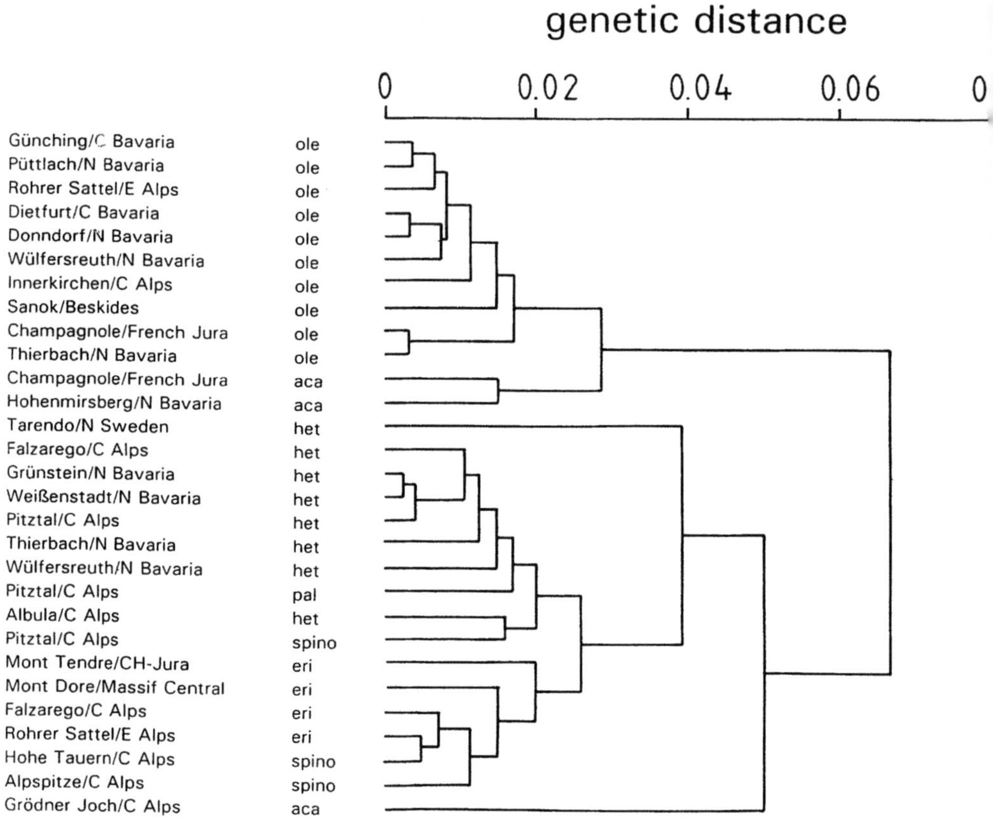

Fig. 2.8. UPGMA cluster of Nei's standard genetic distances (Nei 1972) for 29 *T. conura* populations from different host-plant species and from different geographic areas based on the analysis of 11 enzymes

populations from *C. erisithales* were grouped together with two *C. spinosissimum* populations. In contrast to Komma (1990), I found no diagnostic phospho-gluco-mutase (PGM) allele for *C. erisithales* flies. The Alpine population of *C. palustre* was not separated from *C. heterophyllum* populations. From a number of sampling sites (Thierbach, Wülfersreuth, Champagnole, Pitztal, Falzarego, Rohrer Sattel), I analysed sympatric fly populations from different host plants. *T. conura* populations from a single site which were reared from different host plants showed no more genetic similarity than populations from separated sites with only one available host plant.

2.6 Conclusions

The current host range of *T. conura* is characterized by regional variations in the exploitation of potential host plants. The major factors influencing these differences are host plant related differences like abundancy and flower head size and the escape from natural enemies (Zwölfer and Romstöck-Völkl 1991). For *T. conura*, the key factor for changes in population numbers is the mortality occurring between adult emergence and host-plant recolonization in the following year, i.e. the mortality at the hibernation sites and losses during the recolonization of host-plant habitats (Romstöck-Völkl 1990a). Therefore, high population densities are essential for *T. conura* to compensate for high dispersal losses. Host-plant related factors determining larval densities are host plant abundancy, the number of suitable buds per plant available during the short oviposition period, bud size as a measure of host quality and larval mortality due to bud abortion. A host plant needs a certain density to support a viable *T. conura* population. Small plant populations (e.g. populations of *C. heterophyllum* in the Spanish Pyrenees or populations of *C. acaule* in southern England) would not allow high larval densities. Large buds favour large batches (Fig. 2.1) with less intraspecific competition and thus contribute to higher larval densities. However, host plants with large buds (e.g. *C. heterophyllum*, *C. acaule*) usually produce less flower heads than host plants with comparatively small buds (e.g. *C. oleraceum*, *C. palustre*). Thus, bud size is often counterbalanced by bud number.

These factors can be illustrated by the different exploitation of some host plants within the total range of *T. conura*. In wide parts of the central European highlands (e.g. Fichtelgebirge, Bayerischer Wald, Erzgebirge), *C. heterophyllum* is common and occurs in large dense patches, while the sympatric *C. palustre* – with synchronous bud development (Fig. 2.4) – has a more scattered distribution. In this area, *T. conura* is specialized to *C. heterophyllum* while *C. palustre* is used only exceptionally. In Scotland, the densities of *C. heterophyllum* are much lower and host patches are more isolated than on the continent, while *C. palustre* is widespread and represents a highly predictable resource. Under these conditions, the Scottish population exploits both host plants equally, thereby achieving an increased efficiency of host finding and sufficiently high population densities for larval survival (Romstöck and Arnold 1987).

C. acaule usually grows in a scattered way in dry calcareous grasslands and never reaches high plant densities. Thus, it is unlikely that this host plant might support an independent fly population. A host-plant shift to *C. acaule* in different regions might have occurred independently from different original host plants. This theory is supported by the inconsistent classifications of populations from *C. acaule* in morphological and genetic studies (Figs. 2.7, 2.8). Populations in northern Bavaria and in the French Jura are likely to originate from *C. oleraceum*, a theory supported by the results on host-plant choice (Table 2.2). Central Alpine populations on *C. acaule*, by contrast, might originate from *C. heterophyllum* although the phenology of both host plants differs considerably.

Flower head size is an important host-plant parameter influencing the reproductive success of *T. conura*. Females lay smaller egg batches into smaller flower heads (Fig. 2.1). Additionally, their parasitization rates are higher in small than in large flower heads (Fig. 2.3). Therefore, a female has to lay more egg batches (i.e. attack more buds) to produce the same number of surviving offspring. This means that adult flies forage more intensively between patches, and are exposed to the risk of greater adult mortality (Weisser et al. 1994), e.g. in spider webs (Scheidler 1985). Thus, there should be a selective advantage for flies attacking large heads. Such an escape from natural enemies may be one important selective pressure in the colonization of new host plants and evolution of host races. Feder (1995) demonstrated that the host race of *Rhagoletis pomonella* developing in apples suffered less parasitism than its conspecific developing in hawthorn fruits. Similar differences were found for *T. conura* in *C. heterophyllum* and *C. palustre. T. conura* suffered from very high parasitism in this host plant (Fig. 2.3), while parasitism in *C. heterophyllum* was generally low. Furthermore, the feeding niche of *T. conura* in *C. palustre* is occupied in central Europe – but not in Scotland – by *Tephritis cometa*, and a coexistence of two species occupying the same niche on a common host plant is unlikely (Harris 1989). Thus, both parasitism and interspecific competition may have prevented a successful colonization of *C. palustre* in central Europe, while the disadvantage of high parasitism in Scotland was counterbalanced by the high host plant abundance. In Scotland, *C. heterophyllum* represented an almost enemy-free space which secures a high larval survival rate and may compensate in this way for the relative rarity of *C. heterophyllum*.

T. conura shows similar behavioural, morphological and genetic differentiations between populations as found for host races of the tephritid flies *Rhagoletis pomonella* (McPheron et al. 1988; Feder et al. 1989, 1990, 1994; Luna and Prokopy 1995), *Eurosta solidaginis* (Waring et al. 1990; Craig et al. 1993) and *Tephritis bardanae* (Eber et al. 1991). The current results provide evidence for three *T. conura* host races: a Scottish population and two continental populations. The allopatric Scottish population on *C. palustre/C. heterophyllum* is characterized by morphological (Fig. 2.6) and behavioural differences (Romstöck and Arnold 1987). In contrast to central European populations of *C. heterophyllum*, males and females are frequently found searching on both host plant species in the field, they mate on both hosts, and females accept both plants equally for oviposition.

Populations from the European mainland seem to be separated into two host races with a sympatric and/or parapatric range. Seitz and Komma (1984) described genetic differences between northern Bavarian *T. conura* populations from *C. oleraceum* and *C. heterophyllum*, and the present study confirms this finding for a broader geographic range. *T. conura* populations from other *Cirsium* species show less host plant-related differentiations. They can be either classed with *C. oleraceum* (non-Alpine populations from *C. acaule*) or with *C. heterophyllum* (all other populations; Fig. 2.8). Within these two groups, genetic distances do not exceed 0.025 (exceptions: populations from *C. heterophyllum*/Tarendo-northern Sweden and from *C. acaule*/Grödnertal-central Alps), a distance known as the maximum value between geographic populations in tephritid flies (Berlocher and Bush 1982). The Nei distances of 0.07 between the two main groups provide evidence for a widely reduced gene flow but they are far below the differences found between *Tephritis* species (*T. conura*, *T. arnicae*, *T. cometa*, *T. heiseri*), which range from 0.23 to 0.44 (Komma 1990). The classification of particular populations in morphological studies was somewhat inconsistent with the results from allozyme studies (Fig. 2.7). For example, there were differences between Alpine and northern Bavarian populations from *C. heterophyllum* which are mainly based on differences in oviscapt length (Fig. 2.6). However, such differences between morphological and allozyme data may be not unexpected (Lewontin 1984; Allegrucci et al. 1987).

The separation into two major groups coincides with a number of behavioural differences. First, both overwintered and field-caught males and females preferred their original host plant for searching and resting. Such preferences might be an effective pre-mating barrier between populations (Feder et al. 1994). Second, there are distinct host-plant preferences found for flies originating from *C. heterophyllum* or *C. oleraceum* (see Sect. 2.4.2). Third, an assortative mating - facilitated by the flies' behaviour to mate on the particular host plant - may lead to an inheritance of these differences, although this behaviour has to be confirmed by further experiments with overwintered individuals.

The most important factor influencing the evolution of these two host races in central Europe was certainly the difference in host-plant phenology. Host-plant phenology and synchronization between fly oviposition and host-plant availability differed significantly in sympatric populations on *C. heterophyllum* and *C. oleraceum*, e.g. in northern Bavaria (Fig. 2.5). Both fly and host phenology show only a very slight overlap, a prerequisite for genetic isolation and sympatric speciation (Wood and Guttmann 1982; Zwölfer and Bush 1984; Bush 1992; Payne and Berlocher 1995).

Acknowledgements. First of all, I thank Helmut Zwölfer, who introduced me to insect ecology. His immense knowledge, together with his kindness, produced an excellent atmosphere in his working group. I remember with pleasure the years I had the chance to join his team. Annick Servant and Sabine Eber kindly carried out the allozyme analysis. I also thank G. Bush, G. Bauer, K. Dettner and W. Völkl for their comments on earlier drafts of the manuscript. Financial support was provided by the German Research Council (DFG, Ro 730/1-1).

References

Allegrucci G, Cesaroni D, Sbordoni V (1987) Adaptation and speciation of *Dolichopoda* cave crickets (Orthoptera, Rhaphidophoridae): geographic variation of morphometric indices and allozyme frequencies. Biol J Linn Soc 31:151–160

Berlocher SH, Bush GL (1982) An electrophoretic analysis of *Rhagoletis* (Diptera: Tephritidae) phylogeny. Syst Zool 31:136–155

Berlocher SH, McPheron BA, Feder JL, Bush GH (1993) Genetic differentiation at allozyme loci in the *Rhagoletis pomonella* (Diptera: Tephritidae) species complex. Ann Entomol Soc Am 86:716–727

Bernays E, Graham M (1988) On the evolution of host specifity in phytophagous arthropods. Ecology 69:886–892

Brown JM, Abrahamson WG, Packer RA, Way PA (1995) The role of natural-enemy escape in a gallmaker host-plant shift. Oecologia 104:52–60

Bush GL (1966) The taxonomy, cytology and evolution of the genus *Rhagoletis* in North America (Diptera: Tephritidae). Museum of Comparative Zoology, Cambridge, Massachussetts

Bush GL (1969) Sympatric host race formation and speciation in frugivorous flies of the genus *Rhagoletis* (Diptera: Tephritidae). Evolution 23:237–251

Bush GL (1992) Host race formation and sympatric speciation in *Rhagoletis* fruit flies. Psyche 99:335–357

Craig TP, Itami JK, Abrahamson WG, Horner DH (1993) Behavioral evidence for host-race formation in *Eurosta solidaginis*. Evolution 47:1696–1710

Diehl SR, Bush GL (1984) An evolutionary and applied perspective of insect biotypes. Annu Rev Entomol 29:471–504

Dodson G, George SB (1986) Examination of two morphs of gall-forming *Aciurina* (Diptera: Tephritidae): ecological and genetic evidence for species. Biol J Linn Soc 29:63–79

Eber S, Sturm P, Brandl R (1991) Genetic and morphological variation among biotypes of *Tephritis bardanae*. Biochem Syst Evol 19:549–557

Feder JL (1995) The effects of parasitoids on sympatric host races of *Rhagoletis pomonella* (Diptera: Tephritidae). Ecology 76:801–813

Feder JL, Chilcote CA, Bush GL (1989) Are the apple maggot, *Rhagoletis pomonella*, and blueberry maggot, *R. mendax*, distinct species? Implications for sympatric speciation. Entomol Exp Appl 51:113–123

Feder JL, Chilcote CA, Bush GL (1990) The geographic pattern of genetic differentiation between host associated populations of *Rhagoletis pomonella* (Diptera: Tephritidae) in the eastern United States and Canada. Evolution 44:570–594

Feder JL, Opp S, Wlazlo B, Reynolds K, Go W, Spisak S (1994) Host fidelity is an effective premating barrier between sympatric races of the apple maggot fly, *Rhagoletis pomonella*. Proc Natl Acad Sci USA 91:7990–7994

Feder JL, Reynolds K, Go W, Wang E (1995) Intra- and interspecific competition and host race formation in the apple maggot fly, *Rhagoletis pomonella* (Diptera: Tephritidae). Oecologia 101:416–425

Fox LR, Morrow PA (1981) Specialisation: species property or local phenomenon? Science 211:887–893

Hairston NG, Smith FE, Slobodkin LB (1960) Community structure, population control, and competition. Am Nat 94:421–425

Harris P (1989) Feeding strategy, coexistence and impact of insects in spotted knapweed capitula. In: Delfosse E (ed) Proc VII Int Symp Biol Contr Weeds, Rome. MAF, Rome, pp 39–47

Hsiao TH (1982) Geographic variation and host plant adaptation in the Colorado potato beetle. In: Visser JH, Minks AK (eds) Proc 5th Int Symp Insect-plant relationships, Pudoc, Wageningen, pp 315–324

Jaenike J (1981) Criteria for ascertaining the existence of host races. Am Nat 117:830–834

Katakura H, Shioi M, Kira Y (1989) Reproductive isolation by host specificity in a pair of phytophagous ladybird beetles. Evolution 43:1045–1053

Komma M (1990) Der Pflanzenparasit *Tephritis conura* und die Wirtspflanzengattung *Cirsium*. PhD Thesis, University of Bayreuth, Bayreuth

Lewontin RC (1984) Detecting population differences in quantitative characters as opposed to gene frequencies. Am Nat 123:115–124

Luna IG, Prokopy RJ (1995) Behavioural differences between hawthorn-origin and apple-origin *Rhagoletis pomonella* flies in patches of host trees. Entomol Exp Appl 74:277–282

McPheron BA, Smith DC, Berlocher SH (1988) Genetic differences between host races of *Rhagoletis pomonella*. Nature 336:64–66
Messina FJ (1989) Host preferences of cherry- and hawthorn-infesting populations of *Rhagoletis pomonella* in Utah. Entomol Exp Appl 53:89–95
Möller-Joop H (1988) Biosystematisch-ökologische Untersuchungen an *Urophora solstitialis* L. (Tephritidae): Wirtskreis, Biotypen und Eignung zur biologischen Bekämpfung von *Carduus acanthoides* L. (Compositae) in Kanada. PhD Thesis, University of Bayreuth, Bayreuth
Nei M (1972) Genetic distances among populations. Am Nat 106:283–292
Payne JA, Berlocher SH (1995) Phenological and electrophoretic evidence for new blueberry-infesting species in the *Rhagoletis pomonella* sibling species complex. Entomol Exp Appl 75:183–187
Prokopy RJ, Diehl SR, Cooley SS (1988) Behavioral evidence for host races in *Rhagoletis pomonella* flies. Oecologia 76:138–147
Romstöck M (1984) Zur geographischen Variabilität des mit *Cirsium heterophyllum* Blütenköpfen assoziierten Phytophagenkomplexes. Verh 10 Int Symp Entomofauna Mitteleur, Budapest, pp 123–127
Romstöck M (1987) *Tephritis conura* Loew (Diptera: Tephritidae) and *Cirsium heterophyllum* (L.)Hill: (Cardueae): Struktur- und Funktionsanalyse eines ökologischen Kleinsystems. PhD Thesis, University of Bayreuth, Bayreuth
Romstöck M, Arnold H (1987) Populationsökologie und Wirtswahl bei *Tephritis conura* Loew-Biotypen (Dipt.: Tephritidae). Zool Anz 219:83–102
Romstöck-Völkl M (1990a) Population dynamics of *Tephritis conura* Loew (Diptera: Tephritidae): determinants of density from three trophic levels. J Anim Ecol 59:251–268
Romstöck-Völkl M (1990b) Host refuges and spatial patterns of parasitism in an endophytic host-parasitoid system. Ecol Entomol 15:321–331
Romstöck-Völkl M, Wissel C (1989) Spatial and seasonal patterns in the egg distribution of *Tephritis conura* Loew (Diptera: Tephritidae). Oikos 55:165–174
Scheidler M (1985) Habitatstrukturpräferenzen, Siedlungsdichte und Beutespektrum der Spinnenfauna an Disteln. Diploma Thesis, University of Bayreuth, Bayreuth
Scriber JM (1983) Evolution of feeding specialisation, physiological efficiency, and host races in selected Papilionidae und Saturniidae. In: Denno RF, McClure MS (eds) Variable plants and herbivores in natural and managed systems. Academic Press, New York, pp 373–412
Seitz A, Komma M (1984) Genetic polymorphism and its ecological background in tephritid populations. In: Wöhrmann K, Loeschke V (eds) Population biology and evolution. Springer, Berlin Heidelberg New York, pp 143–158
Tauber CA, Tauber MJ (1989) Sympatric speciation in insects: perception and perspective. In: Otte D, Endler JA (eds) Speciation and its consequences. Sinauer, Sunderland, pp 307–345
Waring GL, Abrahamson WG, Howard DJ (1990) Genetic differentiation among host-associated populations of the gallmaker *Eurosta solidaginis* (Diptera: Tephritidae). Evolution 44:1648–1655
Weisser WW, Houston AI, Völkl W (1994) Foraging strategies in solitary parasitoids: the trade-off between female and offspring mortality. Evol Ecol 8:587–597
Wood TK, Guttmann SI (1982) Ecological and behavioural basis for reproductive isolation in the sympatric *Enchonopa binotata* complex. Evolution 36:233–242
Zwölfer H (1965) A preliminary list of phytophagous insects attacking wild *Cynarea* species (Compositae) in Europe. Tech Bull Commonw Inst Biol Control 6:81–154
Zwölfer H (1982) Patterns and driving forces in the evolution of plant-insect-systems. In: Visser JH, Minks AK (eds) Proc 5th Int Symp on Insect-plant relationships. Pudoc, Wageningen, pp 287–296
Zwölfer H (1986) Insektenkomplexe an Disteln - ein Modell für die Selbstorganisation ökologischer Kleinsysteme. In: Dress A, Hendrichs H, Küppers G (eds) Selbstorganisation. Die Entstehung von Ordnung in Natur und Gesellschaft. Piper, München, pp 181–217
Zwölfer H (1988) Evolutionary and ecological relationships among the insect fauna of thistles. Annu Rev Entomol 33:103–122

Zwölfer H, Bush G (1984) Sympatrische und parapatrische Artbildung. Z Zool Syst Evolutionsforsch 22:211–233
Zwölfer H, Romstöck-Völkl M (1991) Biotypes and the evolution of niches in phytophagous insects on Cardueae hosts. In: Price PW, Lewinsohn TM, Fernandes GW, Benson WW (eds) Evolutionary ecology of plant-animal interactions: tropical and temperate perspectives. Wiley, New York, pp 487–507

3 Symphagy Among Florivorous Fruit Flies (Diptera: Tephritidae) in Southern California

RICHARD D. GOEDEN

3.1 Introduction

Symphagy, as used in this chapter, is the collective sharing of flower heads of a single host-plant species of Asteraceae by more than one species from one or more genera of nonfrugivorous fruit flies (Diptera: Tephritidae) (Goeden 1987, 1989, 1992, 1993, 1994). The manner in which this symphagy is achieved, e.g., whether by spatial or temporal partitionings of the resource that a host-species' flower heads collectively represent (Zwölfer 1987, 1988; Zwölfer and Arnold-Rinehart 1993), is beyond the scope of this chapter and best examined at the individual tephritid and host plant species level (cf. Zwölfer 1965, 1982, 1987, 1988, 1990; Goeden and Ricker 1986, 1987; Headrick and Goeden 1990; Goeden and Headrick 1991, 1992; Goeden et al. 1994). This chapter describes and analyzes 14 years of data on symphagy among 15 genera of florivorous Tephritidae and 12 tribes of Asteraceae in southern California. This is the first comprehensive analysis of symphagy by a multigeneric grouping of geographically defined, nonfrugivorous, Nearctic fruit flies, and was inspired in part by Helmut Zwölfer's comprehensive, long-term field and laboratory studies of the biology, ecology, and evolution of Tephritidae, and other Palearctic insect associates of asteraceous thistles and knapweeds (Zwölfer 1965, 1982, 1988, 1990). I dedicate this analysis of original data as a token of my admiration, respect, and friendship for Helmut on the occasion of his recent retirement from the Faculty of the University of Bayreuth.

3.2 Flora Analyzed

Since 1980, I have sampled the Asteraceae of southern California for associated Tephritidae, and to a lesser degree the Asteraceae of central and northern California. Southern California in the context of this chapter approximates an area of about 777 000 ha and the southern third of California, USA. California is roughly the size of Italy, with a climate and endemic vegetation largely Mediterranean in type, but also including the unlike climates and floras of the Sonoran and Mojave Deserts and several mountain ranges. The flora of southern California is diverse owing to the wide range of topography and influence

Ecological Studies, Vol. 130
Dettner et al. (eds.) Vertical Food Web Interactions

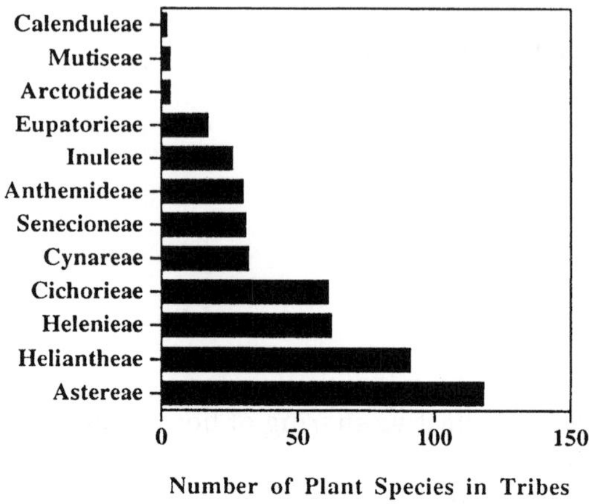

Fig. 3.1. Distribution of plant species among tribes of Asteraceae in southern California

of the Pacific Ocean coast and mountains, and was described by Munz (1974) as comprising 15 distinct types of plant communities.

My database on the florivorous tephritid associates of 489 species and 149 genera of Asteraceae listed by Munz (1974) was begun in 1980, and was updated at the end of each year; consequently, the present work, begun in late 1994, analyzes records for Tephritidae reared during 1980–1993. The 12 tribes that compose the Asteraceae of southern California and their relative sizes are shown in Fig. 3.1. Two tribes, the Arctotideae and Calenduleae, 35 genera, and 80 (16.3%) plant species are adventives in southern California, introduced from elsewhere in North America or from other continents (Robbins 1940; Munz 1974).

3.3 Tephritid Sampling and Rearing Procedures

Approximately 1-l samples of mature flower heads were collected from individual species of Asteraceae at various locations during exploratory collecting trips or as encountered during other field studies. My goal was, and remains, to sample each species of Asteraceae listed by Munz (1974) in southern California at least five times, i.e., at five different locations or at fewer locations over a period of more than 1 year. Heads were bagged along with a whole plant for identification, and a voucher specimen was concurrently pressed if the plant was unrecognized, known to be uncommon, or otherwise thought to be poorly represented among plants previously sampled. Samples were labelled, transported to Riverside in styrofoam cold chests in an air conditioned vehicle, and temporarily stored in a refrigerator at 2–3 °C. After identification of each sampled plant species by a plant systematist, random subsamples of flower heads of hosts of tephritid species under study were dissected, and the remaining flower heads or intact sample were placed in wooden, 35 × 32 × 35 cm, muslin cloth-backed, glass-topped sleeve cages in the insectary of the Department of Entomology, University of California, at 27 ± 1 °C, 14-h photoperiod, and 30–60% RH. Honey (a hygroscopic substance) striped on the underside of the glass top served as a temporary food and water source for the newly emerged flies. Flies were harvested once or twice weekly, 2–3 days after they emerged, and refrigerated until insectary emergence ceased. Next, they were counted, sexed, identified, recorded, and voucher specimens prepared and stored in my research collection. Pressed voucher specimens (ca. 350) of many host-plant species were deposited in the Herbarium of the University of California, Riverside.

3.4 Incidence of Florivorous Tephritidae Among Sampled Asteraceae

Figure 3.2 records the number of times 318 (65% of 489) species and 114 (76.5% of 149) genera of Asteraceae in southern California were sampled as first encountered one to five times during 1980–1993. In addition, many of the more common plant species were sampled more than five times during this period, and 5 genera and 56 species of central and northern California Asteraceae were also sampled, but these data are not included in the present chapter. As this work progresses, fewer new plant species are sampled for the first time each year in southern California, as ever greater proportions of less common species remain unsampled.

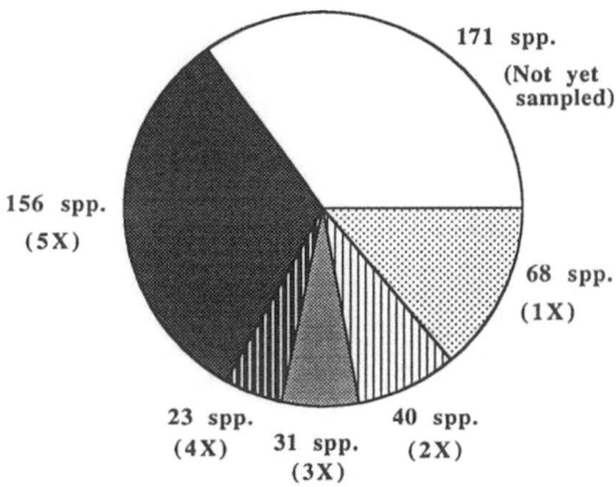

Fig. 3.2. Number of times 489 plant species were sampled during 1980–1993 in southern California

The 318 species of Asteraceae sampled to date (Fig. 3.2) yielded zero to nine species of Tephritidae (Fig. 3.3). Twenty-four percent (75 spp.) yielded no Tephritidae. To date, these plant species appear unattacked and not utilized as a food resource by this most common of all the insect taxa infesting flower heads of Asteraceae in southern California (personal observation). Tephritidae are also the most common insects in the flower heads of Cynareae in Europe (Sobhian and Zwölfer 1985). The Tephritidae is an evolutionarily very young taxon dating back to the Middle Tertiary Period, and most nonfrugivorous species are assocated with Asteraceae, one of the most modern plant families (Zwölfer 1982). Many (110 spp., 35%; Fig. 3.3) plant species sampled in southern California served as host to only a single species of Tephritidae; however, most (113, 41%; Fig. 3.3) plant species were attacked by two to nine species of Tephritidae belonging to one to five genera, which are termed symphagous as defined above. Sobhian and Zwölfer (1985) reported up to 12 phytophagous insect species, including six tephritid species, in flower heads of *Centaurea solstitialis* L. in the Mediterranean Region. Because tephritids have been reared by me from only one (n = 18) or two (n = 19) of 5 samples of 37 different plant species, the 75 species that have been sampled only one to four times (Fig. 3.2), without yielding any flies to date (Fig. 3.3), may yet yield Tephritidae. Because 12.1% of sampled plant species each hosted as many as four to nine tephritid species (Fig. 3.3), but well over half (59.3%) hosted none or only one tephritid species (Fig. 3.3), southern California Asteraceae apparently are not saturated with these phytophagous flies. Therefore, my data clearly support the conclusion recently voiced by Zwölfer and Arnold-Rinehart (1993) that the plant kingdom still offers phytophagous insects many underexploited resources, unattacked taxa, and vacant food niches (Zwölfer 1965; Lawton and Price 1979; Price 1980; Lawton 1984).

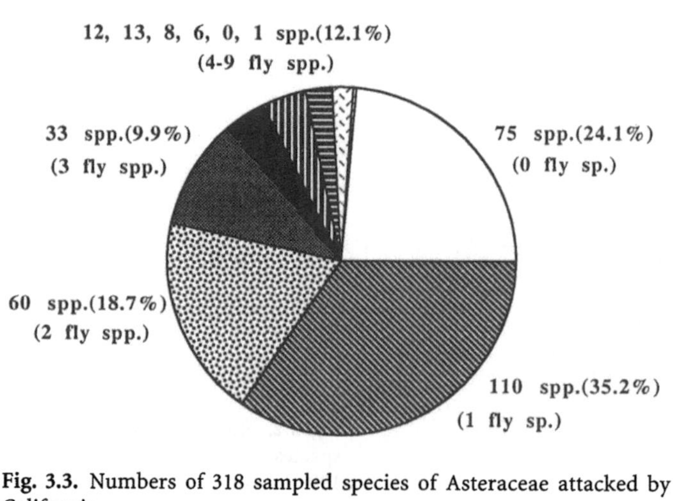

Fig. 3.3. Numbers of 318 sampled species of Asteraceae attacked by 0–9 species in southern California

3.5 Tribal Affinities of Symphagous, Florivorous Tephritidae

The number of plant species that comprise each of the 12 tribes of Asteraceae in southern California showed at least one major deviation from an otherwise direct linear relationship with the number of component plant species lacking Tephritidae ($r^2 = 0.638$; $t = 4.2$, $p = 0.0018$). This deviation involved the tribe Cichorieae (Lactuceae), which has the highest number of tephritid-free plant species in southern California (24 spp.), although it is the fourth largest tribe among the California Asteraceae with 61 spp. (Fig. 3.1). This reduced fruit fly attack probably reflects the better defenses of the Cichorieae against phytophagy, including, perhaps, the milky sap, a triterpene-rich latex, unique to this tribe of Asteraceae (Mabry and Bohlmann 1977). Many florivorous Tephritidae from California are sap-feeders (Goeden et al. 1987, 1994; Headrick and Goeden 1990; Goeden and Headrick 1991, 1992). The other 11 tribes include tephritid-free plant species in direct proportion to their relative numerical sizes in southern California ($r^2 = 0.786$; $t = 5.744$, $p = 0.0003$; Fig. 3.1).

Among those tribes containing species that serve as host plants to only a single species of Tephritidae, the Cichorieae had 17 such host spp., behind both the Heliantheae with 21 spp. and the Astereae with 20 spp.; whereas, the Helenieae had only five species that hosted one species of tephritid (Fig. 3.4), quite disproportionate to its size as third largest tribe in southern California (Fig. 3.1). Regression analysis again showed that the number of plant species in a tribe of Asteraceae infested by a single species of tephritid was significantly and directly related to the size of that tribe in southern California ($r^2 = 0.787$; $t = 6.083$, $p = 0.0001$).

The number of tribes of southern California Asteraceae hosting multiple tephritid species (Fig. 3.5) was significantly and directly related to the numeri-

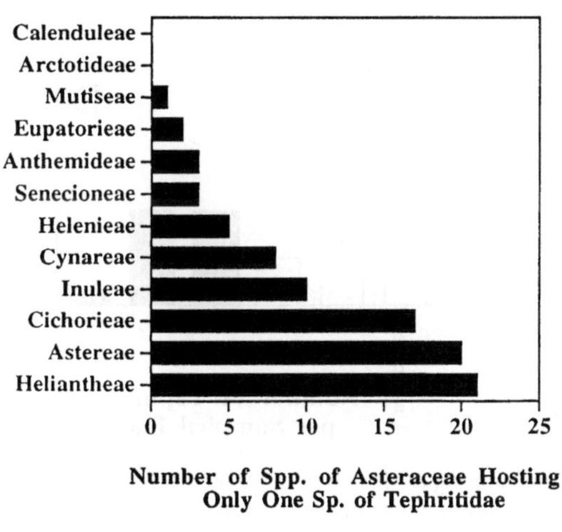

Fig. 3.4. Incidence of species with 12 tribes of Asteraceae hosting only one species of Tephritidae in southern California

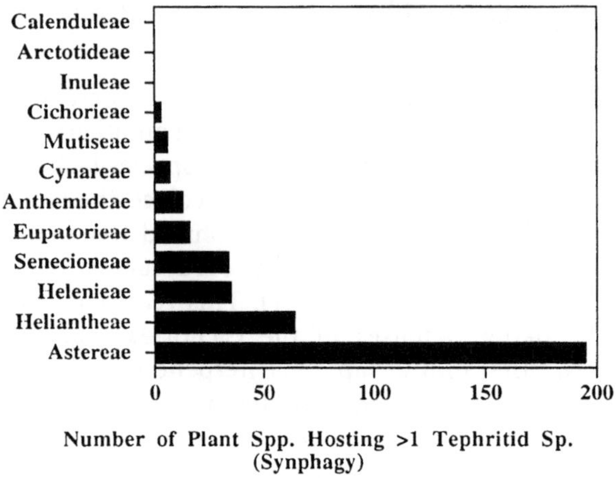

Fig. 3.5. Distribution of symphagy among 12 tribes of Asteraceae in southern California

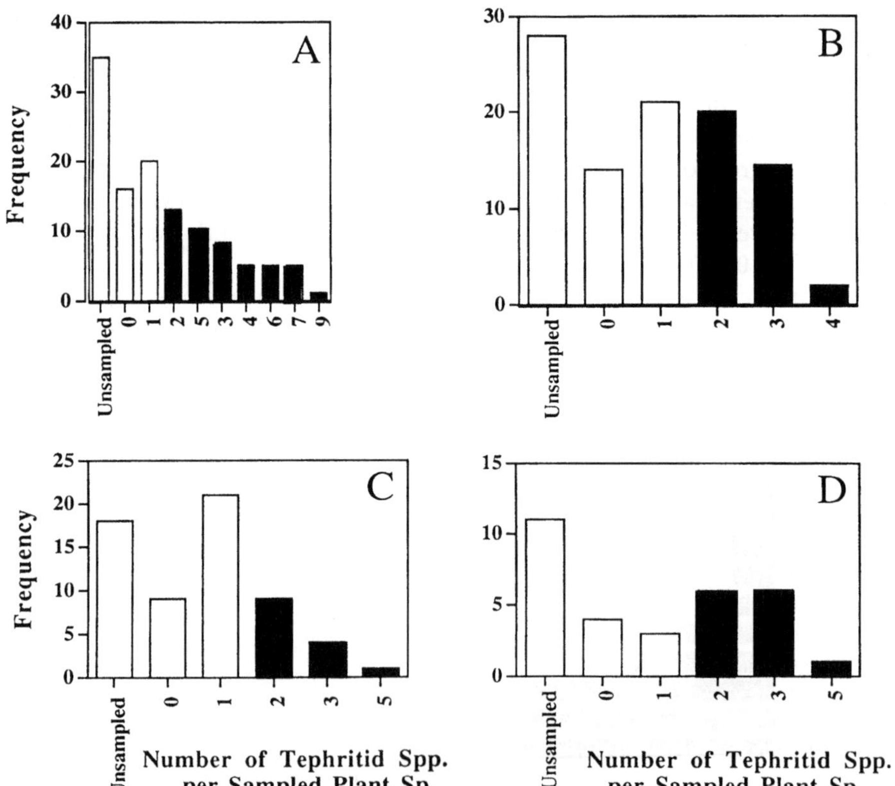

Fig. 3.6A–D. Incidence of symphagy (*dark bars*) among **A** 118 species of Astereae; **B** 91 species of Heliantheae; **C** 62 species of Helenieae; **D** 31 species of Senecioneae in southern California

cal size of these tribes ($r^2 = 0.692$; $t = 4.739$, $p = 0.0008$; Fig. 3.1). However, the Senecioneae contained the fourth largest number of plant species serving as hosts to symphagous Tephritidae (Fig. 3.5), although this is the sixth largest tribe of southern California Asteraceae (Fig. 3.1). The numerical distributions of the host plant species supporting symphagous tephritids among the four largest tribes, the Astereae, Heliantheae, Helenieae, and Senecioneae, are shown in Fig. 3.6. In all but the Helenieae, the numbers of host plants bearing symphagous Tephritidae were greater than the total of those serving as hosts to one or no tephritid species (Fig. 3.6). These empirical data support the relationship demonstrated first by Southwood (1960, 1961) with Hawaiian trees and phytophagous insects, that the abundance of plant species in a geographic area directly determines the richness of their associated insect faunas (Strong et al. 1984). In this regard, Zwölfer (1982) also demonstrated that average species packing, i.e., average number of phytophagous insects species in flower heads of 56 species of European Cardueae thistles, was significantly correlated with the number of species in each of nine host-plant genera.

3.6 Incidence and Coincidence of Genera of Symphagous Tephritidae Within Plant Tribes

The incidence of symphagy among the 15 genera of florivorous Tephritidae and the 12 tribes of Asteraceae in southern California are recorded in Table 3.1. Three tribes yielded no symphagous Tephritidae; the Arctotideae, Calenduleae, and Inuleae. Of these, the first two are adventive and represented by only three and two species, respectively, in southern California. As plant colonizers, they would not be expected to host any species of endophagous phytophages, such as florivorous Tephritidae transferring from other tribes comprising the indigenous flora (Strong et al. 1984; Goeden and Ricker 1986, 1987; Zwölfer 1988). The Inuleae are native to California, but like the Cichorieae noted above, show resistance to tephritid attack by their lack of symphagous tephritid associates (Table 3.1); their main associates are specialized, stenophagous species of *Trupanea* that overcome the otherwise effective defenses of these plants against tephritid attack (Goeden 1992).

Apart from these tribes, symphagous Tephritidae from diverse genera attack the flower heads of all other tribes of southern California Asteraceae to a greater or lesser extent. Among the least widespread genera are native *Urophora* spp., which are restricted to the Astereae (Table 3.1) and, as Goeden (1987) noted, are specific to the subtribe Solidagininae, unlike European *Urophora* spp. that are principally associated with Cynareae (Zwölfer and Arnold-Rinehart 1993). The genus *Neotephritis* is represented in California by the single species, *N. finalis* (Loew), which is restricted to host plants in the tribe Heliantheae, which it shares in symphagy with six other genera of Tephritidae (Goeden et al. 1987). The sole species of *Tomoplagia* in southern

Table 3.1. Incidence of symphagy among florivorous species in 15 genera of Tephritidae and 12 tribes of Asteraceae in southern California during 1980–1993[a]

Genera of Tephritidae	Plant tribes											
	Anthemideae	Arctotideae	Astereae	Calenduleae	Cichoreae	Cynareae	Eupatorieae	Helenieae	Heliantheae	Inuleae	Mutiseae	Senecioneae
Chaetostomella	–	–	–	–	–	2	–	–	–	–	–	–
Dioxyna	–	–	1	–	–	–	–	–	4	–	–	–
Euaresta	–	–	–	–	–	–	–	–	6	–	–	1
Euarestoides	–	–	3	–	–	–	–	2	5	–	–	3
Neaspilota	–	–	33	–	1	–	–	1	–	–	–	–
Neotephritis	–	–	–	–	–	–	–	–	10	–	–	–
Orellia	–	–	–	–	–	2	–	–	–	–	–	–
Paracantha	–	–	–	–	–	3	–	–	1	–	–	–
Campiglossa	1	–	11	–	1	–	–	1	3	–	–	8
Procecidochares	–	–	14	–	–	–	4	–	–	–	–	–
Tephritis	7	–	22	–	–	–	–	1	–	–	–	4
Tomoplagia	–	–	–	–	–	–	–	–	–	–	2	–
Trupanea	5	–	86	–	1	–	12	30	35	–	4	18
Urophora	–	–	25	–	–	–	–	–	–	–	–	–
Xenochaeta	–	–	–	–	–	–	–	–	–	–	–	–

[a] Cumulative total numbers of one to six species in one or more genera of Tephritidae recorded in symphagy from individual species of host plants in each tribe.

California attacks at least two of only three species of Mutiseae extant in southern California (Goeden and Headrick 1991), which incorporates the northwesternmost extension of the distribution of this tribe in North America. According to Jansen and Palmer (1988), this is the most ancient tribe of Asteraceae. The Mutiseae tribe is rich in species, and thus presumably also in florivorous insects (Zwölfer 1982; Jensen and Zwölfer 1994) in the Sonoran Desert regions of Arizona and Mexico (Shreve and Wiggins 1964) and in Central and South America (Aczél 1985).

The evolutionarily young status of the tribe Cynareae in California, and North America, in general, is reflected by the low incidence of symphagy within and among the three genera of Tephritidae, *Chaetostomella*, *Orellia*, and *Paracantha* (Table 3.1), as noted and discussed by Goeden and Ricker (1986, 1987) and Zwölfer (1988). The Cichoreae, too, as demonstrated above, harbor few symphagous Tephritidae in southern California (Fig. 3.5). Here, a single plant species, *Stephanomeria virgata* Bentham, serves as host for all three genera of Tephritidae listed under the Cichorieae in Table 3.1 (Goeden and Blanc 1986; Goeden 1992), including, surprisingly, the specialist stenophages, *Campiglossa sabroskyi* Novak (Goeden and Blanc 1986) and *Neaspilota achilleae* Johnson (Goeden 1989), but neither of the widespread generalists, *Campiglossa genalis* (Thomson) (Goeden et al. 1994) or *N. viridescens* Quisenberry (Goeden and Headrick 1992).

The highest incidence of symphagous Tephritidae occurs in the Astereae, Heliantheae, and Helenieae (Table 3.1, Fig. 3.5), the three largest tribes of Asteraceae, respectively, in southern California. According to Gonzalez (1977), the Astereae and Heliantheae represent the most advanced and most primitive tribes of Asteraceae, respectively [in contrast to the opinion of Jansen and Palmer (1988) cited above, regarding the "most ancient" status of the Mutiseae], confounding any attempt to correlate symphagy with tribal host plant age (Zwölfer 1987). However, the five largest genera of florivorous Tephritidae in southern California in terms of numbers of component species, i.e., *Trupanea*, *Neaspilota*, *Tephritis*, *Campiglossa*, and *Urophora* (Foote et al. 1993), also show the greatest numbers of symphagous associations, i.e., 191, 35, 34, 25, and 25, respectively (Table 3.1). This indicates that, like the European Cynareae, which Zwölfer (1988) suggested provided a platform for species radiation among Tephritidae and other insect taxa, the Astereae, Heliantheae, and Helenieae similarly have provided platforms for species radiation among Tephritidae in California (Table 3.1) and the western USA (Foote et al. 1993). Adoption of a new host plant by transfer from a closely related plant species, followed by reproductive isolation, has been the main mode of speciation among congeneric Tephritidae in Europe (Zwölfer 1982, 1987, 1988). Evidence of this radiation is also provided in Table 3.2, in which 60, 9, 7, 3, and 2 plant species are listed as hosts to congeneric, symphagous Tephritidae in the genera *Trupanea*, *Urophora*, *Tephritis*, *Neaspilota*, and *Campiglossa*, respectively. Indeed, these five largest tephritid genera are the only ones among the 15 florivorous genera analyzed that showed symphagy among congeneric species.

Table 3.2. Number of host-plant species shared by each of 15 genera of florivorous, symphagous Tephritidae in southern California during 1980–1993

	Chaetostomella	*Dioxyna*	*Euaresta*	*Euarestoides*	*Neaspilota*	*Neotephritis*	*Orellia*	*Paracantha*	*Campiglossa*	*Procecidochares*	*Tephritis*	*Tomoplagia*	*Trupanea*	*Urophora*	*Xenochaeta*
Chaetostomella	–	–	–	–	–	–	1	2	–	–	–	–	–	–	–
Dioxyna	–	–	–	–	1	–	–	–	1	1	1	–	4	–	–
Euaresta	–	–	–	5	–	–	–	–	1	–	–	–	3	–	–
Euarestoides	–	–	5	–	–	–	–	–	1	–	2	–	11	–	–
Neaspilota	–	1	–	–	3	–	–	–	10	9	8	–	30	11	–
Neotephritis	–	–	–	–	–	–	–	1	1	–	–	–	10	–	–
Orellia	1	–	–	–	–	–	–	2	–	–	–	–	–	–	–
Paracantha	2	–	–	–	–	1	2	–	–	–	–	–	1	–	–
Campiglossa	–	1	1	1	10	1	–	–	2	4	9	–	22	2	–
Procecidochares	–	1	–	–	9	–	–	–	4	–	7	–	17	4	–
Tephritis	–	1	–	2	8	–	–	–	9	7	7	–	24	5	–
Tomoplagia	–	–	–	–	–	–	–	–	–	–	–	–	2	–	–
Trupanea	–	4	3	11	30	10	–	1	22	17	24	2	60	19	–
Urophora	–	–	–	–	11	–	–	–	2	4	5	–	19	9	–
Xenochaeta	–	–	–	–	–	–	–	–	–	–	–	–	–	–	–

For example, as many as six *Trupanea* spp. utilized the flower heads of each of two different hosts, *Brickellia oblongifolia* Nuttall and *Haplopappus squarrosus* Hooker and Arnott (Goeden 1992). However, 42 (70%) of the 60 plant species attacked by two species of Tephritidae in southern California (Fig. 3.3) each hosted different genera of flies; and 18 (54%), 11 (33%), and 4 (12%) of the 33 plant species attacked by three species of Tephritidae (Fig. 3.3) each hosted two, three, and one genera of flies. Therefore, transfer to new host-plant species apparently largely proceeds independently among tephritid genera.

3.7 Different Levels of Symphagy Among Southern California Tephritidae

Goeden (1992) itemized the numbers of host plants shared by each of 14 species of *Trupanea* in California throughout 1991, which involved the six generalist species in this genus, i.e., *T. actinobola* (Loew), *T. femoralis* (Thomson), *T. jonesi* Curran, *T. nigricornis* (Coquillett), *T. radifera* (Coquillett), and *T. wheeleri* Curran, which shared between 38 and 73% of their reported host-plant species with one or more of five others of these six congeners. *Campiglossa genalis*, the only generalist in its genus in southern California, enters into symphagous associations with a congener in only two species of *Aster* in southern California (Table 3.2; Goeden 1993); otherwise, most of the symphagy listed for *Campiglossa* in Table 3.1 involves *C. genalis* and Tephritidae in other genera. Similarly, the generalist *Neaspilota viridescens* figures in all three symphagous associations with congeners (Table 2, in Goeden 1989), and with noncongeners (Table 3.1). On the other hand, the seven plant species listed in Table 3.2 as hosting congeneric, symphagous *Tephritis* in their flower heads share five generalist species that attack only two or three tribes of Asteraceae (Goeden 1993), not broad generalists, and all five species are variously involved in symphagy with other genera of Tephritidae in the Anthemideae, Astereae, and Senecioneae (Table 3.1).

3.8 Conclusions

Analysis of relations of symphagous, florivorous Tephritidae with flower heads of Asteraceae in southern California showed many parallels with, and provided support for, the seminal findings of Zwölfer and coworkers regarding evolutionary and ecological relations of Tephritidae and the flower heads of Palearctic Cynareae (Cardueae). Symphagy is a common mode of resource sharing among southern California Tephritidae, now recorded from 14 of 15 fruit fly genera and hosts representing nine of 12 tribes of Asteraceae (Tables

3.1 and 3.2). Two tribes lacking hosts of symphagous Tephritidae, the Arctotideae and Calenduleae, are adventive and poorly represented in the southern California flora; whereas native species of Inuleae only serve as hosts to specialized, monophagous, florivorous Tephritidae (Goeden 1992). Individual host-plant species were shown to accommodate as many as nine species of symphagous Tephritidae in their flower heads, although flower heads of Asteraceae as food niches are far from saturated in southern California (Fig. 3.3). The number of host-plant species in tribes of Asteraceae that bear symphagous Tephritidae in their flower heads generally is significantly and directly related to the number of species comprising those tribes. The three largest tribes, Astereae, Heliantheae, Helenieae (Fig. 3.1), bear the largest numbers of symphagous Tephritidae, respectively (Table 3.1, Fig. 3.1); whereas, the Cichorieae and Helenieae, especially the former tribe, contain fewer symphagous Tephritidae in proportion to their sizes in California (Figs. 3.1, 3.5, 3.6C), probably due to their defenses against tephritid attack. Most symphagous associations involve generalist, not stenophagous species of florivorous Tephritidae, primarily belonging to the five largest genera (Tables 3.1 and 3.2). Species packing varied between samples in total numbers and incidences of component tephritids.

Acknowledgements. I thank several colleagues for their helpful criticisms and suggestions for improvement of drafts of my manuscript: Tom Bellows, Louie Blanc, Dan Hare, David Headrick, Andy Sanders, and Jeff Teerink. I and my wife, Joan, also thank Helmut and Uta Zwölfer for the pleasure and privilege of knowing them as friends for so many years.

References

Aczél MI (1955) Fruit flies of the genus *Tomoplagia* Coquillett. Proc US Natl Mus 104:321–411

Foote RH, Blanc FL, Norrbom AL (1993) Handbook of the fruit flies (Diptera: Tephritidae) of America north of Mexico. Cornell University Press, Ithaca

Goeden RD (1987) Host plant relations of native *Urophora* spp. (Diptera: Tephritidae) in southern California. Proc Entomol Soc Wash 89:269–274

Goeden RD (1989) Host plants of *Neaspilota* in California (Diptera: Tephritidae). Proc Entomol Soc Wash 91:164–168

Goeden RD (1992) Analysis of known and new host records for *Trupanea* from California (Diptera: Tephritidae). Proc Entomol Soc Wash 94:107–118

Goeden RD (1993) Analysis of known and new host records for *Tephritis* from California, and description of a new species, *T. joanae* (Diptera: Tephritidae). Proc Entomol Soc Wash 95:425–434

Goeden RD (1994) Analysis of known and new host records for *Paroxyna* from California (Diptera: Tephritidae). Proc Entomol Soc Wash 96:281–287

Goeden RD, Blanc FL (1986) New synonymy, host-, and California records in the genera *Dioxyna* and *Paroxyna*. Pan-Pac Entomol 62:88–90

Goeden RD, Headrick DH (1991) Notes on the biology, hosts, and immature stages of *Tomoplagia cressoni* Aczél in southern California (Diptera: Tephritidae). Proc Entomol Soc Wash 93:549–558H

Goeden RD, Headrick DH (1992) Life history and description of immature stages of *Neaspilota viridescens* Quisenberry (Diptera: Tephritidae) on native Asteraceae in southern California. Proc Entomol Soc Wash 94:59–77

Goeden RD, Ricker DW (1986) Phytophagous insect faunas of the two most common native *Cirsium* thistles, *C. californicum* and *C. proteanum*, in southern California. Ann Entomol Soc Am 79:953–962

Goeden RD, Ricker DW (1987) Phytophagous insect faunas of native *Cirsium* thistles, *C. mohavense*, *C. neomexicanum*, and *C. nidulum*, in the Mojave Desert of southern California. Ann Entomol Soc Am 80:161–175

Goeden RD, Cadatal TD, Cavender GA (1987) Life history of *Neotephritis finalis* (Loew) on native Asteraceae in southern California (Diptera: Tephritidae). Proc Entomol Soc Wash 89:552–558

Goeden RD, Headrick DH, Teerink JA (1994) Life history and description of immature stages of *Paroxyna genalis* (Thomson) (Diptera: Tephritidae) on native Asteraceae in southern California. Proc Entomol Soc Wash 96:612–629

Gonzalez AG (1977) Lactuceae - chemcal review. In: Heywood VH, Harborne JB, Turner BL (eds) The biology and chemistry of the Compositae, vol II. Academic Press, London

Headrick DH, Goeden RD (1990) Resource utilization by larvae of *Paracantha gentilis* (Diptera: Tephritidae) in capitula of *Cirsium californicum* and *C. proteanum* (Asteraceae) in southern California. Proc Entomol Soc Wash 92:512–520

Jansen RK, Palmer JD (1988) Phylogenetic implications of chloroplast DNA restriction site variation in the Mutisieae (Asteraceae). Am J Bot 75:753–766

Jensen U, Zwölfer H (1994) Adjustment of gene flow at the population, species and ecosystem level. In: Schulze ED (ed) Flux control in biological systems: from enzymes to populations and ecosystems. Academic Press, San Diego, pp 447–467

Lawton JH (1984) Non-competitive populations, non-convergent communities, and vacant niches: the herbivores on bracken. In: Strong DR, Simberloff D, Abele LG, Thistle AB (eds) Ecological communities. Conceptual issue and the evidence. Princeton University Press, Princeton, pp 67–99

Lawton JH, Price PW (1979) Species richness of parasites on hosts: agromyzid flies on the British Umbelliferae. J Anim Ecol 48:619–637

Mabry TJ, Bohlmann F (1977) Summary of the chemistry of the Compositae. In: Heywood VH, Harborne JB, Turner BL (eds) The biology and chemistry of the Compositae, vol II. Academic Press, London

Munz PA (1974) A flora of southern California. University of California Press, Berkeley

Price PW (1980) Evolutionary biology of parasites. Princeton University Press, Princeton

Robbins WW (1940) Alien plants growing without cultivation in California. Univ Calif Agric Exp Stn Bull 637:1–128

Shreve F, Wiggins IL (1964) Vegetation and flora of the Sonoran desert. Stanford University Press, Stanford

Sobhian R, Zwölfer H (1985) Phytophagous insect species associated with flower heads of yellow star thistle (*Centaurea solstitialis* L.). Z Angew Entomol 99:301–321

Southwood TRE (1960) The abundance of Hawaiian trees and the number of their associated insect species. Proc Hawaii Entomol Soc 17:299–303

Southwood TRE (1961) The number of species of insects associated with various trees. J Anim Ecol 30:1–8

Strong DR, Lawton JH, Southwood TRE (1984) Insects on plants. Community patterns and mechanisms. Blackwell, Oxford

Zwölfer H (1965) Preliminary list of phytophagous insects attacking wild Cynareae (Compositae) in Europe. Tech Bull Commonw Inst Biol Control 6:81–154

Zwölfer H (1982) Patterns and driving forces in the evolution of plant-insects systems. In: Visser JH, Minks K (eds) Proc 5th Int Symp Insect-plant relationships, Wageningen, Pudoc, Wageningen

Zwölfer H (1987) Species richness, species packing, and evolution in insect-plant systems. Ecological Studies, vol 61. Springer, Berlin Heidelberg New York, pp 301–319

Zwölfer H (1988) Evolutionary and ecological relationships of the insect fauna of thistles. Annu Rev Entomol 33:103–122

Zwölfer H (1990) Disteln und ihre Insektenfauna. In: Streit B (ed) Evolutionsprozesse im Tierreich. Birkhäuser, Basel, pp 255–278

Zwölfer H, Arnold-Rinehart J (1993) The evolution of interactions and diversity in plant-insect systems: The *Urophora-Eurytoma* food web in galls on Palearctic Cardueae. In: Schulze E-D, Mooney HA (eds) Biodiversity and ecosystem function. Ecological studies, vol 99. Springer, Berlin Heidelberg New York, pp 211–233

4 Establishment of *Urophora cardui* (Diptera: Tephritidae) on Canada Thistle, *Cirsium arvense* (Asteraceae), and Colony Development in Relation to Habitat and Parasitoids in Canada

D.P. Peschken and J.L. Derby

4.1 Introduction

Investigations of *the* European *Urophora cardui* (L.) (Diptera: Tephritidae) as a potential biological control agent were initiated three decades ago by Agriculture Canada and the Commonwealth Institute of Biological Control within the Canadian biological control program against introduced thistles (Zwölfer 1965; Peschken 1971). Sparked by this program, the biology of this monophagous stem-gall former on the weed Canada thistle (*Cirsium arvense* (L.) Scop., Asteraceae) was studied extensively. Zwölfer (1967, 1979, 1982, 1988), Zwölfer and Arnold-Rinehart (1993) Zwölfer et al. (1970) and Schlumprecht (1989) reported on its biology, phenology, field host range, parasitism, patchy occurrence within its large general range, and its habitat preferences. Zwölfer et al. (1970) showed that habitat characteristics greatly influence pupation and emergence of adults. Mortality of *U. cardui* larvae from one of its main parasitoids, *Eurytoma robusta* Mayr. (Hymenoptera: Eurytomidae), decreases with an increase in gall size and the number of cells per gall while mortality due to *E. serratulae*, another main parasitoid, is only slightly affected by gall size (Zwölfer and Arnold-Rinehart 1993). The gall is a weak physiological sink when young and rapidly ceases to function as a sink as it ages (Forsyth 1983). In the laboratory, increasing gall size correlated with an increase in the number of main shoots, and with a decrease in plant vigour as measured by several parameters, such as height, and root and shoot fresh weight. Little evidence of a significant reduction in vigour by side-shoot galls was found in the field (Peschken et al. 1982; Forsyth 1983; Peschken and Derby 1992). Galls on the main shoot significantly reduced plant height but the relative frequency of galls on the main shoot is low. Morphology and development of the gall and larvae were investigated in detail be Lalonde and Shorthouse (1982, 1984, 1985) and Shorthouse and Lalonde (1986, 1988). The 20-day gall growth phase is probably the most critical, both in terms of the success of the gall former and stress on the host (Lalonde and Shorthouse 1985). Development of third instar larvae is most rapid during the gall maturation phase and is seriously impaired only if the host is stressed to the point of death and desiccation of nutritive tissue.

Urophora cardui has been released for the biological control of Canada thistle in several countries. It did not establish in New Zealand (Jessep 1989). In

Ecological Studies, Vol. 130
Dettner et al. (eds.) Vertical Food Web Interactions

Montana, USA, the fly failed to become established at two open, relatively dry sites, but survived in low numbers at a moist and well protected site (Story 1985). However, this site was destroyed by housing development (J. Story, pers. comm. 1994). The fly is well established and thriving in Oregon, USA (Coombs 1990; Coombs et al. 1995), and in Washington (Piper and Andres 1995). In Canada, *U. cardui* was released in the provinces of British Columbia, Alberta, Saskatchewan, Ontario, Quebec, New Brunswick (Peschken et al. 1982), Nova Scotia (Sampson and Ingraham 1990), and Prince Edward Island (Diamond et al. 1987). Initially, it became established only in eastern Canada where it is thriving (Laing 1977; Peschken et al. 1982; Forsyth 1983; Sampson and Ingraham 1990; Diamond 1991; M. Betts, P.L. Youwe, pers. comm. 1994).

This chapter reports on the establishment of this gall fly in western Canada (especially Saskatchewan), larval and pupal mortality, habitat preference and resource utilization in the absence of parasitoids in Saskatchewan, and its parasitism in eastern Canada.

4.2 Life Cycle of *Urophora cardui*

Urophora cardui emerges in spring and lays its eggs between the young leaves at the tips of shoots of *C. arvense* in June and July (Peschken and Harris 1975; Peschken et al. 1982). The larvae tunnel into the stem tissue and initiate formation of a multichambered gall. Third instar larvae overwinter. The galls deteriorate during winter and spring and air penetrates the gall which allows the larvae to pupate. Adults exit the gall through callus-filled tunnels which open in spring (Lalonde and shorthouse 1982).

4.3 Release Activities and Experimental Studies in Canada

4.3.1 Release Stock

It was assumed that a population adapted to a colder climate than that of the southern Rhine valley and released into a favourable habitat might succeed in western Canada. *Urophora cardui* galls were obtained from Finland and New Brunswick. The climate in Finland where the fly has recently invaded (Jansson and Lindeberg 1982; Jansson 1992), and in New Brunswick, where *U. cardui* has been established since 1976, is considerably colder than that in the southern Rhine Valley of France and Germany, the source area for the previous release colonies (Walter and Lieth 1960). Winter mortalities of pupae and larvae of *U. cardui* originating from New Brunswick and the Rhine Valley were compared in Saskatchewan during two consecutive winters in 1982/1983 and 1983/1984. In September 1981 and April 1982, *U. cardui* was collected in New Brunswick, and from these collections, one generation was reared in 1982 at the Agriculture Canada Research Station, Regina, Saskatchewan, as part of an unrelated field experiment. Galls produced in this experiment were collected in August 1982 and stored at 4 °C in the laboratory until November. A population was collected in the southern Rhine valley of Germany on 13 November 1982 and airmailed to Regina. On 22 November 1982, galls from both sources were placed outside at the Regina Research Station, in separate screen cages (22 × 27 × 21 cm) and covered lightly with straw. Twenty-five galls from each source were dissected each month in the laboratory until May 1983 inclusive. The exposed larvae were placed in clear plastic petri dishes (diameter 4 cm) at room temperature for pupation and eclosion of adults.

Again in October 1983, galls were collected in New Brunswick, and in the southern Rhine valley of France. On 7 November 1983, galls from both sources were placed outside similarly as described above. Twenty-five galls from New Brunswick and 70 galls from the Rhine Valley were dissected each month until March 1984. More galls from the Rhine Valley were dissected to compensate for the high number of parasitized larvae. The uncovered larvae were treated as above. These winter survival data were analyzed with Genstat 5 software (Lawes Agricultural Trust 1987) using maximum likelihood to estimate parameters of a generalized linear model (McCullagh and Nelder 1989).

The survival of larvae and pupae from a Finnish population was also tested during one winter in Saskatchewan. On 5 November 1985, galls of *U. cardui* which had been collected at Helsinki, Finland in the fall of 1985, were placed outside similarly as described above. Twelve galls were retrieved in November 1985, and in January to April 1986. The uncovered larvae were treated as above. No galls from the Rhine Valley were available for comparison in 1986.

4.3.2 Releases and Population Development in Saskatchewan

The release site (30 × 25 m) (Fig. 4.1a) consisted of a dense thistle patch (15 × 14 m) surrounded by scattered thistles and trees, 60 m south of the shore of Pasqua Lake in Echo Valley Provincial Park, Saskatchewan. From 6 June to 3 July 1984, 3042 flies from galls collected in New Brunswick, and from 11 to 17 June 1986, 287 flies from galls collected in Finland were released. Galls were counted annually in the fall from 1984 to 1994 (except in 1991 and 1992) in three defined areas: the release area (Fig. 4.1a), along a 4-m-wide trail bordered by trees and sloping east for about 600 m to the shore of Lake Pasqua (Fig. 4.1b), and along the shore for 500 m (Fig. 4.1c). In 1993, the extent of the spread of the galls was determined.

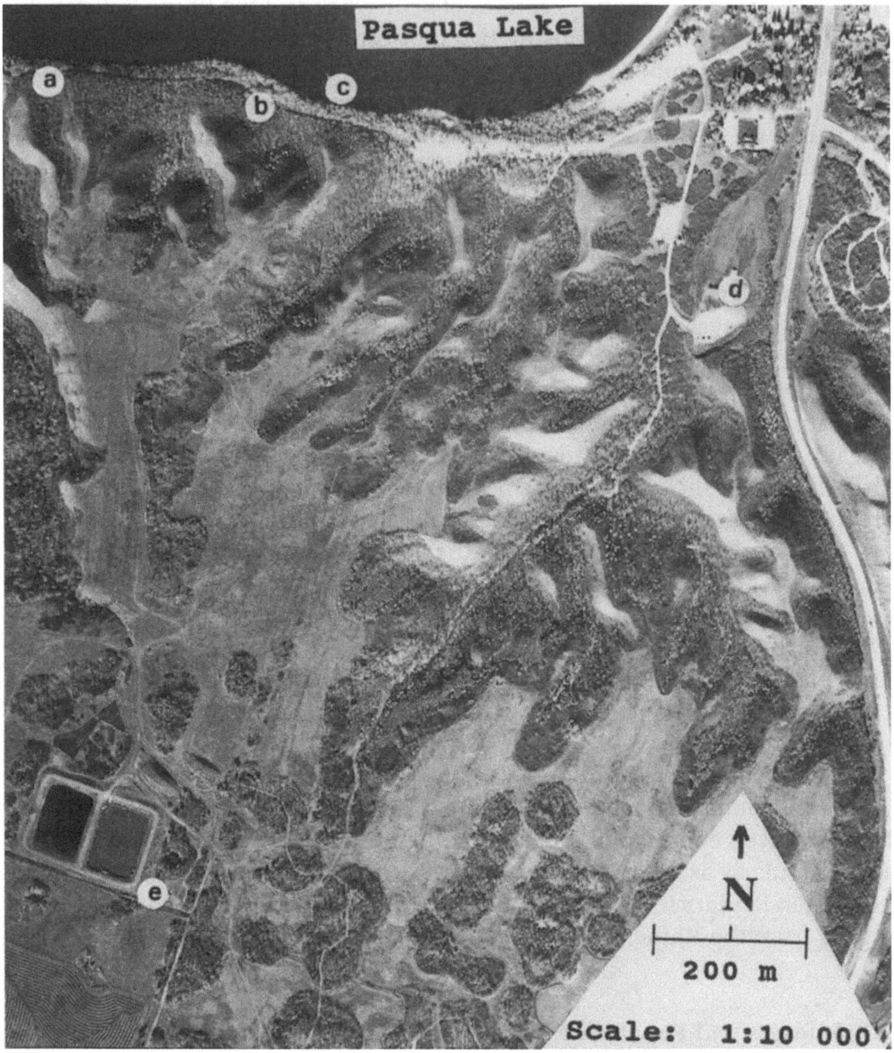

Fig. 4.1. Aerial photograph of the release area: *a* Release site; *b* along narrow trail; *c* along the shore; *d* upland habitat; *e* near small lake and pond

4.3.3 Habitat and Mortality in Saskatchewan

Density of galls was measured in habitats differing in distance to trees and to the lake shore. In 1990, density of galls m^{-2} and the number of galls per attacked stem, were determined at the release site (Fig. 4.1a) and in an area within 5 m of the shore (Fig. 4.1c). Thistles were dense (>20 stems m^{-2}) on both sites. Five 1-m square samples were taken along each of four transects through the thistle patch of the release site (Fig. 4.1a). In an area measuring 4

× 23 m within 5 m of the lake shore, all galls were counted. The number of galls per attacked thistle stem was counted on 20 thistle shoots on both sites and on a third site 120 m south of the lake shore. In 1993, a portion of the lake shore habitat was flooded, but a comparable site along the lane (Fig. 4.1b), was similarly compared to the release site (Fig. 4.1a). In 1993 and 1994, the density of galls was measured in two dense thistle patches in the same upland, dry meadow (Fig. 4.1d), 400 m from the lake shore. The two thistle patches paralleled a grove of trees, one growing 3 to 5 m to the east of the trees, the other 33 m away. In each patch, the galls were counted in 10 1-m squares along each of two transects. Also in 1994, the density of galls was measured similarly in two adjacent but different habitats (Fig. 4.1e). The first was a low, moist area with scattered trees between two small man-made lakes and a pond, the second, an adjacent drier site, starting 30 m away from any trees.

Eclosion of flies was determined for 7 years between 1985 and 1994. A total of 346 galls was collected throughout the survey area in fall to determine summer mortality, and in the spring to determine the total of summer and winter mortality. The galls were dissected, larvae and pupae were placed into petri dishes, and eclosed flies were counted as described in Section 4.3.1.

Summer and winter survival of the immature stages from a total of 60 galls collected at the original release site (Fig. 4.1a) and a total of 72 galls collected at the shore (Fig. 4.1c) were compared for 3 years from 1989/1990 to 1991/1992. These data were also analyzed with Genstat 5 software (Lawes Agricultural Trust 1987) using maximum likelihood to estimate parameters of a generalized linear model (McCullagh and Nelder 1989).

Lalonde and Shorthouse (1982) suggested that "Partial breakdown of the callus plug due to insufficient moisture leads to death of adults or interrupts synchronized pupation and the appearance of adults." Therefore, we collected 80 galls from the release site, along the lane and the lake shore, between 13 June and 12 July 1985, when many adults had emerged. These galls were dissected to determine the number of dead adults.

4.3.4 Recent Releases in British Columbia

Releases of *U. cardui* made since 1982 are shown in Table 4.1. Release sites were moist. The flies originated from collections made in Finland and Echo Valley Provincial Park, Saskatchewan.

4.3.5 Parasitism in Eastern Canada

A colony of *U. cardui*, established in 1975 with stock from the southern Rhine Valley, thrived near Belwood, Ontario. In the fall of 1988, 58 galls, and in 1989, 96 galls were collected and dissected to check for parasitoids. Near Sussex

Table 4.1. Releases of *Urophora cardui* since 1982 in British Columbia

Locality	Year	No. of adults released	Origin of release stock	Year recovered
Vancouver Island	1987	1324	Finland	1987–1988
Kamloops (near Paul Lake)	1991	400[a]	Echo Valley Provincial Park, Saskatchewan	Not recovered
Vancouver	1990	400[a]	Echo Valley Provincial Park, Saskatchewan	1990–1994

[a] 200 galls were released which were estimated to yield 400 flies, based on dissected samples.

Corner in New Brunswick where *U. cardui* was established in 1976, a total of 723 galls, which were produced in the years 1978 and 1980 to 1985, was collected. These galls were dissected and flies and parasitoids allowed to emerge, as described earlier.

4.4 Results

4.4.1 Release Stock

In 1982/1983 and in 1983/1984, the average percentage of adult flies which emerged from unparasitized *U. cardui* larvae of European origin was significantly lower in every month than from larvae of New Brunswick origin (Table 4.2; $p < 0.01$). As winter progressed, the differences became significantly greater ($p < 0.01$). In February, only 23% of the European larvae developed to adults as compared to 85% of those from New Brunswick ($p < 0.1$). There were no parasitoids in the New Brunswick population in 1982/1983, because it had been reared through one generation in Saskatchewan, and it was parasitized at the level of 6% (range 1–8%) in 1983/1984. The European population was parasitized at the average level of 76% (range 69–91%) over 1982/1983 and 1983/1984. Percentage survival of the Finnish larvae was over 90% throughout fall, winter and spring. Parasitism in this population averaged 16% (range 8–35%).

4.4.2 Releases and Population Development in Saskatchewan

In August 1984, 1035 galls were counted, all on the release site (Figs. 4.1a, 4.2) and an adjacent treed area. In 1985, the population of galls decreased sharply to 156 but the colony had spread along a narrow lane with intermittent Canada thistle, leading to the lake shore, 650 m from the release site. The total gall

Table 4.2. Mean percentage emergence (±SE) of *Urophora cardui* flies from unparasitized larvae from galls originating from the southern Rhine Valley in Europe and from New Brunswick, Canada, during the winters of 1982/1983 and 1983/1984, and from Finland in 1985/1986, in Saskatchewan

Month	Emerged flies, % (No. of larvae) (±SE of percent)		
	New Brunswick 1982/1983 and 1983/1984	Rhine Valley 1982/1983 and 1983/1984	Finland 1985/1986
Nov.	79 (306) (±2)	68 (86) (±5)	97 (74) (±2)
Dec.	82 (296) (±2)	51 (78) (±6)	-
Jan.	72 (321) (±2)	68 (70) (±5)	94 (59) (±3)
Feb.	85 (263) (±2)	23 (81) (±5)	91 (56) (±4)
March	76 (285) (±3)	39 (69) (±7)	93 (70) (±3)
April	84 (84) (±4)	34 (12) (±14)	91 (49) (±4)
May	75 (121) (±4)	44 (9) (±16)	-

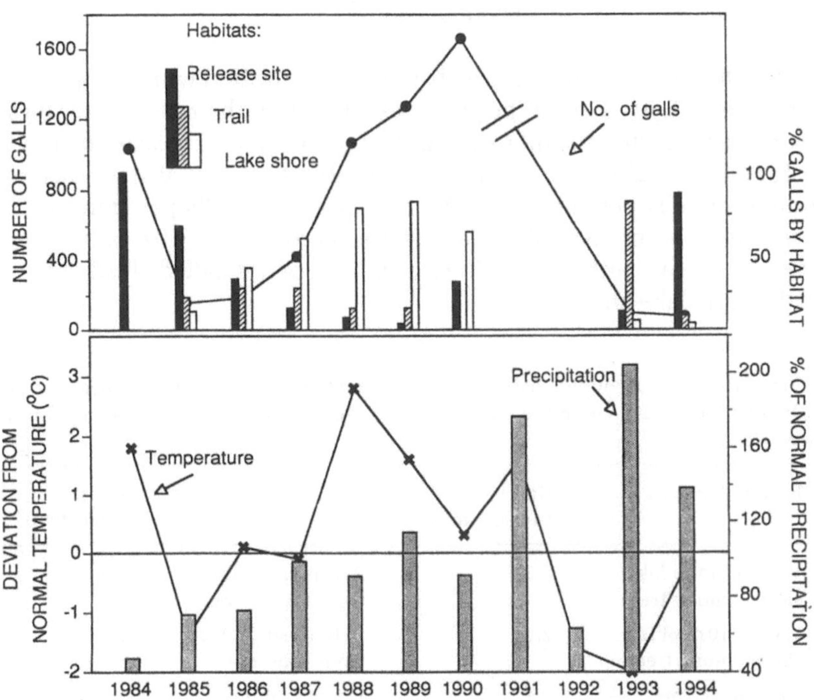

Fig. 4.2. Development of the colony of *Urophora cardui* in three habitats near Pasqua Lake, Saskatchewan, from 1984, the year of the release, to 1994, in relation to deviations from the normal of the average monthly temperatures (°C) and percent of normal precipitations for June, July and August. (NB In 1990, most galls along the trail were destroyed by mowing in the course of trail maintenance)

count peaked in 1990 with 1661 galls. However, in 1993 and 1994, the gall count crashed down to 6 and 5%, respectively, of the 1990 peak (Fig. 4.2). While some of the thistles along the lake shore were flooded due to the unusually high rainfall in 1993, this does not account for the steep population decline. Temperatures during June, July and August, when flies were active and galls matured, were above normal from 1986 to 1990, while rainfall was below normal except in 1989 (Fig. 4.2; Environment Canada Weather Station, Regina, Saskatchewan). The summers of 1992, 1993 and 1994 were characterized by below normal temperatures, and above normal rainfall occurred during 1991, 1993 and 1994. Eclosion of flies from galls collected in fall averaged 83% (range 54–92%), and from those collected in spring averaged 37% (range 21–53%). Peaks in total mortality occurred in galls collected in the spring of 1985 (76%) and 1988 (79%). The spring of 1985 was the driest, and that of 1988 was the hottest of those in which mortality was determined (Fig. 4.2). No parasitoids were found in any year. By 1993, the galls had spread 4 km along the lake shore and 2.5 km inland.

4.4.3 Habitat and Mortality in Saskatchewan

In 1989 and 1990, 82 and 63% of the galls were concentrated within 5 m of the lake shore while in 1994, 87% of the galls were found on the release site (Figs. 4.1c, 4.2). Comparing sites distant to and near trees or water, consistently more galls were found on sites near trees and water (Table 4.3). This also is reflected in the number of galls per stem. In 1990, there were 5.6 galls per stem on 20 galled thistle stems within 5 m of the lake shore, as compared with only 3.2 on the release site (60 m south of the shore), and 1.9 on 20 galled thistle stems 120 m south of the lake shore.

Table 4.3. Comparison of gall count in habitats near and far from the lake shore, and near and far from trees. In 1990 most galls along the trail were destroyed by mowing in the course of trail maintenance

Year	Favourable sites	Gall (density m^{-2})	Unfavourable sites	Gall (density m^{-2})
1990	Within 5 m of lake shore, among trees	6.2	Release site, 60 m from lake shore	2.2
1994	Within 10 m of lake shore among trees	2.4	Release site, 60 m from lake shore	0
1993	Upland, 4–5 m from trees	0.9	Upland, 33 m from trees	0
1994	Same site as above	0.4	Same site as above	0
1994	Moist, treed site, between lake and pond	6.0	30 m from trees	0.1

Analysis of the pooled data showed that significantly more flies emerged from the larvae in the galls which were collected along the shore (82%) than 60 m away from the shore (61%) during the 3 years of the study ($p < 0.01$; Table 4.4). There was a significant interaction between the source and the month of collection of the galls ($p < 0.01$). The mean percentage of emergence from larvae in the galls collected in April along the shore (81%) was significantly higher than that from galls collected in the same month at the release site (28%; $p < 0.01$). However, there was no significant difference in emergence from galls collected in the two locations in September (84 vs. 94%). There also was no significant difference in the number of cells per gall (larvae, pupae and empty cells) between the habitats on the lake (6.3 ± 0.4; n = 499 cells) and 60 m south of the lake (5.6 ± 0.4, n = 436 cells, $p > 0.05$).

In the 80 galls collected from 13 June to 12 July 1985, 61 flies out of a total of 82 healthy pupae had emerged from the galls. No adults were found stuck in the exit tunnels of the galls.

4.4.4 Recent Releases in British Columbia

Urophora cardui failed to establish on Vancouver Island and near Kamloops (M. Betts and P. Youwe, pers. comm. 1995; Table 4.1). However, the colony at Vancouver is thriving with 6 galls found in 1992, 100 in 1993 and 500 in 1994. This site is flat and moist with numerous drainage ditches and trees (D. Ralph, pers. comm. 1995).

4.4.5 Parasitism in Eastern Canada

No parasitoids of *U. cardui* were found in Ontario. In New Brunswick, average total mortality over the seven years when galls were collected was 33%; with 5% of this due to parasitism. By far the most prevalent parasitoid was *Pteromalus*

Table 4.4. Mean percentage emergence (±SE) of *Urophora cardui* flies from galls collected in September and the following April over 3 years (1989/1990 to 1991/1992) at locations 5 and 60 m from Pasqua Lake[a]

Month	Distance to Lake	
	5 m	60 m
September	84 (±2) a	94 (±2) a
April	81 (±3) a	28 (±3) b
September and April pooled	82 (±2) a	61 (±2) b

[a] Means followed by different letters in the same row are different at the 1% level.

elevatus (Walker) (Hymenoptera: Pteromalidae) (94%). In addition, the following parasitoids and/or inquilines were found: six specimens of *Macroneura vesicularis* (Retzius) [syn. *Eupelmella vesicularis* (Retzius)] (Hymemoptera: Eupelmidae) in 1983; one specimen of *Torymus* sp. (Hymenoptera: Torymidae) was found in each of 1982, 1984 and 1985; two of *Eurytoma* sp. (Hymenoptera: Eurytomidae) in 1985, one each of *Tetrastichus* sp. (Hymenoptera: Eulophidae), *Pachyneuron* sp. (Hymenoptera: Pachyneuridae), *Spintherus* sp. (a European genus; identification not certain) (Hymenoptera: Pteromalidae), and *Anthomyza* sp. (Diptera: Anthomyzidae) in 1982; and one *Philonthus* sp. (Coleoptera: Staphylinidae) in 1983.

4.5 Discussion

4.5.1 Release Stock

Urophora cardui became significantly more winter hardy in the seven generations since its release in New Brunswick. Although no European galls were available in 1985/1986 for comparison, it may be assumed that the Finnish larvae, of which over 90% developed into flies, were even more winter hardy than the New Brunswick stock.

Parasitism did not influence winter survival of the unparasitized larvae. This assumption is supported by the fact that survival of the larvae is not correlated with the level of parasitism: the level of parasitism in the Rhine Valley population was highest, but the survival of unparasitized larvae was lowest, and while the level of parasitism of the Finnish population was three times higher than of the New Brunswick population, survival was higher in the Finnish population.

4.5.2 Releases, Population Development and Habitat

The favourable habitat and relatively winter hardy release stock may have combined to achieve establishment in Echo Valley Park. It may be speculated that the release of the Finnish population in 1986 made possible the steady increase in the number of galls until 1990. However, the steep decline of the number of galls, which originated from the release of flies from New Brunswick, from 1035 in 1984 to 156 in 1985, may also have been caused by below average temperatures and precipitation (Fig. 4.2). We believe that in the dry, warm years of 1986 to 1990, *U. cardui* adults found their most favourable habitat along the lake shore which faced north. However, this habitat became too cool during 1992 to 1994, and the release site again became the favoured habitat. These flies are inactive below 20 °C and their activity peaks at 30–34 °C

(Zwölfer et al. 1970). Most flies emerge during June and July (Peschken et al. 1982). From 1992 to 1994, 30 °C was reached on an average of only 1.7 days in June, and 1.3 days in July, while from 1984 to 1990, 30 °C was reached on an average of 5 days in June and 6.3 days in July (Environment Canada Weather Station at Regina, Saskatchewan). Thus, we suggest that in the last 3 years temperature limited fly activity, and that the release area in Saskatchewan is at the margin of climatic tolerance for *U. cardui.*

This may also explain establishment of *U. cardui* near Vancouver, British Columbia, which is slightly warmer than the other two sites in British Columbia (Environment Canada Weather Stations at Victoria, Vancouver and Hefly Creek, British Columbia).

In Europe, *U. cardui* is highly parasitized (average 61% in one area of the southern Rhine Valley; see also Sect. 4.3.1), prefers moist, partially shaded habitats, and predominately colonizes river valleys, giving it a patchy distribution (Zwölfer 1982; Zwölfer et al. 1970). In Saskatchewan, where parasitoids were absent and few galls damaged by predators, consistently more galls were found near trees and on moist sites. Similar observation were made in Oregon, Montana, and Washington (Story 1985; Piper and Andres 1995; E.M. Coombs, pers. comm. 1995). Thus, the patchy European distribution is due to intrinsic factors in the biology of *U. cardui.*

At what stage in its development is *U. cardui* susceptible to dry conditions? In laboratory experiments, Rotheray (1986) found that even though the callus cells of the exit channels had sufficiently decayed, 61% of 59 flies were trapped with their heads against the outer epidermis of the galls which had been kept dry. Even in galls which were kept moist, 23% of 61 flies were trapped. However, we did not find any adults stuck in the escape tunnels in our field collected galls in Saskatchewan, which indicates that callus plugs and the outer epidermis had deteriorated sufficiently to allow escape from the gall. Our data show that larval development and their ability to overwinter may be impaired on dry sites. Swelling of the gall and the amount of nutrient tissue developed in it for larval development are dependent on turgor pressure (Harris 1991). This may be reduced on dry sites by a combination of the high potential for transpiration from the plant's many leaves, and from the stomata of the galls which are stretched and unable to close. Thus, a moist habitat seems to promote larval development in summer resulting in significantly higher winter survival and/or pupation (Lalonde and Shorthouse 1982). This may also explain why almost all failed releases of adult flies produced galls in the first year and colonies declined rapidly in subsequent years (Peschken et al. 1982): the adult flies were able to produce galls but the larvae succumbed during the critical periods of winter and/or pupation in spring under dry conditions.

Urophora cardui is a stenotopic insect which is adapted to warm, moist sites where it builds up dense populations and may exert some pressure on Canada thistle (for a definition of "stenotop" see Schwerdtfeger 1963). There are strong genetic differences between individual subpopulations (Seitz and Komma

1984; Seitz 1995). Thus, races from different habitats may be needed to obtain establishment where it has failed so far.

4.5.3 Parasitism in Eastern Canada

The main parasitoid in New Brunswick, *Pteromalus elevatus*, did not reach detrimental levels. This accidental introduction to North America was first reported from Newfoundland. It attacks tephritid larvae and their parasitoids on certain Asteraceae (Graham 1969). The two main European parasitoids, *Eurytoma serratulae* and *E. robusta* were not found.

4.6 Conclusions

Urophora cardui (L.) (Diptera: Tephritidae) was released in Canada to aid in the biological control of Canada thistle [*Cirsium arvense* (L.) Scop., Asteraceae]. Release stock from the southern Rhine valley in France and Germany became established only in eastern Canada. Populations from New Brunswick and Finland proved to be more winter hardy and became established in Saskatchewan and British Columbia in western Canada after release into a favourable habitat. In Saskatchewan, in years with above normal temperatures and below normal rainfall during June, July and August, the population thrived in a moist but cool habitat. However, in years with below normal temperatures and above normal precipitation, the population declined sharply. Emergence of flies from galls collected in a moist habitat was significantly higher than from those collected in a drier habitat. Parasitism of the flies in New Brunswick averaged only 5%, with *Pteromalus elevatus* (Walker) (Hymenoptera: Pteromalidae) being most prevalent (95%). No parasitoids were found in Ontario or Saskatchewan.

Resource utilization by *U. cardui* is influenced by abiotic factors and the interaction between host plant and gall former, not by parasitoids. Both in Europe, where the larvae are highly parasitized, and in Canada, where levels of parasitism are zero or low, the fly thrives only in moist, warm habitats, especially near bodies of water. The weak physiological sink caused by the larvae draws only slightly on the plant's resources. It may be argued that insects with such narrow ecological requirements should not be used in biological control, but that is hindsight. Being obviously very host specific, and with very few other specialized insects on Canada thistle known at that time (and still today), *U. cardui* was worth a try.

Acknowledgements. We gratefully acknowledge technical assistance by K. Sawchyn in surveys, gall dissection and rearing of flies and parasitoids, and collection of galls in the Rhine Valley by D. Schroeder and H. Zwölfer, in New Brunswick by D.B. Finnamore, and in Finland by A. Jansson. We also thank G. Gibson and J. Huber for identifying the parasitoids, and R. DeClerck-Floate, J.

Hume and J. Soroka for making many useful suggestions and correcting many errors in an earlier draft of the manuscript.

References

Coombs EM (1990) Biological control of weeds. Status Report, Oregon Dep Agric, 12 pp

Coombs EM, Isaacson DL, Hawkes RB (1995) The status of biological control of weeds in Oregon. In: Delfosse ES, Scott RR (eds) Proc 8th Int Symp Biol Contr Weeds, 2–7 Feb 1992, Lincoln University, Canterbury, New Zealand, pp 463–471

Diamond JF (1991) Integrated control of *Cirsium arvense* (L.) Scop. in pastures. MSc Thesis, Plant Science Dep, Macdonald College of McGill University, Ste-Anne-de-Bellevue, PQ

Diamond JF, Sampson MG, Watson AK (1987) Biological control of *Cirsium arvense* (L.) Scop. (Canada thistle) in Prince Edward Island. Res Rep, vol 2. Expert Committee of Weeds, Toronto, Canada, 766 pp

Forsyth SF (1983) Stress physiology and biological control: a case study with Canada thistle (*Cirsium arvense* (L.) Scop.). PhD Thesis, Plant Science Dep, Macdonald College of McGill University, Ste-Anne-de-Bellevue, PQ

Graham MWR de Verre (1969) The Pteromalidae of north-western Europe (Hymenoptera: Chalcidoidea). Bull Br Mus (Nat Hist) Entomol Suppl 16, 352 pp

Harris P (1991) Invitation paper (C.P. Alexander Fund): classical biocontrol of weeds: its definition, selection of effective agents, and administrative-political problems. Can Entomol 123:827–849

Jansson A (1992) Distribution and dispersal of *Urophora cardui* (Diptera: Tephritidae) in Finland in 1985–1991. Entomol Fenn 2:211–216

Jansson A, Lindeberg B (1982) A spectacular Tephritid fly (Diptera) new to Finland. Notulae Entomol 62:151

Jessep CT (1989) *Cirsium arvense* (L.) Scopoli, Californian thistle (Asteraceae). In: Cameron PJ, Hill RL, Bain J, Thomas WP (eds) A review of biological control of invertebrate pests and weeds in New Zealand 1874 to 1987. CAB Int Inst Biol Control Tech Commun 10:343–345

Laing JE (1977) Establishment of *Urophora cardui* L. (Diptera: Tephritidae) on Canada thistle in southern Ontario. Proc Entomol Soc Ont 108:2

Lalonde RG, Shorthouse JD (1982) Exit strategy of *Urophora cardui* (Diptera: Tephritidae) from its gall on Canada thistle. Can Entomol 114:873–878

Lalonde RG, Shorthouse JD (1984) Developmental morphology of the gall of *Urophora cardui* (Diptera, Tephritidae) in the stems of Canada thistle (*Cirsium arvense*). Can J Bot 62:1372–1384

Lalonde RG, Shorthouse JD (1985) Growth and development of larvae and galls of *Urophora cardui* (Diptera, Tephritidae) on *Cirsium arvense* (Compositae). Oecologia 65:161–165

Lawes Agricultural Trust (1987) Genstat 5 reference manual. Genstat 5 Committee, Statistics Department, Rothamstead Experiment Station, Harpenden, Hertfordshire. Clarendon Press, Oxford

McCullagh P, Nelder JA (1989) Generalized linear models. 2nd edn. Chapman and Hall, London

Peschken DP (1971) *Cirsium arvense* (L.) Scop., Canada thistle. In: Biological control programmes against insects and weeds in Canada 1959–1968. Commonw Inst Biol Control Tech Commun 4:79–83

Peschken DP, Derby JL (1992) Effect of *Urophora cardui* L. (Diptera: Tephritidae) and *Ceutorhynchus litura* F. (Coleoptera: Curculionidae) on the weed Canada thistle, *Cirsium arvense* (L.) Scop. Can Entomol 124:145–150

Peschken DP, Harris P (1975) Host specificity and biology of *Urophora cardui* (Diptera: Tephritidae), a biocontrol agent for Canada thistle (*Cirsium arvense*) in Canada. Can Entomol 107:1101–1110

Peschken DP, Finnamore DB, Watson AK (1982) Biocontrol of the weed Canada thistle (*Cirsium arvense*): releases and development of the gall fly *Urophora cardui* (Diptera: Tephritidae) in Canada. Can Entomol 114:349–357

Piper GL, Andres LA (1995) Canada thistle, *Cirsium arvense* (L.) Scop., In: Nechols JA, Andres LA, Beardsley JW, Goeden RD, Jackson CG (eds) Biological control in the U.S. western region: accomplishments and benefits of regional research project W-84 (1964–1989). Univ California, Div Agric and Natural Res Publ, pp 233–236

Rotheray GE (1986) Effect of moisture on emergence of *Urophora cardui* L. (Diptera: Tephritidae) from its gall on *Cirsium arvense* (L.). Entomol Gaz 37:41–44

Sampson MG, Ingraham A (1990) Biological control of weeds in Nova Scotia. Final Report, Project TD38, Nova Scotia Dep Agric and Marketing, 53 pp

Schlumprecht H (1989) Dispersal of the thistle gallfly *Urophora cardui* and its endoparasitoid *Eurytoma serratulae* (Hymenoptera: Eurytomidae). Ecol Entomol 14:341–348

Schwerdtfeger F (1963) Ökologie der Tiere. Autökologie. Paul Parey, Hamburg

Seitz A (1995) Gene flow and the genetic structure of populations of Central European animal species. Verh Dtsch Zool Ges 88:61–76

Seitz A, Komma M (1984) Genetic polymorphism and its ecological background in tephritid populations (Diptera: Tephritidae). In: Wöhrmann K, Loeschke V (eds) Population biology and evolution. Springer, Berlin Heidelberg New York, pp 143–158

Shorthouse JD, Lalonde RG (1986) Formation of flowerhead galls by the Canada thistle gall-fly, *Urophora cardui* (Diptera: Tephritidae), under cage conditions. Can Entomol 118:1199–1203

Shorthouse JD, Lalonde RG (1988) Role of *Urophora cardui* (L.) (Diptera, Tephritidae) in growth and development of its galls on stems of Canada thistle. Can Entomol 20:639–646

Story JM (1985) Status of biological weed control in Montana. In: Delfosse ES (ed) Proc 6th Int Symp Biol Contr Weeds, 19–25 Aug 1984, Vancouver, Canada, pp 837–842

Walter H, Lieth H (1960) Klimadiagramm-Weltatlas. Fischer, Jena

Zwölfer H (1965) Preliminary list of phytophagous insects attacking wild Cynareae (Compositae) species in Europe. Commonw Inst Biol Control Tech Bull 6:81–154

Zwölfer H (1967) Observations on *Urophora cardui* L. (Trypetidae). Weed projects for Canada. Progress Report XIX, Commonw Inst Biol Control, Delémont, 11 pp

Zwölfer H (1979) Strategies and counterstrategies in insect population systems competing for space and food in flower heads and plant galls. Fortschr Zool 25:331–353

Zwölfer H (1982) Das Verbreitungsareal der Bohrfliege *Urophora cardui* L. (Dipt.: Tephritidae) als Hinweis auf die ursprünglichen Habitate der Ackerdistel (*Cirsium arvense* (L.) Scop.). Verh Dtsch Zool Ges 75:298

Zwölfer H (1988) Evolutionary and ecological relationships of the insect fauna of thistles. Annu Rev Entomol 33:103–122

Zwölfer H, Arnold-Rinehart J (1993) The evolution of interactions and diversity in plant-insect systems. The *Urophora-Eurytoma* food web in galls on palearctic Cardueae. Ecological Studies, vol 99. Springer, Berlin Heidelberg New York, pp 211–233

Zwölfer H, Englert W, Pattullo W (1970) Investigations on the biology, population ecology and the distribution of *Urophora cardui* L. Weed projects for Canada. Progress Report XXVII, Commonw Inst Biol Control, Delémont, 17 pp

Part B

Host-Parasite Interactions

5 Host Relationships at Reversed Generation Times: *Margaritifera* (Bivalvia) and Salmonids

G. Bauer

5.1 Introduction

Generation time is supposed to be an important factor for the evolution of host-parasite relationships. The shorter an organism's generation time, the faster its potential rate of evolution. Thus, short-lived hosts (the generation time is equal to or only slightly longer than that of the parasite) can be thought as being involved in an arms race where new host defences select for new parasite offences and vice versa. On the other hand, long-lived hosts are unlikely to match the faster evolving attack possibilities of parasites. Therefore the evolution of "covenants" may account for the continued persistence of such host-parasite systems (Freeland 1986).

But how about host-parasite systems in which the generation time of the parasite is longer? Such systems exist among the Unionoidea; freshwater mussels whose larvae (glochidia) parasitize on gills or fins of fish. During the parasitic phase, which lasts from 2–3 weeks to up to 10 months (Bauer 1994), the glochidia are completely encysted by host tissue, implying a close physical intimacy. An extreme example for long generation times occurs in freshwater pearl mussels (Margaritiferidae), whose glochidia are gill parasites. This group contains species with the longest life span of invertebrates on earth (Comfort 1957; Hutchinson 1979). The maximum life span of *M. margaritifera* of up to 140 years (Grundelius 1987) considerably exceeds the life span of most fish species. Thus, one might expect a selective disadvantage for the parasite as its long generation time should lead to low rates of evolution compared to its host. On the other hand, there are strong indications for considerable evolutionary success, such as the high phylogenetic age of the genus *Margaritifera*, its large distribution areas and the high population densities of the three *Margaritifera* species sensu stricto (Smith 1980; Chesney et al. 1993). Therefore their host-parasite systems are analyzed here with respect to evolutionary mechanisms and outcomes.

5.2 The Three *Margaritifera* Species Sensu Stricto

M. margaritifera occurs on both sides of the northern Atlantic (Fig. 5.1). *M. falcta* and *M. laevis* are distributed around the northern Pacific. According to

Ecological Studies, Vol. 130
Dettner et al. (eds.) Vertical Food Web Interactions

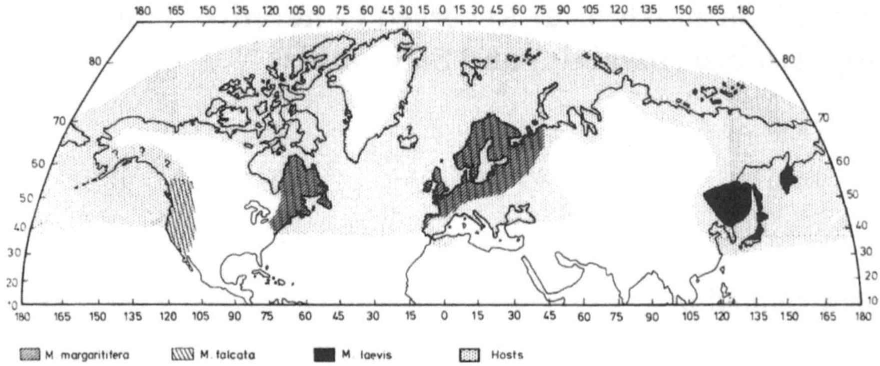

Fig. 5.1. Approximate distribution area of the three *Margaritifera* species sensu stricto and natural distribution of their hosts. (Taylor and Uyeno 1966; Scott and Crossman 1973; Smith 1976; Clarke 1981; Banaresku 1992; Ziuganov et al. 1994)

Taylor and Uyeno (1966), they are more closely related to each other than to *M. margaritifera*. These three species live, with very few exceptions, in fast flowing streams which are poor in lime and nutrients (Murphy 1942; Bauer et al. 1992).

5.3 Characteristics of the Host-Parasite Relationship

5.3.1 The Parasitic Stage

Throughout its distribution area, only one spawning season per year is reported for *Margaritifera* (Murphy 1942; Awakura 1968; Karna and Millemann 1978; Smith 1978; Young and Williams 1984a; Bauer 1987a; Ziuganov et al. 1994).

Glochidial morphology as described for *M. margaritifera* (Harms 1907; Young and Williams 1984b), *M. falcata* (Murphy 1942; Karna and Millemann 1978) and *M. laevis* (Awakura 1968), seems to be identical. Glochidia of *Margaritifera* are amongst the smallest within the Unionoidea (Bauer 1994). They require a parasitic phase of around 1500 day-degrees to complete development into the young mussel. Whereas glochidia of most mussel species metamorphose on the host without any significant growth being involved (Surber 1912), the length of *Margaritifera* glochidia increases more than five times during parasitism. There is little if any size difference in the parasitic stage at the beginning and completion of metamorphosis either:

1. within the distribution area of a species (see in particular *M. margaritifera*) or
2. among the three species (Table 5.1).

Table 5.1. Size of glochidia and size of young mussels when leaving the host fish. (After Murphy 1942; Awakura 1968; Smith 1976; Karna and Millemann 1978; Young and Williams 1984b; Cunjak and McGladdery 1990; Buddensieck 1991; Ziuganov et al. 1994 and own data)

Species	Locality	Length of:	
		Glochidia (mm)	Young mussel (mm)
M. margaritifera	Germany	0.07	0.4–0.5
	Scotland	0.07	0.33–0.48
	Karelia	0.05–0.08	0.4–0.5
	Massachusetts	0.06	0.38
	Nova Scotia	?	0.38
M. falcata	Oregon	0.07	0.36
	California	0.05–0.06	0.39–0.42
M. laevis	Hokkaido	0.07–0.09	0.42–0.44

5.3.2 Host Range and Host Specificity

Glochidia of the three species develop exclusively on members of the subfamily Salmoninae (Fig. 5.2). The genus *Salvelinus* has a circumpolar distribution, but its susceptibility has only been tested for *M. margaritifera*. The genus *Hucho* is distributed in Europe and Asia, but its susceptibility for *Margaritifera* is only known for the European *Hucho hucho*.

Apparently, the fish species serving as hosts in the Atlantic region are resistant to pearl mussel glochidia from the Pacific region and vice versa (Fig. 5.2). For example, *Oncorhynchus thsawytscha* and *O. keta* are excellent hosts for *M. falcata* or *M. laevis*, respectively. However, when they were experimentally infected with glochidia of *M. margaritifera* the parasites were sloughed within 5 days. Conversely, the Atlantic salmon (*Salmo salar*) and the brook trout (*Salvelinus fontinalis*) are resistant to *M. falcata* (Murphy 1942; Meyers and Millemann 1977). Within their respective host ranges, however, there is little if any evidence for phylogenetic relationships (Fig. 5.3). For example, *M. falcata* develops on the comparatively primitive *O. clarki* as well as on the highly derived *O. nerka* and *M. margaritifera* may even use *H. hucho* which splits very early from the *Salmo* lineage.

5.3.3 Relation Between the Generation Time of Parasite and Host

In northern Bavaria the life span of *M. margaritifera* is between 60 and 110 years (Bauer 1992), which contrasts sharply with that of its host, the brown trout, which attains a maximum age of not much more than 4 years (Hänfling

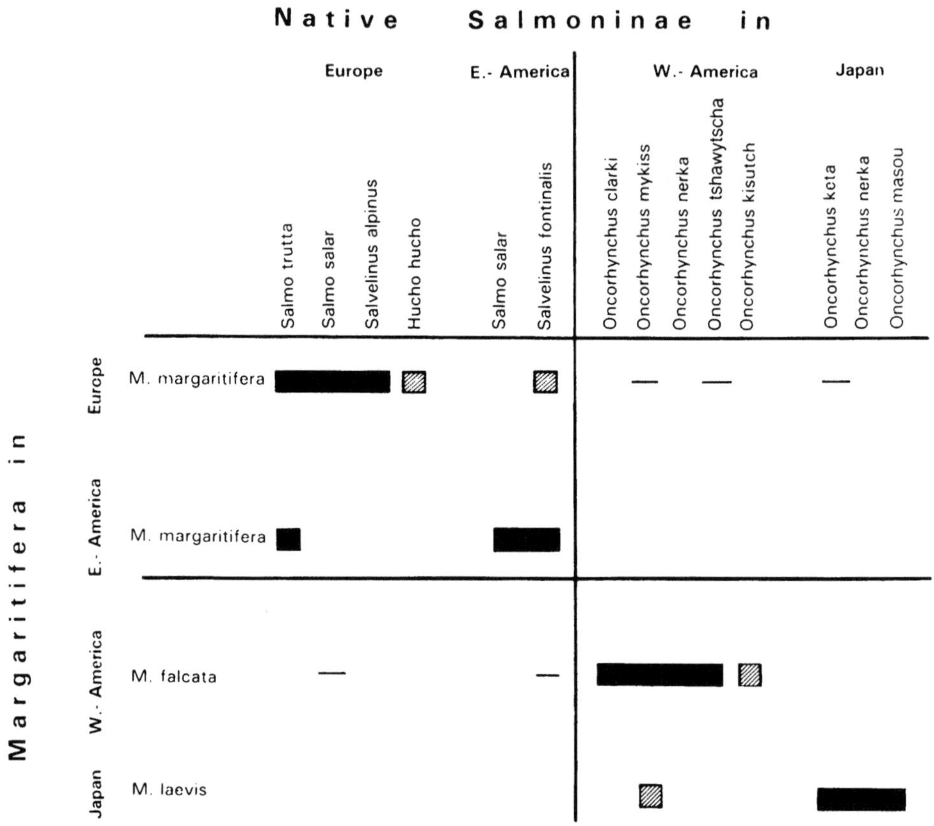

Fig. 5.2. The available data on the host range of *M. margaritifera*, *M. falcata* and *M. laevis*. *Black bars* highly susceptible host; *hatched bars* less susceptible host (some glochidia complete development, but many are sloughed); *minus signs* completely resistant; *no symbols* no data available. (Data from Murphy 1942; Awakura 1968; Utermark 1973; Smith 1976; Meyers and Millemann 1977; Bauer 1987c; Cunjak and McGladdery 1990 and own recent investigations on the relationship *M. margaritifera*, *H. hucho*, *O. tshawythscha* and *O. keta*)

1993). Further to the north, the Atlantic salmon becomes increasingly important as a host and the relationship is 40–140 years parasite life span (Grundelius 1987; Ziuganov et al. 1994) to 6–8 years for the life span of the host (Beverton and Holt 1959). If the northernmost mussel populations utilize mainly the Arctic charr (*Salvelinus alpinus*), the relationship could be around 100–130 years parasite life span to 20–30 years for the Arctic charr (Beverton and Holt 1959). (It is not known whether such populations exist.)

The *Oncorhynchus* species attain an age of 3–7 years (Scott and Crossmann 1973), compared to a life span of 60+ years for *M. falcata* (Clarke 1981) and 30–40 years for *M. laevis* (Ziuganov et al. 1994).

As far as we know, the survivorship curve of adult pearl mussels is convex (Bauer 1987a), indicating not only a high maximum life span but also a high

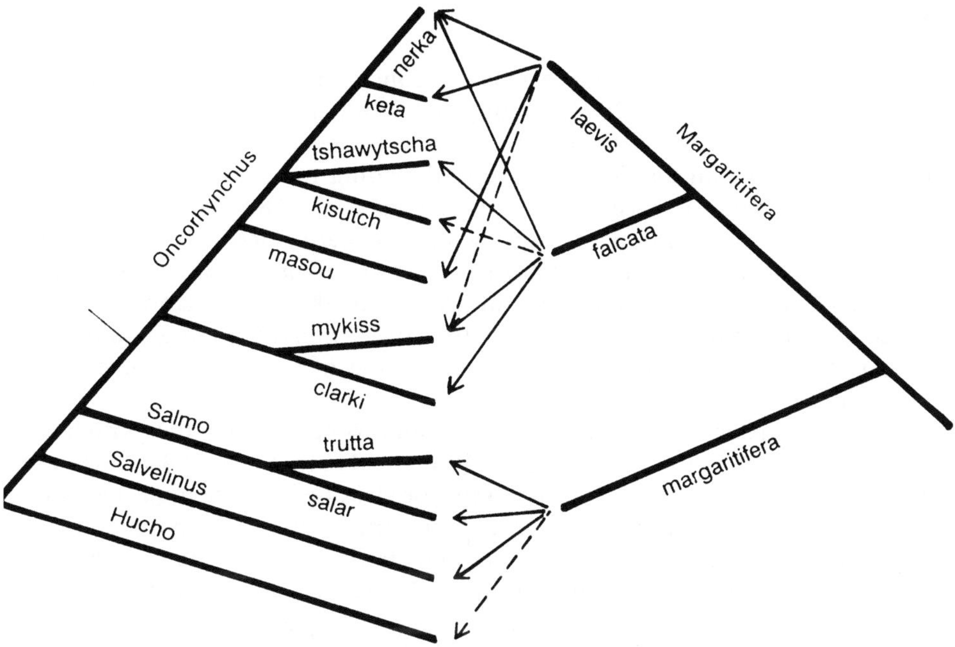

Fig. 5.3. Phylogenetic relationships within the Salmoninae hosts (Smith and Stearley 1989) and their *Margaritifera* parasites. *Arrows* Parasitic relations; *dashed arrows* point to less susceptible hosts

life expectancy. The available data suggest a juvenile period of around 10 years (Young and Williams 1984a). After this, pearl mussels reproduce every 1 or 2 years until they die (Bauer 1987a). Thus, it seems reasonable to assume that in most populations the generation time of *Margaritifera* exceeds that of its host by a factor of between 10 and 20.

5.4 The Framework for the Evolution of the Host-Parasite Relationship

5.4.1 Condition 1: Mode of Host Infection and Parasitic Association

There is no active host selection among the Unionoidea (Kat 1984). *Margaritifera* glochidia reach their host passively in the water current and if they attach to unsusceptible fish species they die (Murphy 1942; Awakura 1968; Young and Williams 1984b; Bauer 1987b). Furthermore they are obligate gill parasites, which means that they are in close contact with the circulatory system and thus to the immune system of the host (Neves et al. 1985).

5.4.2 Condition 2: Mutual Impact

Even on susceptible hosts there may be considerable glochidial mortality (Fustish and Millemann 1978; Young and Williams 1984b; Bauer 1987b) which must be attributed to the host response. Besides a nonspecific tissue response, specific antibodies are produced leading to acquired immunity of the host after two or three infections (Meyers et al. 1980; Bauer and Vogel 1987). On the other hand, parasitism by *Margaritifera* glochidia can affect the fitness of its host. Experimentally, lethal infection intensities can be generated (Murphy 1942; Meyers and Millemann 1977). Cunjak and McGladdery (1990) suggest an increasingly detrimental impact of *M. margaritifera* glochidia on young-of-the-year Atlantic salmon as a function of time and the degree of infestation.

However, the dense populations of salmonids occurring together with large *Margaritifera* populations (Young and Williams 1984b; Bauer 1988; Ziuganov et al. 1994) suggest that *Margaritifera* must be classified as a "benign" parasite.

A further factor limiting the impact of the parasite on its host populations is its smaller distribution area (Fig. 5.1) and its degree of habitat specificity, which is much higher than that of its hosts. Thus, even within the distribution area of *Margaritifera* its occurrence is highly patchy and comparatively few populations of the hosts are in contact with the parasite (only those occurring in rivers extremely poor in nutrients and lime).

In summary, the selection pressures exerted between host and parasite are highly asymmetric. The parasite completely depends on the native Salmoninae, but the impact on its hosts seems to be rather low.

5.4.3 Condition 3: Phylogenetic History

The salmonids probably originated during the Tertiary (Frost and Brown 1967). *Hucho*, which split early from the *Salmo* lineage (Fig. 5.3), remained landlocked in the Old World. *Salvelinus* developed anadromous forms, preferred low temperatures and attained a circumpolar distribution. The genus *Salmo* at first remained restricted to the Atlantic basin, as it preferred higher temperatures. At the end of the Tertiary there was a comprehensive connection between the cold-temperate fauna of the Atlantic and Pacific. *Salmo* reached the Pacific where it underwent considerable radiation leading to its sister group. *Oncorhynchus* now includes eight anadromous species and inland forms (Naeve 1958; Smith and Stearley 1989).

The life-history data of the three *Margaritifera* species considered here suggest that they are monophyletic. In particular, the close adaptation to the Salmoninae, the identical tiny glochidia and large young mussels (Table 5.1) are features which must have been inherited from a common ancestor. According to the phylogeny of the Salmoninae (Fig. 5.3) and the distribution of *Margaritifera* (Fig. 5.1), *M. margaritifera* should be the oldest species. It must have originated before the Atlantic split the American from the European populations, i.e. before the early Tertiary (60 million years ago) when the last connection via the Thule Bridge existed.

The present distribution of the three margaritiferids is in the southern range of the salmonid's distribution area (Fig. 5.1), indicating a temperature preference comparable to *Salmo* (and not *Salvelinus*). Therefore *Margaritifera* presumably did not reach the Pacific before the end of the Tertiary, i.e. not before *Salmo* invaded the Pacific basin. Here, it developed *M. falcata* and spread along the west coast of North America. When it was closed it crossed the Bering bridge, invaded Asia and developed into *M. laevis*.

5.4.4 Condition 4: Rates of Evolution

The rate of evolution is high among the Salmoninae. Most of the present species are supposed to have originated postglacially (Naeve 1958; Behnke 1972). In spite of this short period of time, a great variety of life histories evolved. The phylogenetic history supports a stepwise transition from freshwater forms (like *Hucho*) to anadromy and to increasing loss of freshwater stages (as in *O. gorbuscha*, Scott and Crossmann 1973). However, isolated trends in the other direction also occurred in most groups (Smith and Stearley 1989).

On the other hand, the rates of evolution are extremely low among the Margaritiferidae. They are the most ancient and anatomically conservative family of the Unionoidea (Chesney et al. 1993). The genus *Margaritifera* probably originated during the upper Cretaceous period and underwent very little radiation since then. Less than ten recent species are currently recognized. Thus, the relation between the generation times of *Margaritifera* and salmonids is reflected in their rates of evolution.

5.5 Evolutionary Mechanisms

Neither model of the evolution of host-parasite systems accounts for the salmonid-*Margaritifera* relationship. Arguing against coevolution or diffuse coevolution is the highly asymmetric selection pressures between host and parasite. This situation is similar to the "sequential evolution" model for plant-insect systems (Jermy 1984). If one considers Atlantic and Pacific Salmoninae as two lineages, then this model seems to be valid: the pearl mussels follow these two lineages without major evolutionary feedback, leading to one Atlantic and two closely related Pacific species. However, the diversity of the hosts' life histories does not serve as a major factor in mussel evolution (as required for the sequential evolution model). This is particularly evident at the hosts' species level: the new niches which are rapidly provided by the hosts do not result in radiation of the parasite. Presumably, the evolution of this host-parasite relationship is shaped by two opposing forces.

5.5.1 Selective Pressure for a Host Range as Broad as Possible

The passive mode of host infection (condition 1) strongly selects for a host range encompassing as many individual fish as possible in pearl mussel rivers, a habitat where the fish fauna is dominated by various salmonids (Awakura 1968; Karna and Millemann 1978; Young and Williams 1984a; Cunjak and McGladdery 1990; Bauer et al. 1992).

The parasite's impact on the hosts is apparently very weak (condition 2). The hosts evolve quickly and continuously, independent of the parasite, which is not able to follow their evolution (condition 4). Therefore any *Margaritifera* specialized on one particular host species is in danger of "losing" its host during evolution. This danger vanishes when it is independent of its hosts' evolution, a status which may be attained by adapting to a pool of Salmoninae species living as parr in pearl mussel rivers. Accordingly, the phylogenetic relationships, as suggested in Fig. 5.3 (old parasite species - old host species and vice versa) can be attributed to the phylogenetic history in two different distribution areas as outlined in condition 3: *M. margaritifera* adapted to the pool of Atlantic species and the closely related *M. laevis* and *M. falcata* to the pool of Pacific species.

5.5.2 Selective Pressure for a High Degree of Adaptation

The margaritiferids, which are "evolutionarily sluggish", depend on hosts which evolve quickly (conditions 3, 4) due to selection pressures not related to their *Margaritifera* parasites (condition 2). Thus, the mussels would be at a considerable disadvantage against their hosts, should the fight between host and parasite escalate, especially as the glochidia have to cope with a well-developed host response (conditions 1, 2). Therefore the parasite is forced to adapt biochemically as closely as possible to its host, in order to minimize the effectiveness of the hosts' defence mechanisms. There are some indications that masking may be an important process in this context (Arey 1932; Baer 1951). It seems reasonable that such a highly specific adaptation can only be achieved for a few related host species to which the protective mechanisms of the glochidia could adapt.

This close adaptation is evident in three features of the *Margaritifera*-salmonid relationship:

1. The tiny glochidia indicate a narrow host range since within the Unionoidea there is a positive correlation between the size of the host range and glochidial size. Also, the considerable growth of glochidia during parasitism may be related to the high degree of adaptation since larger

glochidia, which are able to exploit a larger host range usually do not grow (Bauer 1994).
2. The degree of host specificity is very high. As shown in Fig. 5.2, Salmoninae native in the Atlantic basin are unsusceptible in the Pacific basin and vice versa.
3. Whereas many life-history traits of *Margaritifera* are considerably plastic (Bauer 1992), the traits closely associated with the host relationship (glochidial and young mussel size, Salmoninae hosts) are strongly canalized. They appear identical in a large and disjunct distribution area (Table 5.1), indicating that they are of fundamental importance for fitness and therefore under strong selective control (Berven and Gill 1983; Smith-Gill 1983).

Besides the positive relationship between the size of the host range and glochidial size, there is a trade-off between size and number of glochidia among the Unionoidea (Bauer 1994). Thus, the narrow host range of pearl mussels allows a reduction in glochidial size and therefore a considerable increase in fertility.

5.6 Evolutionary Outcome

The compromise between these opposing selection pressures results in a parasite whose host relationship seems at the first sight paradoxical.

On the one hand, it is highly specific as the host range is exclusively restricted to the native Salmoninae (Fig. 5.2). Furthermore, glochidial size is at the lowest boundary of the size spectrum among the Unionoidea (Bauer 1994), indicating that it may have reached a "design barrier" (Stearns 1977). Thus, an even higher degree of adaptation to the hosts could not be accompanied by a further reduction in glochidial size and consequently a further increase in fertility.

On the other hand, the parasite is unspecific in that it is adapted to most native Salmoninae, largely independently of their phylogeny (Fig. 5.3).

This suggests that, on the one hand, the degree of adaptation is always stabilized on the same optimum (or even maximum). On the other hand, the niche dimension "host range" is held constant by adapting to the pool of native Salmoninae species. Both factors might explain the constancy and persistence of the host-parasite relationship in time (more than 60 million years) and space.

Acknowledgements. This chapter is dedicated to my honoured teacher Helmut Zwölfer. I am grateful to the members of Tierökologie I, University of Bayreuth, for the excellent atmosphere, to Dr. T. Petney for valuable comments, to Mrs. A. Servant-Miosga, who did the drawings, and to the Fachberatung für Fischerei, Oberfranken, for providing fish.

References

Arey LB (1932) The formation and structure of the glochidial cyst. Biol Bull 53:217–227
Awakura T (1968) The ecology of parasitic glochidia of the freshwater pearl mussel *Margaritifera laevis*. Sci Rep Hokkaido Fish Hatchery 23:1–17
Baer JG (1951) The ecology of animal parasites. Univ Illinois Press, Urbana
Banaresku (1992) Distribution and dispersal of freshwater animals in North America and Eurasia. Aula, Wiesbaden
Bauer G (1987a) Reproductive strategy of the freshwater pearl mussel. J Anim Ecol 56:691–704
Bauer G (1987b) The parasitic stage of the freshwater pearl mussel. II. Susceptibility of brown trout. Arch Hydrobiol 76:393–402
Bauer G (1987c) The parasitic stage of the freshwater pearl mussel. III. Host relationships. Arch Hydrobiol 76: 413–423
Bauer G (1988) Threats to the freshwater pearl mussel in central Europe. Biol Conserv 45:239–253
Bauer G (1992) Variation in the life span and size of the freshwater pearl mussel. J Anim Ecol 61:425–436
Bauer G (1994) The adaptive value of offspring size among freshwater mussels (Bivalvia; Unionoidea). J Anim Ecol 63:933–944
Bauer G, Vogel C (1987) The parasitic stage of the freshwater pearl mussel. I. Host response to glochidiosis. Arch Hydrobiol 76:393–402
Bauer G, Höchwald S, Silkenat W (1992) Spatial distribution of freshwater mussels. The role of host fish and metabolic rate. Freshwater Biol 26:377–386
Behuke RJ (1972) The systematics of salmonid fishes of recently glaciated lakes. J Fish Res Board Can 29:639–677
Berven KA, Gill DE (1983) Interpreting geographic variation in life history traits. Am Zool 23:85–97
Beverton RJH, Holt SJ (1959) A review of the life spans and mortality rates of fish in nature, and their relation to growth and other physiological characteristics. CIBA Found Colloq Ageing 5:142–180
Buddensiek V (1991) Untersuchungen zu den Aufwuchsbedingungen der Flußperlmuschel *Margaritifera margaritifera* in ihrer früher postparasitären Phase. PhD Thesis, Tierärztliche Hochschule, Hannover
Chesney HCG, Oliver PG, Davis GM (1993) *Margaritifera durrovensis*: taxonomic status, ecology and conservation. J Conchol Lond 34:267–299
Clarke AH (1981) The freshwater molluscs of the Canadian Interior Basin. Malacologia 13:1–495
Comfort A (1957) The duration of life in molluscs. Proc Malacol Soc Lond 32:219–241
Cunjak RA, McGladdery SE (1990) The parsite-host relationship of glochidia on the gills of young-of-the-year Atlantic salmon. Can J Zool 69:353–358
Freeland B (1986) Arms races and covenants: the evolution of parasite communities. In: Kikkawa J, Anderson DJ (eds) Community ecology: pattern and process. Blackwell, Oxford, pp 289–303
Frost WE, Brown ME (1967) The trout. The New Naturalist. Collins, London
Fustish CA, Millemann RE (1978) Glochidiosis of salmonid fishes II. J Parasitol 64:155–157
Grundelius E (1987) Flodpaerlmusslans Tillbakagang i Dalarna. Information fran Soetvattenslaboratoriet Drottningholm 4
Hänfling B (1993) Wanderverhalten, Wachstumsparameter und Mortalität der Bachforelle *Salmo trutta*. Ms Thesis, University of Bayreuth, Bayreuth
Harms W (1907) Zur Biologie und Entwicklungsgeschichte der Flußperlmuschel. Zool Anz 31:814–824
Hutchinson GE (1979) An introduction to population ecology. Yale University Press, London
Jermy T (1984) Evolution of insect/host plant relationships. Am Nat 124:609–630
Karna DW, Millemann RE (1978) Glochidiosis of salmonid fishes III. J Parasitol 63:728–733
Kat PW (1984) Parasitism and the Unionacea. J Parasitol 59:189–207

Meyers TR, Millemann RE (1977) Glochidiosis of Salmonid fishes I. Comparative susceptibility to experimental infection with *Margaritifera margaritifera.* J Parasitol 63:728–733
Meyers TR, Millemann RE, Fustish CA (1980) Glochidiosis of Salmonid fishes IV. Humoral and tissue response of Coho and Chinook salmon to experimental infection with *Margaritifera margaritifera.* J Parasitol 66:274–281
Murphy G (1942) Relationship of the freshwater pearl mussel to trout in the Truckee River. Calif Fish Game 28:89–102
Naeve F (1958) The origin and speciation of *Oncorhynchus.* Trans R Soc Can 52:25–39
Neves RJ, Weaver LR, Zale AV (1985) An evaluation of host fish suitability for glochidia of *Villosa vanuxemi* and *V. nebulosa.* Am Midl Nat 113:13–19
Scott WB, Crossman EJ (1973) Freshwater fishes of Canada. Fisheries Research Board of Canada, Ottawa
Smith DG (1976) Notes on the biology of *Margaritifera margaritifera* in central Massachusetts. Am Midl Nat 1:252–256
Smith DG (1978) Biannual gametogenesis in *Margaritifera margaritifera* in northeastern North America. Bull Am Malacol Union Inc:49–53
Smith DG (1980) Anatomical studies on *Margaritifera margaritifera* and *Cumberlandia monodonta.* Zool J Linn Soc 69:257–270
Smith G, Stearley R (1989) The classification and scientific names of rainbow and cutthroat trout. Fisheries 14:4–10
Smith-Gill SJ (1983) Developmental plasticity: developmental conversion versus phenotypic modulation. Am Zool 23:47–55
Stearns SC (1977) The evolution of life history traits; a critique of the theory and a review of the data. Annu Rev Ecol Syst 8:145–171
Surber T (1912) Identification of the glochidia of freshwater mussels. Bur Fish Doc 771:3–13
Taylor DW, Uyeno T (1966) Evolution of host specificity of freshwater salmonid fishes and mussels in the north Pacific region. Venus 24:199–209
Utermark W (1973) Untersuchungen über die Wirtsfischfrage für die Glochidien der Flußperlmushel. MS Thesis. Tierärztliche Hochschule, Hannover
Young M, Williams J (1984a) The reproductive biology of the pearl mussel in Scotland. I. Field studies. Arch Hydrobiol 99:405–422
Young M, Williams J (1984b) The reproductive biology of the freshwater pearl mussel in Scotland. II. Laboratory studies. Arch Hydrobiol 100:29–43
Ziuganov V, Zotin A, Nezlin L, Tretiakov V (1994) The freshwater pearl mussels and their relationships with salmonid fish. Vniro Publishing House, Moscow

6 The Community Structure of Ticks on Kudu, *Tragelaphus strepsiceros*, in the Eastern Cape Province of South Africa

T.N. PETNEY and I.G. HORAK

6.1 Introduction

Arthropod and helminth parasites are classified together as macroparasites on the basis of ecological similarities (Anderson and May 1979). The relationship between each group and its host species for such parameters as ratio of body size, intrinsic population growth rate, life span and the pathological significance of their interaction are comparable but significantly different from microparasites (viruses, bacteria, protozoans) and parasitoids.

There are also major differences between arthropod parasites and helminths. Helminths, as endoparasites, live in a relatively constant internal environment once they have found a host. In such an environment, where they can remain for considerable times, they are not subject to seasonal influence and as a consequence their host is not subject to rapid changes in the structure of its helminth "infracommunity". Thus data on helminth communities are often restricted to a certain season (Bush and Holms 1986; Esch et al. 1988) or include the tacit assumption of constancy through time (Goater et al. 1987).

With few exceptions arthropod parasites are ectoparasitic and in many cases they are only temporary parasites with one or more free-living life history stages (mosquitoes, fleas) or with varying periods without contact with the host during the parasitic stage (mosquitoes, ticks). Such species are subject to seasonal changes in their environment which can significantly influence their presence or abundance (Demarais et al. 1987) which in turn is likely to subject each individual host to a changing spectrum of parasites.

There is a wealth of data available on the community structure of helminths which has been used to build the hypotheses dealing with the factors determining community structure for macroparasites (Holmes 1987; Esch et al. 1990). In comparison with the information available on helminths, that available on the community structure of ectoparasites is rudimentary. In this contribution, we offer an analysis of the influence of climate on the structure of infracommunities of ticks on a large African antelope, the greater kudu (*Tragelaphus strepsiceros*).

Ticks represent part of the guild of blood-feeding arthropods which includes mites, sucking lice and fleas as well as a variety of diptera. Ixodid ticks have three developmental stages which are parasitic although most of their duration is free living. The eggs hatch into larvae on the ground. These must

Ecological Studies, Vol. 130
Dettner et al. (eds.) Vertical Food Web Interactions

then find a host on which they engorge. In most species the engorged larvae then detach from the host before moulting into nymphs on the ground. The nymphs must then find a new host before engorging, usually detaching, and then moulting into adults. Adults then find a new host and after mating, which, except in the genus *Ixodes*, takes place on the host, the female engorges, drops from the host and lays eggs before dying. Members of some genera use the same host for each developmental stage without detaching (one-host ticks) and in other genera the larvae and nymphs may share a common host (two-host ticks) while in the majority of genera larvae, nymphs and adults each have a separate host (three-host ticks). Although host specificity is common within some genera of ticks, many species feed on a wide range of hosts, in some cases ranging from reptiles through birds to mammals (Hoogstraal and Aeschlimann 1982). This is particularly true for many of the species infesting wild and domestic ungulates in Africa.

The greater kudu is a large antelope, with males reaching 1.4 m at the shoulder and weighing up to 250 kg. It prefers savannah woodland vegetation, is gregarious, and feeds by browsing (Smithers 1983).

6.2 Methods and Materials

Andries Vosloo Kudu Reserve (33°06′S; 26°41′E) comprises an area of 6497 ha of Valley Bushveld vegetation in the eastern Cape Province of South Africa. At the time of our study it was stocked with approximately 450 greater kudu, 140 eland (*Taurotragus oryx*), 100 African buffalo (*Syncerus caffer*), 54 red lartebeeste (*Alcelaphus buselaphus*), 80 bushbuck (*Tragelaphus scriptus*) and 18 springbok (*Antidorcas marsupialis*) (Petney and Horak 1987). A single greater kudu was shot on the reserve at monthly intervals from February 1985 until January 1986 and then at three monthly intervals from March to December 1986. All samples, except those for December, were collected in the last half of the month. Dead animals were transported to the laboratory as soon as possible and processed as indicated in MacIvor et al. (1987) to remove ticks.

Non-parametric Spearman-rank correlations were used to compare species richness and intensities of infestation for each sampling time. The Shannon Index was used to measure the diversity of ticks on each kudu sampled. This index was calculated for each developmental stage. Similarity in infestation patterns was determined by cluster analysis using the infestation intensities for all developmental stages of all tick species for all months excluding March 1986. This month was excluded from the analysis as it showed an exceptional intensity of infestation of larval *Amblyomma hebraeum*. This analysis was carried out using SPSS (Statistical Package for the Social Sciences).

The relationship between ln (intensity of infestation), species richness and diversity for each developmental stage and climate was investigated using

Table 6.1. The number and life history stage of each tick species collected on all of the kudu sampled

Species	*A. hebraeum*	*A. marmoreum*	*B. decoloratus*	*Haem. silacea*	*Hy. rufipes*	*Hy. truncatum*	*I. pilosus*	*R. appendiculatus*	*R. e. evertsi*	*R. exophthalmus*	*R. glabroscutatum*	*R. simus*
Larvae	32 009	2 112	14	10 003	0	0	0	4 391	296	0	13 596	0
Nymphs	3 408	0	36	2 305	0	0	14	775	145	0	7 069	0
Adults	682	0	54	1 825	3	2	14	514	28	145	1 597	6
Number of engorged females (%)	22/252 (8.7)	na	0/20	106/415 (25.5)	na	na	0/14	24/190 (12.6)	0/10	14/145 (9.6)	20/544 (3.7)	0/4

The nearly complete engorgement of females implies that mating has occurred and that the host is suitable.
na, not applicable.

multiple linear regressions. Both data for the month in which sampling occurred and for the previous month ("lag" prefix) were used in the analysis. The natural logarithm of the intensity of infestation was used to overcome problems associated with the usual aggregated dispersion pattern of ticks (Petney et al. 1989).

Climatic data were collected in Port Elizabeth, approximately 150 km to the southwest of the Andries Vosloo Kudu Reserve (Schulze 1972). Variables used for the analyses were: mean maximum monthly temperature (MAT), mean minimum monthly temperature (MIT), mean relative humidity (RH), mean rainfall (RAI), mean radiation/day (RAD) and evaporation (EVA). The climate of Port Elizabeth, with cool, sometimes moist winters and hot, often dry summers is qualitatively similar to that of the Kudu Reserve except that Port Elizabeth is situated on the coast. These analyses were carried out using Student SYSTAT.

6.3 Results

Of the 12 species of ticks collected only one, *Boophilus decoloratus*, has a one-host life cycle. *Rhipicephalus evertsi evertsi*, *Rhipicephalus glabroscutatum* and the two *Hyalomma* species are 2-host species with larvae and nymphs occurring on the same host individual. Of these four two-host species the two *Hyalomma* species were represented only by adults while both *R. e. evertsi* and *R. glabroscutatum* were represented by all stages of development All other species are three-host ticks (Table 6.1). Of these seven three-host species, three were represented by all developmental stages, two species only by adults, while *A. marmoreum* was represented only by larvae and *Ixodes pilosus* by nymphs and females. Females nearing complete engorgement were collected from five species. In four species, each represented by 20 or less females, no near complete engorgement was observed. Four species were represented by less than 20 individuals. All other species were represented by over 100 individuals ranging up to 36099 *A. hebraeum*. Of these, 18323 larvae came from a single kudu examined in March 1986.

Species richness ranged from 3 to 7 for larvae, 2 to 6 for nymphs, 4 to 8 for adults and 6 to 9 for all developmental stages (Table 6.2). The species richnesses were not significantly correlated for any pair of developmental stages. The intensity of infestation of adults was significantly negatively correlated with the intensities of both larvae ($r_s = -0.449$, $p < 0.05$) and nymphs ($r_s = -0.653$, $p < 0.05$), but there was no significant correlation between the intensities of infestation of larvae and nymphs. For the different developmental stages species richness was not significantly correlated with intensity of infestation for any stage except in larvae ($r_s = 0.790$, $p < 0.01$). Overall species richness was significantly correlated with that of adults ($r_s = 0.835$, $p < 0.01$) but not with that of either larvae or nymphs.

Table 6.2. Mean species richness (**A**; ±SE) and mean Shannon diversity (**B**; ±SE) for each developmental stage of tick and months of maximum and minimum infestation

A	Species richness					
	Max.	Month	Min.	Month	x̄	SE
Larvae	7	May	3	Sept., Oct., Dec. 1986	4.88	0.32
Nymphs	6	April	2	Feb.	4.38	0.24
Adults	8	April	4	May, Aug., March 1986	5.56	0.29
B	Shannon diversity (H)					
	Max.	Month	Min.	Month	x̄	SE
Larvae	1.80	Mar. 1986	0.25	Sept. 1985	4.88	0.32
Nymphs	1.24	April	0.26	Dec. 1985	4.38	0.24
Adults	1.48	Mar. 1986	0.61	Aug. 1985	5.56	0.29

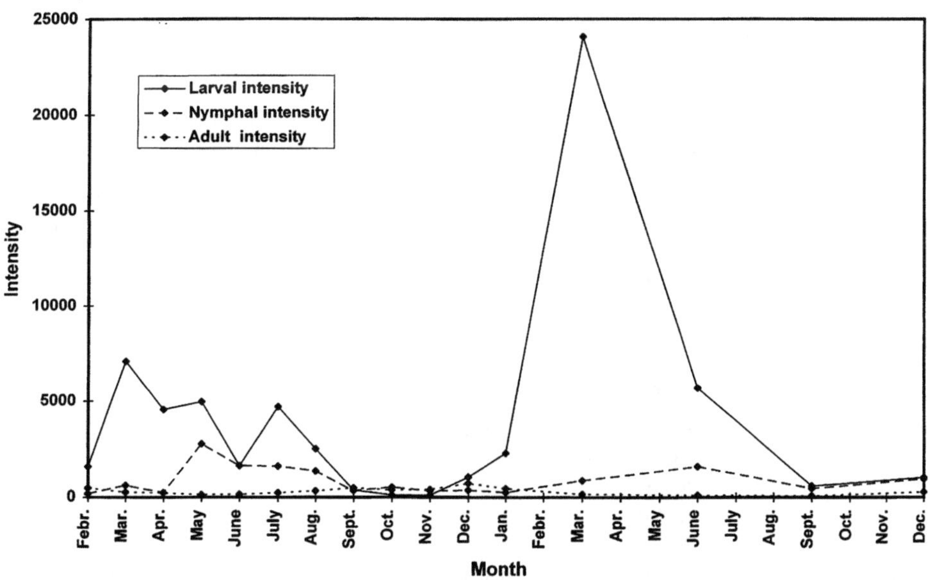

Fig. 6.1. The intensity of infestation for each developmental stage of ticks of each sampling month

Total tick burdens were seasonal with September to December (spring/early summer) 1985 and September to December 1986 showing low infestation intensities (Fig. 6.1). During this time the intensities of infestation of nymphs and larvae were low while adult intensities were high. High larval intensities of infestation occurred between March and August (autumn/winter) and high nymphal intensities between May and August. Ranks of total intensity of

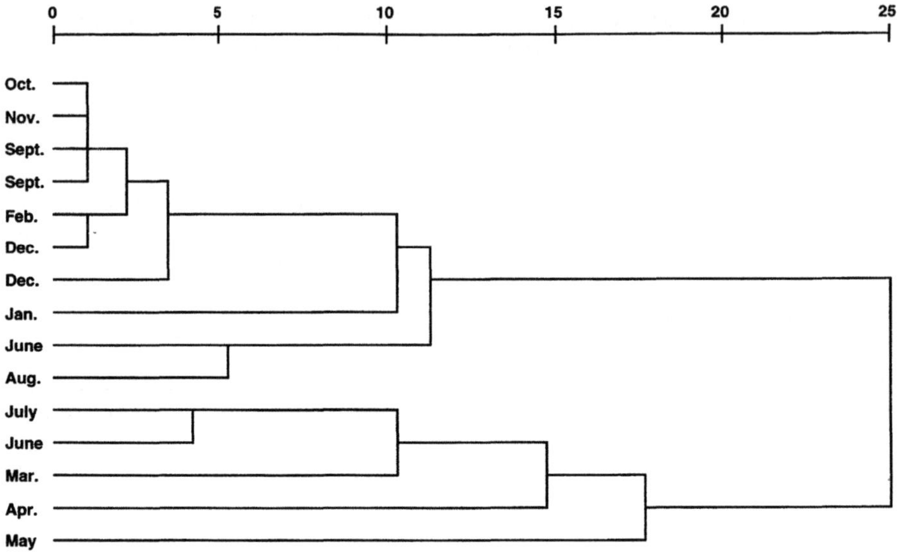

Fig. 6.2. The relationship between pattern of infestation and month of sampling

Table 6.3. Multiple regression analyses of species richness, intensity of infestation and Shannon diversity of larvae, nymphs and adults against climatic variables. No significant correlation was found for adult species richness

Variable	Predictors	Adj. multiple R^2	*F*-value	Probability
Larval species	lagMAT RAD	0.809	24.54	<0.001
Larval ln	lagRH lagRAD RAD	0.894	33.77	<0.001
H larvae	lagMAT MAT	0.512	6.83	<0.01
Nymphal species	lagMAT lagMIT MAT	0.589	5.74	<0.05
Nymphal ln	RAD	0.503	14.16	<0.01
H nymphs	lagRAI RAI lagRAI lagEVA	0.850	15.58	<0.001
Adult ln	RAD	0.434	10.75	<0.01
H adults	lagMAT	0.397	9.21	<0.01

H, Shannon diversity.

infestation for all collecting months show a strong correspondence between years (March 15, 16; June 9, 12; September 3, 4; and December 5, 7).

Two distinct clusters separated autumn and early winter samples (excepting June) from the spring/summer samples in relation to the overall intensity of infestation of ticks (Fig. 6.2). Considerable variation was present in the former group while the latter could be divided into two subgroups. The first of these subgroups, incorporating spring and summer, showed strong similarity between months; the second included January, June 1985 and August.

Regression analyses show significant relationships between climatic factors and the species richness, intensity of infestation and Shannon diversity for each developmental stage except for species richness of adults (Table 6.3). For the species richness and the intensity of infestation of larvae as well as for the Shannon diversity of nymphs over 80% of variability could be accounted for by climatic parameters. Total species richness was not significantly correlated with any climatic parameter.

6.4 Discussion

Ticks, particularly larvae which hatch together from an egg batch, usually have an aggregated dispersion pattern (Petney et al. 1989). Thus accurate estimates of mean intensities of infestation are difficult to determine even with large sample sizes. Our data, with a single kudu sampled each month, are clearly inadequate to give more than a rough approximation of reality. Unfortunately it is equally clear that sample sizes sufficient for an accurate determination of mean infestation intensity can rarely if ever be collected over a sufficient period of time and within a uniform habitat for large and/or uncommon hosts.

Host preferences account for much of the variation in the contribution to the tick community between adult and immature ticks of the same species. *A. marmoreum* adults occur almost exclusively on large reptiles, particularly the tortoise *Geochelone pardalis* (Dower et al. 1988). Immatures of the two *Hyalomma* species and *Rh. exophthalmus* feed predominantly on birds and/or lagomorphs (Walker 1991; Keirans et al. 1993). *Rh. simus* immatures parasitize various species of rodents (Walker 1991). All stages of *I. pilosus* can occur on the same host.

Each of the species uncommon on kudu has been recorded on a variety of other wild ungulates although all are consistently rare on this host (Horak et al. 1983). The concept of core/satellite species is only applicable for all developmental stages of some of the species examined. In other species only larvae (*A. marmoreum*) or adults (*R. exophthalmus*) can be considered as core "species".

The community structure of ticks on kudu is not static but varies greatly with season. Species richness for any developmental stage changes by a factor of 2 and intensity of infestation by an order of magnitude. The distinction between late autumn/winter and spring/summer samples corresponds broadly with findings for other hosts sampled in the eastern Cape Province (Horak et al. 1991a,b).

The significance of climatic conditions for ticks is determined by the sensitivity of the free-living stages to high temperature and low humidity (Norval 1977). The larval stage is the most sensitive, desiccating rapidly at high temperatures and low humidities. This can account for the large degree of variability in species number and intensity of infestation of larvae explained by (lag) MAT, RH and RAD in the multilinear regressions. The nymphal and adult stages appear to show less sensitivity to climate although there are still significant climatic effects. The variability in the sensitivity of Shannon diversity to climatic variables can probably be at least partially explained by the strongly aggregated dispersion pattern to ticks which would tend to decrease evenness.

Climatic variables are only able to explain part of the dynamics of community structure in these ticks. The detailed interpretation of our results must await more finely tuned comparisons of potential direct and indirect interactions between the different tick species and between ticks and host. A variety of potentially important factors are already known. Different tick species or developmental stages within a species can occupy different anatomical sites on the host (niche segregation; Kaiser et al. 1982). Similar species which occupy the same or overlapping sites can compete for space (Andrews and Petney 1981; Norval and Short 1984) or interfere with one another's reproductive behaviour (Andrews et al. 1982). Some species are known to have an immunosuppressant effect on the host (Ribeiro et al. 1985) and in other cases infestation by one tick species can induce a cross immunity to another species (Bull et al. 1981). Pheromones can attract individuals of the same species to attach to an already infested host (Norval et al. 1989).

Acknowledgements. We would like to thank Prof. G. Bauer (Freiburg) for his interest, advice and for running the cluster analysis and Drs. L. Erdinger and M. Maiwald (Heidelberg) for help with the figures.

References

Anderson RM, May RM (1979) Population biology of infectious diseases: part 1. Nature 280:361–367

Andrews RH, Petney TN (1981) Competition for sites of attachment to hosts in three parapatric species of reptile tick. Oecologia 51:227–232

Andrews RH, Petney TN, Bull CM (1982) Reproductive interference between three parapatric species of reptile tick. Oecologia 52:281–286

Bull CM, Sharrad RD, Petney TN (1981) Parapatric boundaries between Australian reptile ticks. Proc Ecol Soc Aust 11:95–107

Bush AO, Holms JC (1986) Intestinal helminths of lesser scaup ducks: an interactive community. Can J Zool 64:142–152

Demarais S, Jacobson HA, Guynn DA (1987) Effects of season and area on ectoparasites of white-tailed deer (*Odocoileus virginianus*) in Mississippi. J Wilde Dis 23:261–266
Dower KD, Petney TN, Horak IG (1988) The relative success of natural infestations of the ticks *Amblyomma hebraeum* and *Amblyomma marmoreum* on Leopard Tortoise, *Geochelone pardalis*. Onderstepoort J Vet Res 55:11–13
Esch GW, Kennedy CR, Bush AO, Aho JM (1988) Patterns in helminth communities in freshwater fish in Great Britain: alternative strategies for colonization. Parasitology 96:519–532
Esch GW, Shostak AW, Marcogliase DJ, Goater TM (1990) Patterns and processes in helminth parasite communities: an overview. In: Esch GW, Bush AO, Aho JM (eds) Parasite communities: patterns and processes. Chapman and Hall, London, pp 1–19
Goater TM, Esch GW, Bush AO (1987) Helminth parasites of sympatric salamanders: ecological concepts at infra community, component and compound community levels. Am Midl Nat 118:289–300
Holmes JC (1987) The structure of helminth communities. Int J Parasitol 17:203–208
Hoogstraal H, Aeschlimann A (1982) Tick-host specificity. Mitt Schweiz Entomol Ges 55:5–32
Horak IG, Potgieter FT, Walker JB, de Vos V, Boomker J (1983) The ixodid tick burdens of various large ruminant species in South African nature reserves. Onderstepoort J Vet Res 50:221–228
Horak IG, Knight MM, Williams EJ (1991a) Parasites of domestic and wild animals in South Africa. XXVIII. Helminth and arthropod parasites of angora goats and kids in valley bushveld. Onderstepoort J Vet Res 58:253–260
Horak IG, Spickett AM, Braack LEO, Williams EJ (1991b) Parasites of domestic and wild animals in South Africa. XXVII. Ticks on helmeted guinea fowl in the eastern Cape Province and eastern Transvaal lowveld. Onderstepoort J Vet Res 58:137–143
Kaiser MN, Sutherst RW, Bourne AS (1982) Relationship between ticks and zebu cattle in southern Uganda. Trop Anim Health Prod 14:63–74
Keirans JE, Walker JB, Horak IG, Heyne H (1993) *Rhipicephalus exophthalmus* sp. nov., a new tick species from southern Africa, and a redescription of *Rhipicephalus oculatus* Neumann, 1901, with which it has hitherto been confused (Acari: Ixodida: Ixodidae). Onderstepoort J Vet Res 60:229–246
MacIvor KM, Horak IG, Holton KC, Petney TN (1987) An evaluation of live and destructive sampling techniques to determine parasitic tick populations. Exp Appl Acarol 3:131–143
Norval RAI (1977) Studies on the ecology of *Amblyomma hebraeum* Koch in the eastern Cape Province of South Africa. II. Survival and development. J Parasitol 63:740–747
Norval RAI, Short N (1984) Interspecific competition between *Boophilus decoloratus* and *Boophilus microplus* in southern Africa. In: Griffiths DA, Bowman CE (eds) Acarology IV, vol 2. Ellis Horwood, Chichester, pp 1242–1246
Norval RAI, Andrew HR, Yunker CE (1989) Pheromone mediation of host selection in the bont tick (*Amblyomma hebraeum* Koch). Science 243:364–365
Petney TN, Horak IG (1987) The effect of dipping on parasitic and free-living populations of *Amblyomma hebraeum* on a farm and on an adjacent nature reserve. Onderstepoort J Vet Res 54:529–533
Petney TN, van Ark H, Spickett AM (1989) On sampling tick populations: the problem of overdispersion. Onderstepoort J Vet Res 57:123–127
Ribeiro JMC, Makoul GT, Levine L, Robinson DR, Spielman A (1985) Antihemostatic, antiinflammatory and immunosuppressive properties of the saliva of the tick, *Ixodes dammini*. J Exp Med 161:332–344
Schulze BR (1972) South Africa. In: Griffiths JF (ed) Climates of Africa, vol 10. World survey of climatology. Elsevier, Amsterdam, pp 501–586
Smithers RHN (1983) The mammals of the southern African subregion. University of Pretoria, Pretoria, xxii + 736 pp
Walker JB (1991) A review of the ixodid ticks (Acari, Ixodidae) occurring in southern Africa. Onderstepoort J Vet Res 58:81–105

7 The Epidemiology of Parasitic Diseases in *Daphnia*

D. Ebert, R.J.H. Payne, and W.W. Weisser

7.1 Introduction

Parasites and pathogens may be directly or indirectly involved in the ecology and evolution of a broad range of phenomena: population dynamics and extinctions, maintenance of genetic diversity and sexual selection, to name just a few. Certainly parasites - here broadly defined to include viruses, bacteria, protozoa and helminths - possess features which make them very attractive as explanatory factors in the evolution and ecology of their host. These features include their typically narrow host range, the adverse effects parasites have on host fecundity and survival, and the density dependence of transmission (Hassell and May 1973; Anderson and May 1979, 1991; May and Anderson 1979). However, the bulk of available information stems from theoretical and laboratory studies, while studies in natural populations are scarce. For example, experimental approaches give clear support for density dependence of transmission (Blower and Roughgarden 1989; Ebert 1995; D'Amico et al. 1996; Knell et al. 1996), but there exists very little data showing density-dependent transmission in natural populations (Dobson and Hudson 1986; Scott and Dobson 1989). Similarly, although laboratory studies have demonstrated a clear effect of parasites on host density (Sait et al. 1994; Mangin et al. 1995), reports of parasite-mediated reduction of host density in the field are rare (Dobson and Hudson 1986; Scott and Dobson 1989). It is essential for our understanding of host-parasite interactions to compliment the results of laboratory work with data from natural populations.

7.1.1 Parasites in Zooplankton Populations

Traditionally, zooplankton ecology has focused on the effects of intra- and interspecific competition and on predation, and plankton dynamics have been claimed to be internally generated (McCauley and Murdoch 1987; McCauley 1993). Feedback loops between short-term changes in zooplankton density and predators have been dismissed as unlikely because such feedback is contingent upon the rate of change in the density of a particular predator species being a function of the prey's current density (McCauley and Murdoch 1987; McCauley et al. 1988). Microparasites, however, do have exactly those features which can lead to feedback loops. They are often host specific and have a very

Ecological Studies, Vol. 130
Dettner et al. (eds.) Vertical Food Web Interactions

short generation time, which enables them to respond very quickly to changes in host density. Here, we investigate the role that microparasites play in the ecology of their *Daphnia* hosts. First, we summarize the results of field surveys investigating the abundance of parasites in natural *Daphnia* populations. We then outline some of the biological features of these microparasites, and finally we present a model of their epidemiology.

7.2 The Abundance of *Daphnia* Microparasites in Natural Populations

The first step in understanding the relevance of parasites to their natural host populations is to assess their abundance. To quantify microparasite prevalences (here defined as the proportion of adult hosts that are infected with micro-endoparasites) of *Daphnia* in natural populations field samples were collected from sites in England and Sweden. First, three English ponds were studied over a period of 1 year (about 10–12 *Daphnia* generations, 65 samples in total) and host density and fecundity were assessed, together with parasite prevalence, richness, diversity and host specificity (Stirnadel 1994; Stirnadel and Ebert 1997). Parasite prevalences were high throughout the year, averaging 84% in adult *D. magna*, 53% in *D. pulex* and 36% in *D. longispina*. In all three host species clutch size in parasitized females was significantly lower than in uninfected females (>20% reduction in *D. magna*, >25% reduction in *D. pulex* and >7% reduction in *D. longispina*). Whether reduced fecundity was a result of parasitism or whether the infections were a consequence of the hosts being weakened for some other reason is difficult to assess. Nonetheless, laboratory experiments confirmed the adverse effect of various parasites on host fecundity (e.g. Ebert 1994a,b, 1995; Mangin et al. 1995; Ebert and Mangin, in prep.). Only two of the 11 common micro-endoparasites found in these three ponds (17 species in total) showed no specificity within the three *Daphnia* host species; the other nine common parasites infected either only one or two of the three sympatric host species, or differed in their host specificity across the three ponds. The latter observation might indicate specialization of parasites to the currently, or formerly, predominant host community in a pond.

To assess the possibility of extrapolating our findings from these three strongly infected ponds to other populations, single samples were analysed from 43 *Daphnia* populations in southern England. Of all populations 91% suffered from endoparasite infections (mainly Microsporidia), with mean prevalences of 43% in *D. magna*, 69% in *D. pulex* and 43% in *D. longispina* (Brünner 1996; Brunner and Ebert, in prep.). A further, similar survey was conducted in rockpools along the Swedish east coast near Uppsala (Bengtsson and Ebert, in prep.). In 24 of 50 (48%) *D. pulex* populations and 9 of 25 (36%) *D. longispina* populations investigated micro-endoparasite infections were found. Across all ponds average microparasite prevalences were 15.5% for *D.*

pulex and 9.1% for *D. longispina*. Although average prevalences were lower in the Swedish rockpools than in English ponds, Swedish infections were primarily due to a single, extremely virulent microsporidium species. This undescribed microsporidium infected ovaries and fat cells of its hosts and reduced clutch size by 98%.

Given that many microparasite infections are only detectable after visible signs of infection have developed, our observations are probably conservative. The results of these field surveys clearly indicate that *Daphnia* parasites are abundant and are likely to play an important role in both the ecology and evolution of their hosts.

It should be mentioned here that vertebrate predators (fishes and newts) were absent in most of the populations surveyed and therefore we must be careful in extrapolating our results to habitats with high levels of vertebrate predation on *Daphnia*. There are two factors suggesting that levels of parasitism in *Daphnia* populations with strong vertebrate predation might be lower than in populations with less predation. Firstly, the likelihood of being infected increases with body size (Vidtmann 1993; Stirnadel 1994), which is probably a result of both higher filtration rates (and thus parasite spore uptake rates) and an accumulation effect with age. In ponds with high adult mortality – as is typical for ponds with fish and newt predators – parasite prevalence might therefore be reduced. Secondly, choice test in the laboratory show that some diseases make their hosts more conspicuous through a reduction in transparency and increase the likelihood of predation strongly (Lee 1994; Lee and Ebert, in prep.). Similarly, increased susceptibility to predation was reported for hosts carrying large loads of epibionts (Willey et al. 1990; Allen et al. 1993; Chiavelli et al. 1993; Threlkeld and Willey 1993).

Thus, parasite mortality in *Daphnia* populations with fish predators is high because of the lower life expectancy of their hosts, and additionally due to the preferences of visually hunting predators for conspicuous *Daphnia*. Therefore, *Daphnia* parasites might not be able to persist in the presence of strong fish predation (see also our mathematical model in Sect. 7.5.2). Consistent with this hypothesis is the observation that large *Daphnia* species are typically found in fishless ponds, and that most described *Daphnia* parasites are reported from large *Daphnia* species such as *D. magna* and *D. pulex*. Very few parasites have been reported from small *Daphnia* species, which co-occur with fish, such as *D. cucullata* or *D. galeata* (Green 1974).

7.3 The Biology of Transmission in Aquatic Systems

7.3.1 Waterborne Transmission

The most important aspect of epidemiology is the mode of transmission. To our knowledge, the first description of a plankton parasite life cycle, including a test of the mode of transmission, was Chatton's (1925) description of the amoeba *Pansporella perplexa* in *D. pulex*. This parasite is transmitted between hosts via waterborne infective stages, which are ingested by the filter-feeding hosts. It was later shown for many other plankton parasites that waterborne transmission is a common route of dispersal. In laboratory transmission experiments we have confirmed the existence of waterborne spores for several *Daphnia* microparasites, including the bacteria White Bacterial Disease and *Pasteuria ramosa*, the amoeba *Pansporella perplexa*, the yeast *Metschnikowiella biscuspidata*, and the microsporidium *Glugoides intestinalis* (formerly *Pleistophora intestinalis*). Waterborne transmission is also common for many described epibionts of the Cladocera (Green 1974; Threlkeld et al. 1993). There are, however, also vertically transmitted (often transovarian) parasites of *Daphnia* (Mangin et al. 1995) as well as some which need a secondary host to complete development (Green 1974). Here, we concentrate on plankton parasites with direct, waterborne and horizontal transmission.

7.3.2 Survival of Transmission Stages Outside the Host

Planktonic populations typically undergo tremendous fluctuations in density, often over several orders of magnitude. Some plankton organisms might even temporarily disappear from their habitat and survive in the form of resting stages. Since these bottlenecks in host density pose a problem for horizontally transmitted parasites, Green (1974) suggested that plankton parasites should have persisting transmission stages which can endure phases of low host density. He suggested that pond sediments form spore banks for these infective stages, similar to the way that they harbour resting stages of many plankton organisms.

To test this hypothesis mud samples were collected from different ponds harbouring parasitized populations of *D. magna*. Subsamples of these sediments were placed in beakers and uninfected *D. magna* added. When the hosts were dissected after 24 days, infections with three different microparasites were found: the bacterium *P. ramosa* and the yeast *M. biscuspidata* were found in the haemolymph and the microsporidium *G. intestinalis* was found in the host gut (Ebert 1995). Using a mud sample that had been kept at 4 °C for 4 years, it was still possible to infect the *Daphnia* with *P. ramosa*. Similarly, mud samples containing spores of *G. intestinalis* remained infectious after storage at 4 °C for 3 months and in a different study for 5 months at 12 °C (D. Ebert unpubl.). These estimates of spore durability should be interpreted with caution until a systematic investigation to estimate spore durability appropriately (e.g. their half-lifetime) is completed. Nonetheless, the results clearly confirm Green's (1974) hypothesis that pond sediments can serve as "parasite spore banks" and that parasites can survive periods of low host density in a "sit-and-wait" stage.

The uptake of spores from the sediments is a consequence of poor feeding conditions for the hosts. Some cladocerans change their behaviour when feeding conditions deteriorate and switch from normal filter feeding in the free water to a browsing behaviour on bottom sediments. This behaviour serves to stir up particles from the sediments, which are then ingested by filter feeding (Horton et al. 1979; Freyer 1991). It is important to note here that spore uptake from the pond sediments is a density-independent form of transmission.

In considering the survival of transmission stages, it would appear that parasites in aquatic systems face less problems than their terrestrial counterparts, as many common sources of spore mortality present in terrestrial parasites do not exist for waterborne transmission stages. For example, desiccation, one of the main threats to the survival of air- or soilborne spores, is irrelevant in the aquatic environment. In addition, water not only provides protection from UV radiation, but its high heat capacity also buffers the effect of rapid changes in temperature and prevents overheating. Since the protection of spores from adverse environmental effects can be considered to be costly, one might speculate that aquatic parasites should be able to shift the trade-off between offspring quantity and quality towards production of more offspring.

7.4 The Spread of Microparasites

After a parasite appears for the first time in a new host population it can only persist if on average each infection causes at least one secondary infection – that is, the basic reproductive rate of the parasite, R_0, must be larger than 1. There has been much discussion concerning which factors are responsible for the spread of a parasite in a plankton population.

7.4.1 Parasite Transmission Is Density Dependent

The standard assumption of epidemiological theory that parasite transmission is density-dependent has often been discussed with regard to plankton parasites (e.g. Canter and Lund 1951, 1953; Miracle 1977; Brambilla 1983). Certainly the most convincing data are that presented by Canter and Lund (1953), who observed strong fluctuations of the diatom *Fragilaria crotonensis* in an English lake. Whenever the density of this algae reached more than about 100 cells/ml, a fungal parasite (*Rhizophidium fragilariae*) spread rapidly and host density dropped by two orders of magnitude. For *Daphnia* no such example exists, although the reported data do not contradict density dependence (see Sect. 7.6). We have used an experimental approach to test for density dependence.

The microspordian gut parasite *G. intestinalis* in *D. magna* proved to be an ideal system for experimental epidemiology. The life cycle of *G. intestinalis* is

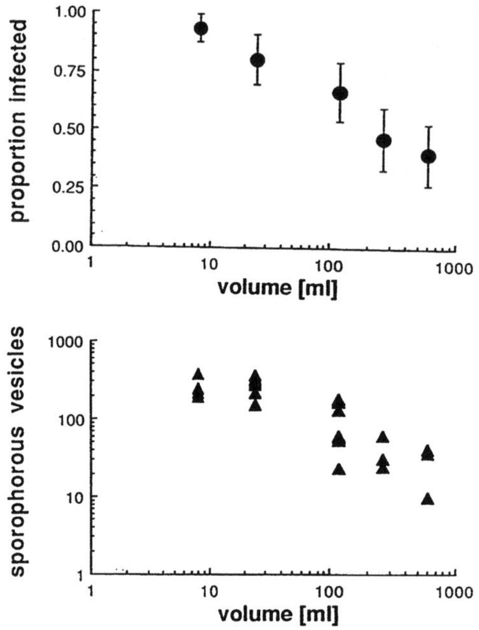

Fig. 7.1. Density-dependent transmission of the microsporidium *G. intestinalis*. *Above* Relation between beaker size and parasite transmission (±SE). The proportion of hosts which became infected is shown on the y-axis (n = 15 per beaker size). *Below* The impact of beaker size on spore load within the host. Each point gives the mean of five females. The mean number of sporophorous vesicles per replicate is shown. (From Ebert 1995 with permission)

direct, and transmission to new hosts occurs only 3 days after infection (Ebert 1994a,b). The waterborne spores of this parasite are transmitted with the faeces. Laboratory experiments showed that transmission of *G. intestinalis* is strongly density dependent and that the intensity of infection (parasite load per host) increased more rapidly when hosts were more crowded (Fig. 7.1). The upper graph shows the results of an experiment conducted by placing one infected and one uninfected host for 2 days in breakers of different sizes. The previously uninfected hosts were latter tested for the presence of the parasite. The lower graph shows the results when equally infected young females were placed in groups of five in breakers of different size. After 12 days the intensity of infection was determined (Fig. 7.1).

Density-dependent transmission was also tested for *M. biscuspidata* and *P. ramosa* (D. Ebert, in prep.). These parasites are transmitted only after the infected host has died and the spores are released from the cadaver. The appropriate way to test for density dependence is therefore to produce spore suspensions and to test the infectivity of different spore concentrations. In both species of parasite, transmission probability was reduced when spore density was low, but transmissibility reached 100% when the spore density was high.

From these experiments we conclude that density dependence is indeed a real phenomenon in the spread of parasitic infections in *Daphnia* populations. However, the finding of a density-dependent mechanism of transmission does not reveal the practical importance of it for the epidemiology of natural *Daphnia*-parasite populations.

7.4.2 Parasite Transmission Can Be Limited by Low Temperatures

It has often been observed that plankton epidemics are predominantly found during the warm summer months. Ruttner-Kolisko (1977) proposed that transmission of a microsporidian parasite in rotifers is impaired at low temperatures. We tested this hypothesis with *G. intestinalis* in *D. magna*. It was found that parasite transmission was strongly impaired below 12 °C. This is consistent with the observation that *G. intestinalis* decreased in late autumn in *D. magna* populations in south England (Stirnadel 1994). Poor transmissibility at temperatures below 25 °C were reported for *P. ramosa*, parasitizing the cladoceran *Moina rectirostris* (Sayre et al. 1979). In contrast, *P. ramosa* in *D. magna* can be transmitted at 15, 20 and 25 °C in the laboratory (Ebert et al. 1996). Thus it appears that the temperature criterion is species- and strain-dependent.

7.4.3 Host Stress Might Facilitate Parasite Spread

It has been claimed that *Daphnia* cultures kept under poor conditions are more susceptible to infections (Seymour et al. 1984; Stazi et al. 1994), and France and Graham (1985) have observed higher rates of microsporidiosis in crayfish in acidified lakes. We could not find any experimental evidence to support the stress hypothesis. Transmission of *G. intestinalis* appeared to be largely independent of the feeding conditions of *D. magna* and did not differ among age groups or host sexes (Ebert 1995). Other forms of stress and other parasites have not yet been evaluated.

7.5 Epidemiology of *Daphnia* Microparasites

The results discussed thus far indicate that the invasion, spread and persistence of a microparasite in *Daphnia* populations cannot be attributed to one single factor. Rather, the relevant factors might vary temporally, or act synergistically. In the following sections we develop a framework to encompass the different facets of microparasite epidemiology.

7.5.1 A Mechanism for Invasion, Spread and Decline of Parasites in Cladocerans

For parasites in the temperate zone the epidemiology of most microparasites follows a similar pattern (Green 1974; Ruttner-Kolisko 1977; Redfield and Vincent 1979; Brambilla 1983; Yan and Larsson 1988; Vidtmann 1993). Prevalences are typically low in winter and in early spring. After host plankton densities peak in spring parasite prevalence increases. Prevalence fluctuates throughout the summer, decreases in the autumn, and in many cases parasites disappear in winter completely. Green (1974) suggested that the epidemics of some microparasites (e.g. the bacterium *Spirobacillus cienkowskii*) start when a benthic feeding host acquires a parasite from the mud. Once the cycle is started other cladocerans that are partially benthic and partially free water foragers become infected and transmit the parasite to those cladocerans which live in the free water. The parasites disappear from the pond when the hosts go into diapause at the end of the season.

Ebert (1995) proposed a single species version of this model. Following diapause *Daphnia* hatch from their ephippia and re-colonize a pond. Under good feeding conditions in spring the population increases rapidly, until a severe food shortage leads to a population crash. Starving *Daphnia* change their behaviour and switch from filter feeding in the free water to a browsing behaviour on the bottom sediments, which serves to stir up food particles from the sediments (Horton et al. 1979; Freyer 1991). During this browsing, the animals pick up the long-lasting parasite spores from the bottom of the pond. Once the first hosts are infected, the disease can spread further through density-dependent horizontal transmission. The epidemic ends when either environmental conditions deteriorate (e.g. low temperature prevents transmission) or host density falls under the critical value necessary for parasite persistence.

7.5.2 A Mathematical Model for the Epidemiology of Plankton Parasites

We now develop an epidemiological model of an aquatic host-parasite system, which takes into account the features of *Daphnia* microparasites. In particular, we look into the combined effects of density-dependent host-to-host transmission (waterborne transmission), density-independent transmission from the spore bank in the pond (sediment-borne transmission) and the long durability of spores in the sediments. We model the case of the planktonic crustacean *Daphnia* and a horizontally transmitted microparasite, developed from a mathematical model of this system by Weisser et al. (in prep.). Our model is not intended for the analysis of long-term dynamics of host-parasite interactions. This allows us to make one further simplification. We assume that the number of spores in the spore bank effectively remains constant over a single season, not being significantly affected by spore uptake or by the sinking of spores into the sediment.

The dynamics of the microparasite infection of a single *Daphnia* consumer population is described by following the temporal variation in numbers of algae cells, A, uninfected, X, and infected, Y, *Daphnia* individuals, and of free-floating spores, Z. The model takes the following form:

$$
\begin{aligned}
\frac{dA}{dt} &= rA\left(1-\frac{A}{K}\right)-f(A)(X+Y)\\
\frac{dX}{dt} &= \left[\psi_X(A)-\mu\right]X+\psi_Y(A)Y-\beta XZ-H(A)X\\
\frac{dY}{dt} &= \beta XZ+H(A)X-\mu Y\\
\frac{dZ}{dt} &= sY-mZ
\end{aligned}
\qquad (1\text{–}4)
$$

The rationale underlying this formulation can be described as follows: Algae, A, on their own are self-sustaining and their growth is described by a simple logistic equation with growth rate r and carrying capacity K. The function $f(A)$ determines how the *Daphnia* feeding rate depends on the algae density A (Lampert 1987). The functions $\psi_X(A)$ and $\psi_Y(A)$ describe the algal-dependent birth rates of unparasitized and parasitized individuals respectively. Since we exclude vertical transmission, both infected and uninfected females produce uninfected offspring. The functions $f(A)$ and $\psi_Y(A)$ are monotonically increasing with saturation at high algal densities (Lampert 1987). The feeding function, $f(A)$, can be caricatured by the form $f(A) = dA/a_2$ for $A < a_2$ and $(A) = d$ for $A > a_2$, where a_2 is the algal density above which consumer feeding is no longer limited by resource density (Fig. 7.2). For $\psi_X(A)$ we assume that there is a minimum algal density a_1 below which reproduction is zero (e.g. Lampert 1987, Fig. 7.2). The caricature form for the reproduction function $\psi(A)$ is given by $\psi_i(A) = 0$ for $0 < A < a_1$, $\psi_i(A) = b_i\ (A - a_1)/(a_2 - a_1)$ for $a_1 < A < a_2$, and $\psi_i(A) = b_i$ for $A > a_1$, where $i = X$ for unparasitized individuals and $i = Y$ for parasitized *Daphnia* individuals. We assume that the effect of the parasite is to lower the birth rate of its host so that $b_X > b_Y$. Both infected and uninfected *Daphnia* individuals die at rate μ.

To model the infection process we distinguish between density-dependent uptake of free-floating spores and density-independent transmission via spore uptake from the sediment. The rate at which uninfected consumers are infected by free-floating spores in proportional to the number of encounters between spores, Z, and uninfected hosts, X, with proportionality constant, β. If algal density falls below a_3 *Daphnia* individuals feed on the ground and take up spores at rate $H(A)$ which increases for smaller A (Fig. 7.2). Infected *Daphnia* release spores into the water at rate s, and the spores die at a rate m. All parameters are positive.

7.5.3 Analysis of the Model

In the absence of density-independent transmission ($H = 0$), one can investigate the ability of a parasite to invade an uninfected *Daphnia* population by

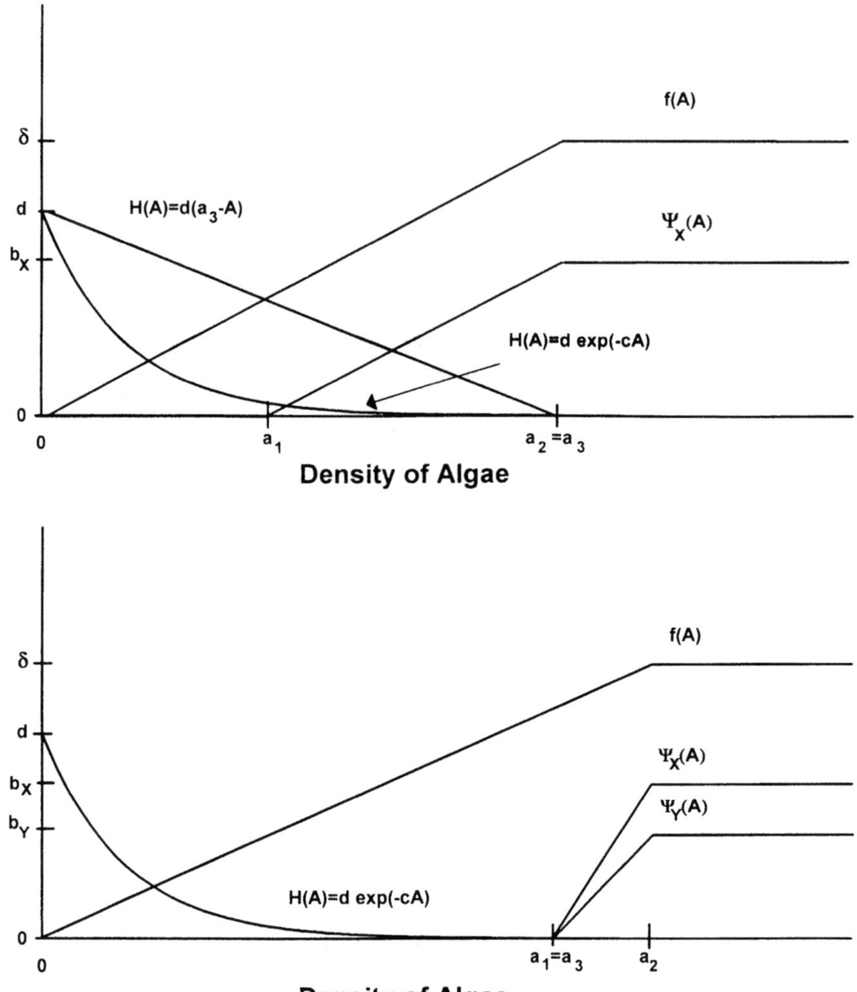

Fig. 7.2. The functions $f(A)$, *Daphnia* feeding, $\psi(A)$, reproduction and $H(A)$, spore uptake rate from the sediment. *Upper graph:* $a_3 = a_2$ $(A^* < a_3)$, $H(A) = d(a_3 - A)$ for $A < a_3$, $H(A) = 0$ otherwise (linearly decreasing); and $H(A) = d\,e^{-cA}$ for $A < a_3$, $H(A) = 0$ otherwise (exponentially decreasing). *Lower graph:* $a_3 = a_1$ $(A^* > a_3)$ for exponentially decreasing $H(A)$. (See text for further explanations)

calculating the basic reproductive rate of the parasite, R_0. This gives the number of secondary infections a single infected individual will produce during its lifetime, if introduced into a population of uninfected hosts at equilibrium (Anderson and May 1979). In the present context, if an infected individual of the consumer species is introduced into an uninfected population at equilibrium, it will produce spores at rate s until it dies. Since the life-expectancy of a consumer is $1/\mu$, the lifetime production of spores of an infected individual can be calculated as s/μ. The probability that a single spore infects a host during its lifetime depends on the life expectancy of a spore (=1/ m), and on the equilibrium density of *Daphnia* individuals in the *uninfected* population, $X^*_{\text{uninfected}}$. Thus, the basic reproductive rate of the parasite is given by:

$$R_o = \frac{s\beta}{\mu m} X^*_{\text{uninfected}} \tag{5}$$

An infection will only spread in the host population if $R_0 > 1$. Equation (5) shows that a parasite can increase R_0 by increasing the transmission efficiency, β, the rate at which spores are released by infected host individuals, s, or by increasing spore longevity, $1/m$. R_0 also increases if the *Daphnia* equilibrium density of an uninfected host population, $X^*_{\text{uninfected}}$, is increased. Increasing *Daphnia* birth rate, bx, increases $X^*_{\text{uninfected}}$. Also, $X^*_{\text{uninfected}}$ can be increased by increasing the algal growth rate, r, or increasing the carrying capacity of the resource, K (Weisser et al., in prep.). In contrast, increased host mortality μ can reduce R_0. Thus, in a situation in which *Daphnia* is heavily predated upon, for example by fish, the basic reproductive rate of a parasite is reduced. This implies that some parasites might be able to persist in a *Daphnia* population only in the absence of predators. The Appendix gives the equilibrium values and stability properties of the model.

Suppose *Daphnia* feed on the sediment and take up spores from it whenever algal density falls below a_3. If $a_3 > A^*$ (the algal equilibrium density calculated in the absence of density-independent transmission), then there will always be uptake of spores by *Daphnia* from the ground. Under these conditions the parasite will persist in the population even though the infection might not be able to spread via waterborne spores alone (i.e. $R_0 < 1$). If, however, $a_3 < A^*$, then *Daphnia* will not feed on the sediment once the population densities become close to the equilibrium densities, and initial parasite infection can only persist if $R_0 > 1$. Figure 7.3 illustrates the effect of a_3.

It has been acknowledged that the combination of a saturating consumer functional response with a high nutrient supply can be very destabilizing in resource-consumer models (Rosenzweig 1971; DeAngelis 1992). This is also true in *Daphnia* populations (Murdoch and McCauley 1985; McCauley and Murdoch 1987). In our case, if the algal carrying capacity is close to the algal density, which is limiting host feeding and reproduction, $K < a_2$, the system returns to a locally stable equilibrium point. But if K is high relative to a_2 then pronounced fluctuations can develop in the system. Unfortunately, the simple expression for R_0 given in Eq. (2) applies only if the pre-infection population has a constant size. Weisser et al. (in prep.) describe and provide computer simulations of the expected parasite invasion behaviour when the pre-infection population exhibits stable limit cycles.

Weisser et al. (in prep.) also describe the way in which the presence of the parasite is expected to modulate the nature, and occurrence, of any periodic

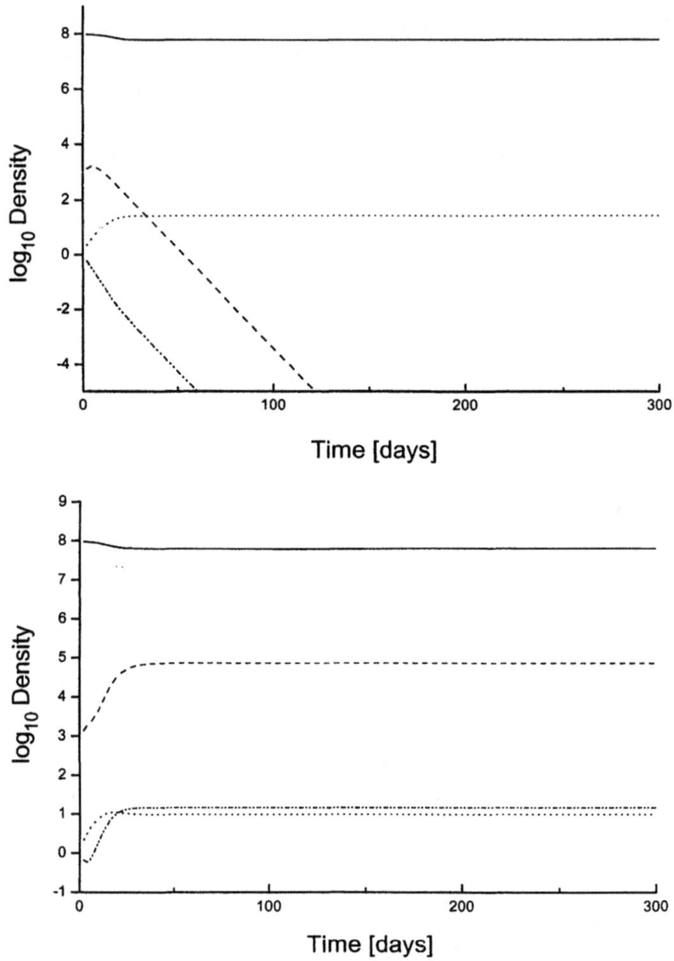

Fig. 7.3. Temporal changes in algae *Daphnia* and parasite spore densities. The effect of changing a_3 on the persistence of the parasite. *Upper graph*: $R_o < 1$, $a_3 = a_1$. *Lower graph*: $R_o < 1$, $a_3 = a_2$. Note that the parasite is able to persist, although $R_o < 1$. $A = A/a_2$, $X = Xd/a_2$, $Y = Yd/a_2$, $Z = Zb$, $K = a_2 = 10^8$, $a_1 = 10^7$, $b_X = 0.45$, $b_Y = 0.225$, $\mu = m = 0.1$, $\delta = 1.5 \times 10^6$, $b = 10^{-6}$, $d = 10^{-8}$. (–) Algae, *A*; (. . .) uninfected *Daphnia*, *X*; (–..–) infected *Daphnia*, *Y*; (– –) free-floating spores, *Z*

cycles, showing that it enhances the oscillatory properties of the system. Briefly, when R_o is very high or the birth rate of infected individuals is very low positive equilibria are more likely to become unstable and bifurcate into stable periodic cycles (Fig. 7.4). Put rather simplistically, this is because both the increasing of virulence and/or the increasing of R_o decrease total *Daphnia* density at equilibrium up to a point when it is no longer stable. The analytical predictions are summarized in the Appendix, in which we also derive an expression predicting that uptake of spores from the pond sediments will have

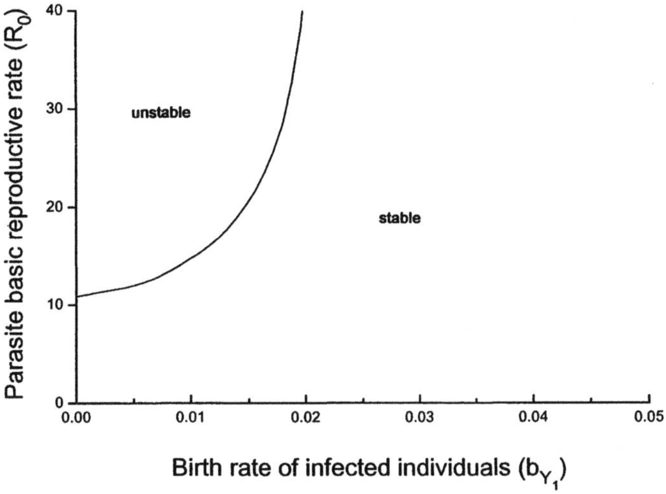

Fig. 7.4. Stability diagram for the mathematical model when $H(A) = 0$. To the left of the line the (positive) equilibrium is unstable, to the right of the line the equilbrium is stable. If $R_0 < 1$, the parasite cannot invade the community (not shown in the diagram). $K = a_2 = 10^8$, $a_1 = 10^7$, $b_X = 0.45$, $\mu = m = 0.1$, $\delta = 1.5 \times 10^6$, $b = 10^{-6}$

a dampening effect, serving to mitigate the influence of the free-floating spore-derived infection. This is illustrated in Fig. 7.5: because spores are regularly taken up from the sediment, infection is a more consistent feature of the system and less dependent upon the spore density in the water. The effect is strongest in that range of parameter values where the transition from local stability to stable periodic cycles occurs.

Buffering of the community fluctuations is not the only possible consequence of spore uptake from the ground. A more interesting, and potentially more important, function of the sediment spores is their possible role in re-initiating infection of *Daphnia* on a seasonal basis. Assume that the parasite is initially absent from a *Daphnia* population in early spring. With increasing temperature and sunshine the carrying capacity of algae increases. These increasing levels of algae cause the system to start to oscillate, and thus – because the fluctuations can lead to algal density regularly falling below a_3 – there will be a repeated uptake of spores from the sediment; and the parasite thereby re-infects the host population (Fig. 7.6). This scenario provides a mechanism by which *Daphnia* populations can be newly infected each spring.

7.6 Discussion

Over the last three decades freshwater zooplankton population dynamics were believed to be shaped by the effects of inter- and intraspecific competition and

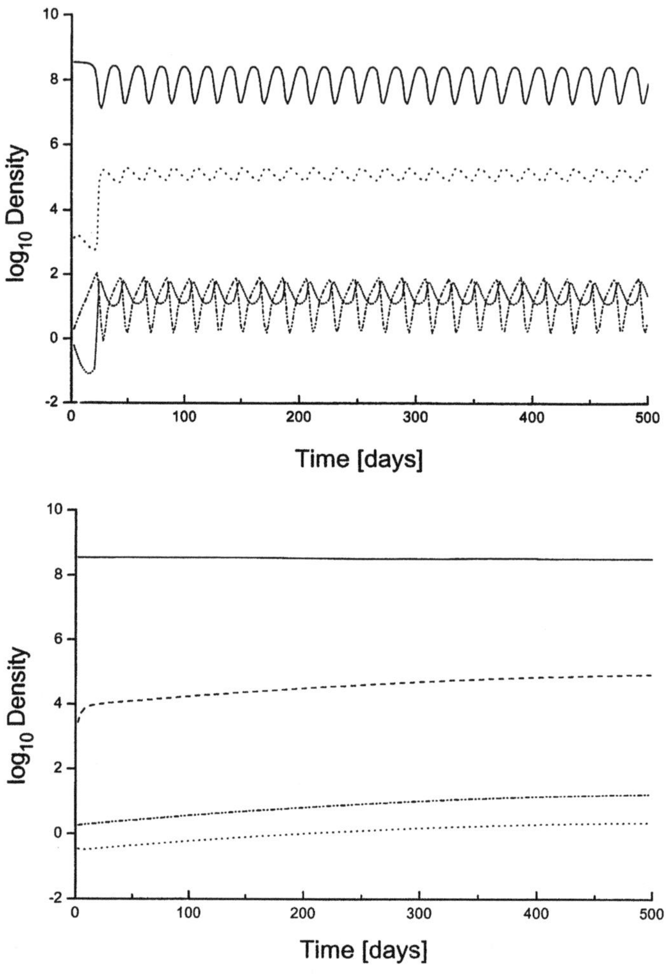

Fig. 7.5. The dampening influence of spore uptake from the pond sediments on the population dynamics of algae, *Daphnia* and parasite spores. $H(A) = d(a_3 - A)$. *Upper graph*: Oscillations, $a_3 = 10^8$. *Lower graph*: Monotonic damping, $a_3 = 5 \times 10^8$. Parameters for both graphs: $K = 3.5 \times 10^8$, $a_2 = 10^8$, $a_1 = 10^7$, $b_X = 0.45$, $b_Y = 0.225$, $\mu = 25$, $m = 0.2$, $d = 1.5 \times 10^6$, $b = 10^{-6}$, $s = 1000$, $d = 10^{-8}$. (–) Algae, *A*; (. . .) uninfected *Daphnia*, *X*; (–..–) infected *Daphnia*, *Y*; (– –) free-floating spores, *Z*

predation. Subsequently, in the 1990s, parasites and epibionts were added to this list (e.g. Threlkeld et al. 1993; Ebert 1995), yet their relative importance in natural populations remains to be investigated. The results of our field surveys confirm that microparasites are indeed very abundant in natural *Daphnia* populations and it seems likely that they play a significant role in population dynamics, competition and life history of their hosts. Others have established the role of epibionts in natural populations (for a review see Threlkeld et al. 1993).

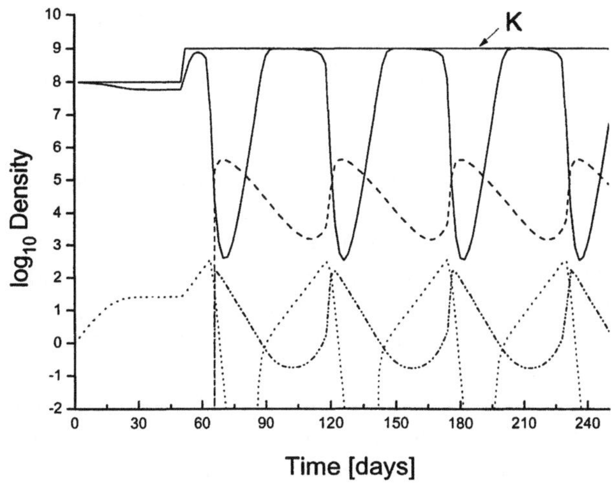

Fig. 7.6. The effect of a spring increase in the algae carrying capacity K on the *Daphnia* population oscillations and the uptake of spores from the sediment over the time course of a season. $H(A) = d(a_3 - A)$, $\psi_i(A) = b_i A/(a_4 + A)$, $i = X, Y$. $a_3 = 5 \times 10^6 < A^*$, $a_2 = 10^8$, $a_4 = 7 \times 10^7$, $b_X = 0.45$, $b_Y = 0.225$, $\mu = 0.25$, m = 0.2, $\delta = 1.5 \times 10^6$, b = 10^{-6}, s = 1000, d = 10^{-8}. (–) Algae, A; (. . .) uninfected *Daphnia*, X; (–..–) infected *Daphnia*, Y; (– –) free-floating spores, Z

Density-dependent transmission is undoubtedly an important factor for the epidemiology of microparasites, including *Daphnia*; nevertheless, the existence of additional density-independent transmission creates an interesting twist to the dynamics of parasite invasion. First, if a parasite infects hosts while they browse on the pond substrate it can persist in that host population even when each primary infection produces less than one secondary infection, that is even when $R_0 < 1$. Second, interactions with feeding conditions generate a complex scenario. Under poor feeding conditions, browsing on pond sediments continually causes new infections, but not under good feeding conditions; whereas good feeding conditions not only increase host numbers but also increase host-to-host transmission, and thereby the parasite's reproductive rate. Stress due to poor feeding conditions seems not to increase host susceptibility, but the consequent behavioural alteration does, by exposing the hosts to the sporebank (Ebert 1995).

This complexity is reflected in the results of various field studies on microparasite epidemiology in zooplankton. Brambilla (1983) observed microsporidian epidemics in *D. pulex* in three successive years, with peak prevalences close to 100% in adult females. Parasites were generally present whenever the host density was above ten animals/l, but in one year the parasite suddenly disappeared in mid-summer despite high host densities. Vidtmann (1993) observed that the microsporidium *Larssonia daphniae* was present only when host density was high, and yet was often absent during periods of high host density. Similar results were reported by Yan and Larsson (1988).

Ruttner-Kolisko (1977) described a significant relationship between host density and prevalence, and even attributed a strong population decline of the rotifer *Conochilus unicornis* to a microsporidian epidemic: "... *Plistophora* finally terminates its host species". Stirnadel (1994) was not able to detect density-dependent interactions between any of three *Daphnia* species and their microparasites. Moreover, there was no detectable food effect on parasite prevalence, but for some parasites a seasonal pattern was detected. Green (1974) found seasonal patterns in parasite prevalence but did not relate these to changes in host density. Despite this paucity of published evidence in support of a critical role for density dependence in *Daphnia* epidemiology, most studies do note that there is a minimum host density for parasite persistence, although the behaviour at high densities has yet to be determined.

The mathematical model we present highlights some of the features peculiar to plankton parasites. In a previous model, Weisser et al. (in prep.) showed that host-specific microparasites are able to promote coexistence among *Daphnia* species which would, in the absence of parasites, competitively exclude one another from a community. It was also shown that the basic reproductive rate of the parasite, R_0, as used in classical epidemiology models inspired by terrestrial systems, must be adapted when the uninfected system is cyclic rather than steady. Here we describe why the existence of a density-independent transmission term – caused by uptake of sediment spores under conditions of poor *Daphnia* feeding – significantly modifies the conditions for parasite persistence. The basic reproductive rate [as defined in our model, Eq. (5)] can become redundant altogether as a means of predicting parasite persistence when there is such a large, non-depleting spore bank in the sediment. Instead it is both the feeding behaviour of *Daphnia* and the properties of the resource that determine parasite invasions. We describe the dynamics of a mechanism by which infection could occur on a seasonal basis: rising ambient temperatures or sunlight levels, say, in the spring lead to an increase in algal carrying capacity, and consequently to an increasing likelihood of oscillatory behaviour; if large oscillations develop the algal density can temporarily dip low enough to provoke bottom-feeding by the *Daphnia*, leading to uptake of spores from the pond sediment, and thus to the initiation of a new infective epidemic.

Although, our model was developed for planktonic organisms, our results concerning the role of a density-independent mode of infection could also have relevance to a number of soilborne diseases. Fleming and coworkers (1986) investigated the density-dependent transmission of a virus in different populations of the soil-dwelling pasture pest *Wiscana* sp. (Lepidoptera: Hepialidae). Evidence for density-dependent transmission was found only in young pastures, not in old pastures. The lack of density dependence in old pastures could be a result of transmission occurring mainly from a spore pool, which had been accumulated over several generations. In laboratory populations of a virus-insect system, Sait and coworkers (1994) attributed their failure to detect density dependence to the rapid accumulation and long persistence of virus transmission stages within the cages. Viral contamination of the soil has been repeatedly claimed to be the source of various viral infections (e.g. granulosis virus and nuclear polyhedrosis virus infecting lepidopterans (Kellen and Hoffmann 1987; Young 1990; Woods et al. 1991). Similarly, the insect pathogen *Bacillus thuringiensis* is often found to accumulate in the soil (Dai et al. 1996). Thus it appears that durable transmission stages and their accumulation in pond sediments or soil might be a widespread phenomenon in natural host-parasite systems.

Daphnia and its microparasites are one of the few systems where both host and parasites have generation times short enough to allow experimental ecological studies to be carried out in the laboratory. The wide range of parasites available (bacteria, fungi, protozoa) allows for the testing and comparison of epidemiological, evolutionary and genetic models of infectious diseases. Our model shows that interactions of only a few ecological factors can produce complex epidemiological patterns. Only laboratory-based experimental tests can lead to the disentanglement of the factors acting simultaneously on interacting species. Nevertheless, field studies remain essential in determining the relative importance of these factors in the real world.

Acknowledgements. We thank D. N. Racey, T. Little, A. Carius and D. Brunner for reading earlier version of the manuscript. This work was supported by the Swiss Nationalfond Grant No. 3100-43093.95 (DE) and the British Natural Environment Research Council (DE, RJHP and WWW).

References

Allen YC, De Stasio BT, Ramcharan CW (1993) Individual and population level consequences of an algal epibiont on *Daphnia*. Limnol Oceanogr 38:592–601

Anderson RM, May RM (1979) Population biology of infectious diseases: part I. Nature 280:361–367

Anderson RM, May RM (1991) Infectious diseases of humans. Oxford University, Oxford

Blower SM, Roughgarden J (1989) Parasites detect host spatial pattern and density: a field experimental analysis. Oecologia (Berl) 78:138–141

Brambilla DJ (1983) Microsporidiosis in a *Daphnia pulex* population. Hydrobiologia 99:175–188

Brunner DU (1996) The population structure of *Daphnia magna*: genetics, parasites, competitors and environmental variables. Diploma Thesis, University of Basel, Basel, Switzerland

Canter HM, Lund JWG (1951) Studies on plankton parasites III. Examples of the interaction between parasitism and other factors determining the growth of diatoms. Ann Bot 15:359–372

Canter HM, Lund JWG (1953) Studies on plankton parasites II. The parasitism of diatoms with special reference to lakes in the English Lake District. Transact Br Mycol Soc 36:13–37

Chatton E (1925) *Pansporella perplexa* amœbien a spores protégées parasite des daphnies. Ann Sci Nat Zool Biol Anim 8:5–85

Chiavelli DA, Mills EL, Threlkeld ST (1993) Host preference, seasonality, and community interactions of zooplankton epibionts. Limnol Oceanogr 38:574–583

Dai JY, Yu L, Wang B, Luo XX, Yu ZN, Lecadet MM (1996) *Bacillus thuringiensis* subspecies *huazhongensis*, serotype H40, isolated from soils in the People's Republic of China. Lett Appl Microbiol 22:42–45

D'Amico v, Elkinton JS, Dwyer G, Burand JP (1996) Virus transmission in gypsy moth is not a simple mass action process. Ecology 77:201–206
DeAngelis DL (1992) Dynamics of nutrient cycling and food webs. Chapman and Hall, London
Dobson AP, Hudson PJ (1986) Parasites: disease and the structure of ecological communities. TREE 1:11–15
Ebert D (1994a) Genetic differences in the interactions of a microsporidian parasite and four clones of its cyclically parthenogenetic host. Parasitology 108:11–16
Ebert D (1994b) Virulence and local adaptation of a horizontally transmitted parasite. Science 265:1084–1086
Ebert D (1995) The ecological interactions between a microsporidian parasite and its host *Daphnia magna*. J Anim Ecol 64:361–369
Ebert D, Mangin K (1995) The evolution of virulence: when familiarity breeds death. Biologist 42:154–156
Ebert D, Rainey P, Embley TM, Scholz D (1996) Development, life cycle, ultrastructure and phylogenetic position of *Pasteuria ramosa* Metchnikoff 1888: rediscovery of an obligate endoparasite of *Daphnia magna* Straus. Philos Trans R Soc B 351:1689–1701
Fleming SB, Kalmakoff J, Archibald RD, Stewart KM (1986) Density-dependent virus mortality in populations of *Wisecana* (Lepidoptera: Hepialidae). J Invert Pathol 48:193–198
France RL, Graham L (1985) Increased microsporidian parasitism of the crayfish *Orconectes virilis* in an experimentally acidified lake. Water Air Soil Pollut 26:129–136
Freyer G (1991) Functional morphology and the adaptive radiation of the Daphniidae (Branchiopoda: Anomopoda). Philos Trans R Soc B 331:1–99
Green J (1974) Parasites and epibionts of Cladocera. Trans Zool Soc Lond 32:417–515
Hassell MP, May RM (1973) Stability in insect host-parasite models. J Anim Ecol 42:693–726
Horton PA, Rowan M, Webster KE, Peters RH (1979) Browsing and grazing by cladoceran filter feeders. Can J Zool 57: 206–212
Kellen WR, Hoffmann DF (1987) Laboratory studies on the dissemination of a granulosis virus by healthy adults of the Indian meal moth *Plodia interpunctella* (Lepidoptera: Pyralidae). Environ Entomol 16:1231–1234
Knell RJ, Begon M, Thompson DJ (1996) Transmission dynamics of *Bacillus thuringiensis* infecting *Plodia interpunctella*: a test of the mass action assumption with an insect pathogen. Proc R Soc Lond B 263:75–81
Lampert W (1987) Feeding and nutrition in *Daphnia*. Mem Ist Ital Idrobiol 45:143–192
Lee VA (1994) Parasitically-induced behavioural changes in zooplankton (*Daphnia magna*). Master Thesis, University of Oxford, Oxford
Mangin KL, Lipsitch M, Ebert D (1995) Virulence and transmission modes of two microsporidia in *Daphnia magna*. Parasitology 111:133–142
May RM, Anderson RM (1979) Population biology of infectious diseases: part II. Nature 280:455–461
McCauley E (1993) Internal versus external causes of dynamics in a freshwater plant-herbivore system. Am Nat 141:428–439
McCauley E, Murdoch WW (1987) Cyclic and stable populations: plankton as paradigm. Am Nat 129:97–121
McCauley E, Murdoch WW, Watson S (1988) Simple models and variation in plankton densities among lakes. Am Nat 132:383–403
Miracle MR (1977) Epidemiology in rotifers. Arch Hydrobiol Beih Ergeb Limnol 8:138–141
Murdoch WW, McCauley E (1985) Three distinct types of dynamic behaviour shown by a single planktonic system. Nature 316:628–630
Redfield GW, Vincent WF (1979) Stages of infection and ecological effects of a fungal epidemic on the eggs of a limnetic copepod. Freshwater Biol 9:503–510
Rosenzweig ML (1971) Paradox of enrichment: destabilization of exploitation ecosystems in ecological time. Science 171:385–387
Ruttner-Kolisko A (1977) The effect of the microsporid *Plistophora asperospora* on *Conochilus unicornis* in Lunzer Untersee (LUS). Arch Hydrobiol Beih Ergeb Limnol 8:135–137

Sait SM, Begon M, Thompson DJ (1994) Long-term population dynamics of the Indian meal moth *Plodia interpunctella* and its granulosis virus. J Anim Ecol 63:861–870

Sayre RM, Adams JR, Wergin WP (1979) Bacterial parasite of a cladoceran: morphology, development in vivo and taxonomic relationship with *Pasteuria ramosa* Metchnikoff 1888. Int J Syst Bacteriol 29:252–262

Scott ME, Dobson A (1989) The role of parasites in regulating host abundance. Para Today 5:176–183

Seymour R, Cowgill UM, Klecka GM, Gersich FM, Mayes MA (1984) Occurrence of *Aphanomyces daphniae* infection in laboratory cultures of Daphnia magna. J Invertebr Pathol 43:109–113

Stazi AV, Mantovani A, Fuglieni F, Dojmi di Delupis GL (1994) Observations on fungal infection of the ovary of laboratory-cultured *Daphnia magna*. Bull Environ Contam Toxicol 53:699–703

Stirnadel HA (1994) The ecology of three *Daphnia* species – their microparasites and epibionts. Diploma Thesis, University of Basel, Busel, Switzerland

Stirnadel HA, Ebert D (1997) Prevalence, host specificity and impact on host fecundity of microparasites and epibionts in three sympatric *Daphnia* species. 66:212–222

Threlkeld ST, Willey RL (1993) Colonization, interaction, and organization of cladoceran epibiont communities. Limnol Oceanogr 38:584–591

Threlkeld ST, Chiavelli DA, Willey RL (1993) The organisation of zooplankton epibiont communities. TREE 8:317–321

Vidtmann S (1993) The peculiarities of prevalence of microsporidium *Larssonia daphniae* in the natural *Daphnia pulex* population. Ekologija 1:61–69

Willey RL, Cantrell RL, Threlkeld ST (1990) Epibiotic euglenoid flagellates increase the susceptibility of some zooplankton to fish predation. Limnol Oceanogr 35:952–959

Woods SA, Elkington JS, Murray KD, Liebhold AM, Gould JR, Podgwaite JD (1991) Transmission dynamics of a nuclear polyhedrosis virus and predicting mortality in gypsy moth (Lepidoptera: Lymantriidae) populations. J Econ Entomol 84:423–430

Yan ND, Larsson JIR (1988) Prevalence and inferred effects of microsporidia of *Holopedium gibberum* (Crustacea: Cladocera) in a Canadian Shield lake. J Plankton Res 10:875–886

Young SY (1990) Effects of nuclear polyhedrosis virus infections in *Spodoptera ornithogalli* larvae on post larval stages and dissemination by adults. J Invertebr Pathol 55:69–75

Appendix

Equilibria and Stability

The model described by Eqs. (1)–(4) has four possible equilibria:

- E_i:No species present. This state is always vulnerable to invasion by the algae.
- E_{ii}: Only algae present ($A^* = K$). The *Daphnia* can invade provided that $\psi_X(K) > \mu + H(K)$. If $H(K) > 0$ then the *Daphnia* and its parasite invade simultaneously.
- E_{iii}: Algae and *Daphnia*, without parasite. The equilibrium levels of algae and *Daphnia* are given by the solutions of $\psi_X(A^*) - \mu = 0$ and $rA^*(1 - A^*/K) - f(A^*)X^* = 0$. This state exists only if $\psi_X(K) > \mu + H(K)$ (i.e. if *Eii* is unstable), and can go unstable in two ways: by bifurcating into stable oscillations (see below), or by being invaded by the parasite. If $H(A^*) = 0$ then the parasite invades when the "basic reproductive rate", $R_0 = s\beta X^*/m\mu$, is greater than unity (i.e. the eigenvalue $\lambda = s\beta X^*/m - \mu$ is positive). If $H(A^*) > 0$ then the parasite always invades, and knowledge of R_0 is superfluous.
- E_{iv}: Algae, *Daphnia* and parasite. This equilibrium exists only if state E_{iii} is unstable. It can go unstable by bifurcating into limit cycles, as described below. The equilibrium conditions are

$$
\begin{aligned}
rA^*(1-A^*/K) &= f(A^*)(X^*+Y^*) \\
Y^*(\mu - s\beta/m) &= X^* H(A^*) \\
X^*(\psi_X(A^*)-\mu) &= -Y^*(\psi_Y(A^*)-\mu)
\end{aligned}
\tag{A1}
$$

Limit Cycles

For this analysis we follow Weisser et al. (in prep.) and take the rate of change in the number of free-floating spores to be fast in comparison with the other variables, so that a pseudo-steady state hypothesis may be applied to the spore dynamics such that $Z \approx Ys/m$. The general Jacobian for small perturbations about steady state E_{iv} is

$$
\begin{pmatrix}
Q^* & -f(A^*) & -f(A^*) \\
P^* - X^* H'(A^*) & -\psi_Y(A^*)Y^*/X^* & \psi_Y(A^*) - X^* s\beta/m\mu \\
X^* H'(A^*) & \mu Y^*/X^* & -H(A^*)X^*/Y^*
\end{pmatrix}
\tag{A2}
$$

where $Q = r(1-2A/K) - f'(A)(X+Y)$, $P = \psi'_X(A)X + \psi'_Y(A)Y$, and dashes indicate differentiation with respect to A. This leads to the cubic eigen-equation

$$
\lambda^3 + g_1{}^*\lambda^2 + g_2{}^*\lambda + g_3{}^* = 0
\tag{A3}
$$

where

$$
\begin{aligned}
g_1 &= -Q + H\,X/Y + \psi_Y\,Y/X \\
g_2 &= -Q\{H\,X/Y + \psi_Y\,Y/X\} + (\mu - \psi_Y)Ys\beta/m + fP \\
g_3 &= (\mu - \psi_Y)[-QYs\beta/m - fH'(X+Y)] + fP\{H\,X/Y + \psi_Y\,Y/X\}
\end{aligned}
$$

The Routh-Hurwitz conditions tell us that this state is stable only if $g_1 > 0$, $g_3 > 0$ and $\xi = g_3 - g_1 g_2 < 0$. If the first two of these conditions are satisfied then there is the possibility of a Hopf bifurcation occurring at $\xi = 0$, going unstable as ξ increases. In the vicinity of such a bifurcation the oscillations will have period $2\pi/\omega$, where $\omega^2 = g_2$. In Weisser et al. (in prep.) we prove analytically for a system the same as here, except with $H(A) = 0$ for all A, that the limit cycles associated with this bifurcation – both in the absence and presence of parasites – must be stable (i.e. the bifurcation is supercritical). Using basic continuity arguments (differentiability w.r.t. parameters: Arnold 1973), it is possible to show that the equivalent limit cycles in the case $H(A) > 0$ must likewise be stable.

It is important to note that if the uninfected system exhibits stable oscillations then the use of R_0 for prediction of parasite invasion requires it to be modified as an average over the whole cycle (including taking account of any periods when A drops low enough that $H > 0$), and this becomes less reliable the larger the magnitude of the oscillations.

Influence of Parasitism and of Spore Uptake

It is shown in Weisser et al. (in prep.) that the limit cycles are more likely to occur the larger K, the carrying capacity of the algae. Furthermore, the presence of the parasite ($\beta > 0$) is expected to enhance the magnitude and/or likelihood of oscillations. Here we indicate how the uptake of spores from the pond floor ($\beta > 0$ and $H > 0$) is likely to influence the situation. For understanding the effect of $H(A)$ on the Hopf bifurcation it is useful to rewrite ξ in the form

$$
\xi = Q(g_1 G_1 + fP) + G_3
\tag{A4}
$$

where

$$G_1 = HX/Y + \psi_Y Y/X$$

$$G_3 = (\mu - \psi_Y)[fPY/X - fH'(X+Y) - Ys\beta/m(HX/Y + \psi_Y Y/X)]$$

Thus the Hopf bifurcation occurs when

$$Q = Q_3 = -G_3/(g_1 G_1 + fP) \tag{A5}$$

going unstable for increasing Q. It is easily shown that in the absence of infection $Q_3 = 0$. Although Q occurs on both the right-hand side and left-hand side of Eq. (A5), this form is useful because it allows insight into the role of H: if we assume that changing β and/or H does not significantly alter the size of Q (in Weisser et al. (in prep.) it is shown that when using realistic parameter values the magnitude of Q is predominantly controlled by the relative sizes of K and a_2, and only insignificantly by the indirect effects of β and H), then the effect of changing H can be understood simply by examination of Q_3. Thus we see that increasing H decreases G_3 but increases g_1, so that Q_3 becomes smaller. What this means is that whatever the intrinsic effect of the parasite upon the Hopf bifurcation, the effect of added benthic spore uptake is to act in opposition. We therefore predict, from this somewhat heuristic analysis, that uptake of spores from the pond bottom will have a dampening effect upon the oscillations.

Part C

Aspects of Chemical Ecology in Different Food Chains

8 Inter- and Intraspecific Transfer of Toxic Insect Compound Cantharidin

K. DETTNER

8.1 Introduction

Many arthropods are known to take up and to sequester toxic secondary compounds without being damaged (Bowers 1990). On the contrary, they use these plant-produced defensive chemicals for their own purposes and may increase their individual fitness. Very peculiar associations are observed between secondary chemicals from plants and so-called pharmacophagous insects. To these organisms special exogenous secondary compounds are highly attractive and are subsequently taken orally, detoxified and sequestered. These compounds drug-like may also increase survivorship of the pharmacophagous organism, because they are used as pheromone precursors or have morphogenetic activities (Boppré 1986). Moreover, these biologically active substances may be intra- and interspecifically transferred which may even result in a transfer of a special compound through trophic levels of an ecosystem.

Obviously, such pharmacophagously used compounds must not necessarily originate from plants. The toxic insect compound cantharidin may have the same activities. For a long time the extraordinary attractancy of this terpene anhydride for various arthropods remained a mystery, as illustrated by important contributions to the understanding of the biology of canthariphilous insects by Görnitz (1937) and Young (1984c,d). Since then, several aspects on the biology of canthariphilous insects have been published which led to a mosaic-like, sometimes blurred picture. It is the aim of this chapter to compile all these data on canthariphily in order to understand this unusual phenomenon. Certainly, the ingestion of a drug does not represent an uptake of nutritive material. However, many canthariphilous species must take up considerable amounts of prey tissue in order to liberate the bound cantharidin from the biomatrix. Therefore canthariphily can be interpreted as a true trophic interrelation.

Ecological Studies, Vol. 130
Dettner et al. (eds.) Vertical Food Web Interactions

8.2 Toxic Cantharidin: Biological Activity, Mode of Action, Occurrence and Ecological Significance

Cantharidin, a formal monoterpene anhydride (2-endo, 3-endo-dimethyl-7-oxabicyclo[2.2.1]heptane-2-exo, 3-exo-dicarboxylic anhydride; Fig. 8.3,1), represents an unusual natural compound with various pronounced physiological activities. It is very toxic to most animals including man (lethal dose for humans: 10–60 mg; intraperitoneal mouse LD_{50}: 1.0 mg/kg), induces serum-filled blisters if applied externally to the skin, acts as an antitumour agent, diuretic, abortifacient and may induce priapism in humans (McCormick and Carrel 1987). In spite of its toxicity, the last mentioned effect was responsible for its disuse as an aphrodisiac during the centuries. Submillimolar concentrations of cantharidin act as a feeding deterrent against predaceous arthropods and vertebrates (Carrel and Eisner 1974). Probably its extraordinary high toxicity to mammals has prevented the contact and systemic insecticide cantharidin and its analogues from being used for pest control.

According to Graziano et al. (1988), the toxicity of cantharidin and related herbicidal oxabicycloheptanes (e.g. endothal which inhibits root growth in higher plants) at the cellular level is attributable to a specific binding site in cytosol of mouse liver and other tissues (blood, brain, heart, kidney, skin or stomach). Experimental data on structure-activity relationships of endothal and cantharidin analogues with respect to their herbicidal activity and toxicity to mice indicate that a related oxabicycloheptane target site involving both the carboxylate groups, bridged oxygen of cantharidin and a metal ion might exist in plants and mammals (Matsuzawa et al. 1987). In 1992 Li and Casida were successful in identifying this 100-kDa cantharidin-binding protein (CBP) from mouse liver as protein phosphatase 2A (PP2A). PP2As are most highly conserved proteins (Shenolikar 1994) which are present in different organisms such as invertebrates, vertebrates, yeast and higher plants. As antagonists of proteinkinases these PP2As dephosphorylate proteins and therefore represent regulatory key enzymes in the cytosol. It is remarkable that PP2A is also present in olfactory cilia of mammals (Kroner et al. 1996), where it plays a regulatory role in governing the responsiveness of olfactory neurons. In order to activate the phosphorylated and thus inactivated odorant receptor proteins, the phosphate groups must be rapidly removed by PP2A (Kroner et al. 1996).

In nature cantharidin is only known to be produced by blister beetles (Meloidae) and the smaller oedemerid beetles (Oedemeridae) where it is found in haemolymph and various tissues (Fig. 8.1A). From alkalinized extracts of *Mylabris* and *Lytta* beetles the cantharidinimide could also be isolated (Purevsuren et al. 1987). Meloids with about 3000 species (120 genera) occur in warm dry regions and the aposematically coloured beetles are often found on flowers. Some groups are phytophagous, whereas larvae feed on eggs of grasshoppers or eggs and food stores in bee nests.

In meloid species, mostly males are capable of biosynthesizing cantharidin from a methyl farnesoate intermediate and may collect the anhydrid in their

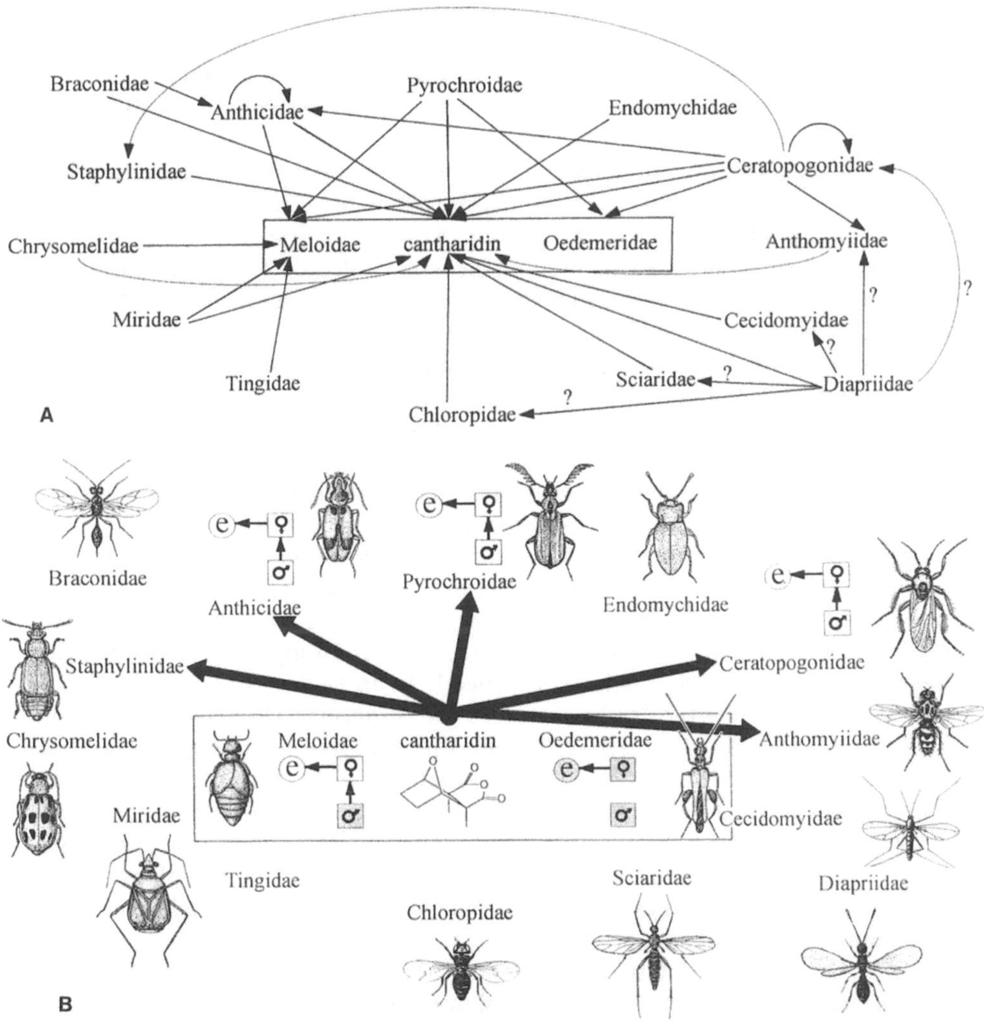

Fig. 8.1. **A** Attractancy of authentic cantharidin, cantharidin-containing producers (*centre* Meloidae, Oedemeridae) and canthariphilous species (*periphery*) for canthariphilous species as observed in the laboratory and in the field (see Sect. 8.5.2). Usually, attractancies may result in cantharidin uptake. *Arrows* within anthicid beetles and ceratopogonid flies indicate feeding on dead conspecifics (Anthicidae) and on other species of Ceratopogonidae. **B** Arrangement of cantharidin producers and canthariphilous taxa as in **A**. *Thick arrows* illustrate uptake and sequestration of cantharidin insofar as determined by gas chromatography – mass spectrometry. *Boxes* (males, females) and *circles* (*e* eggs) are connected by *arrows* if flows of cantharidin from males to females and to eggs (*e*) and larval stages have been proved. *Hatched boxes or circles* indicate capacity of individual sex or stage to produce cantharidin. *Boxes or circles without hatching* illustrate that individuals or stages are not capable of producing cantharidin

accessory glands which represent "cantharidin kidneys" (McCormick and Carrel 1987). The beetles usually contain several milligrams of cantharidin (maximally 11.1 mg/beetle in an *Epicauta* species) per individual (Frenzel and Dettner 1994); the biosynthesis of the toxic compound can be inhibited by systemic administration of 6-fluoromevalonate (McCormick and Carrel 1987). Since field-collected meloid specimens of both sexes contain cantharidin, this compound must be transferred from males to females (Fig. 8.1B). During their continuous matings, males of the genera *Lytta* and *Epicauta* transfer sperms and their cantharidin load from their accessory glands to females. Indeed, synthesis of cantharidin was also observed in both sexes of *Epicauta* and *Meloe* larvae (Meyer et al. 1968; Carrel et al. 1993).

In males both body-accumulated cantharidin and increased cantharidin biosynthesis during copulation may represent processes in order to maximize intersexual transfer of this compound. Subsequently, female meloids may deposit several hundreds of micrograms of cantharidin in each egg mass. By this paternal chemical defence, eggs and larvae are therefore effectively protected against many predators (McCormick and Carrel 1987).

Also, adults of Oedemeridae, the second natural source for cantharidin, are often brightly coloured (100 genera/1000 species). During the day, they feed on pollen and nectar on flowers, their larvae are pulp-feeders within umbellifer stalks or take up their food from rotten fungus-containing wood. In oedemerid beetles, both sexes (maximally 38.5 μg/beetle in an *Oedemera* species) and all developmental stages biosynthesize this toxic compound (Frenzel and Dettner 1994; Holz et al. 1994; Fig. 8.1A). The finding that males fed with deuterated cantharidin do not transfer this labelled material to females is thus conceivable. However, freshly hatched individual larvae of *Oedemera femorata* may actively increase their cantharidin titre. The first individual of a first stage larva, after hatching from its egg batch which is deposited within plant stalks, feeds on neighbouring eggs, maybe in order to increase its cantharidin load by egg cannibalism (Holz et al. 1994).

As a characteristic defensive behaviour meloid and oedemerid beetles may liberate cantharidin-containing haemolymph by reflex bleeding if they are disturbed. In plants cantharidin has not yet been discovered; however, insecticidal seeds of the Indian tree *Butea frondosa* (= *B. monosperma*; Leguminosae) contain palasonin (demethylcantharidin), where the 3-methyl group of cantharidin is missing (Bochis and Fisher 1968). Contrary to *Butea* seeds, extracts of roots and leaves are not toxic (Feuell 1965).

8.3 Attractivity and Significance of Natural and Synthetic Cantharidin to Canthariphilous Insects

For most organisms and especially insects, toxic cantharidin represents an effective feeding deterrent. However, apart from cantharidin producers meloid and oedemerid beetles, cantharidin acts as a potent attractant, arrestant and feeding stimulant to minute fractions within various insect taxa. Living and especially dead meloids and oedemerids and even their cantharidin-containing faces, and all other sources of cantharidin such as remnants and faces from cantharidin-tolerant predators which previously fed on cantharidin sources (bird droppings, spider webs with remnants of cantharidin-containing insects), are highly attactive to these so-called canthariphilous insects (Figs. 8.1, 8.2). Sometimes it has been observed that canthariphilous insects attack living or dead specimens of other canthariphilous species (Figs. 8.1A, 8.2F,G) which may even result in cannibalism. Canthariphilous insects are found in various families of Heteroptera, Coleoptera, Diptera and Hymenoptera (Fig. 8.1). By using cantharidin traps the attractancy to vagile adults was reported. In most cases, however, it is unknown whether the terpene also attracts larval stages of canthariphilous insects. Before compiling canthariphilous taxa and discussing their bionomy in Section 8.5.2, this section summarizes general data on attractancies and significance of cantharidin to these insects.

If filter papers of traps are impregnated with different amounts of cantharidin (as acetonic solutions) the attracted canthariphilous insects within a given time interval differ qualitatively and quantitatively (Fig. 8.2A,D). Görnitz (1937) could attract three beetle specimens of *Notoxus monoceros* with 0.1 μg cantharidin during 1.5 h. Similar low threshold values for attractancies were reported for canthariphilous Diptera such as *Atrichopogon oedemerarum* (Ceratopogonidae) or *Anthomyia pluvialis* (Anthomyiidae; Mayer 1962; Frenzel et al. 1992). In order to attract 12 specimens of the pyrochroid beetle *Neopyrochroa flabellata* within 24 h, nearly 400 μg cantharidin was necessary (Young 1984c). The data indicate that the chemoreceptors of canthariphilous insects may be highly sensitive to this solid terpene anhydride which is characterized by a very low vapour pressure (1.17×10^{-4} Pa at 20 °C).

By covering cantharidin-impregnated filter papers of traps with gauze, the trap allows attraction of canthariphilous insects without access to the bait. Therefore these unfed individuals only contain cantharidin which was previously collected and sequestered in the field. The capacity of these pharmacophagous insects to enrich orally taken cantharidin can be subsequently demonstrated in the lab. Field-collected females of the fungus gnat *Atrichopogon oedemerarum* from northern Bavaria contained significantly more total cantharidin than males. If given access to the synthetic terpene in the laboratory, females may increase cantharidin in their bodies by about 43-fold (conc.: 29-fold) and males by about 96-fold (conc.: 74-fold). In contrast, both sexes of field-collected specimens of *Atrichopogon trifasciatus* and *Anthomyia pluvialis* contained similar cantharidin titres. Fed specimens contained 2- to 4-fold (males) and 8- to 11-fold more of the substance (Frenzel and Dettner 1994). Similar rates of sequestration were also observed in various pyrochroid and anthicid species (Schütz and Dettner 1992; Holz 1995; Fig. 8.1B). If sequestration of cantharidin is analysed in different body compart-

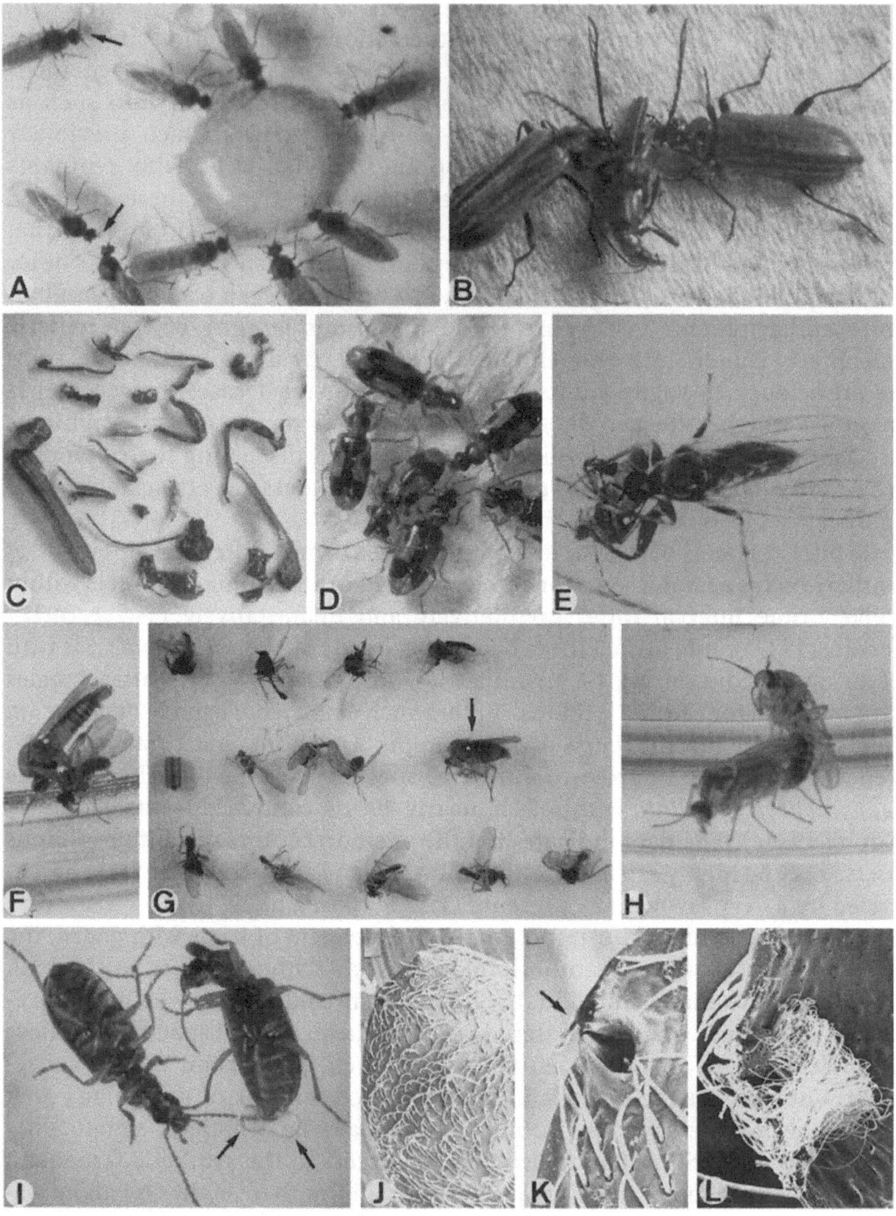

Fig. 8.2

ments there can be shown distinctly increased rates of incorporation for internal male genital organs (Schütz and Dettner 1992; Holz 1995). In contrast to canthariphilous beetles, sequestered cantharidin seems to be distributed thoughout all body-parts of ceratopogonid flies (Frenzel and Dettner 1994).

As reported above, many canthariphilous insects (beetles, ceratopogonids, anthomyiid flies) are directly attracted to cantharidin traps in order to feed on and to sequester the toxin. However, other canthariphilous taxa exist (e.g. sciarid flies, diapriid Hymenoptera) which are attracted towards cantharidin sources, but most specimens only stay one to several metres in the vicinity of the traps (K. Dettner, unpubl.). Obviously, other unknown signals seem necessary to evoke a close-range attraction.

Among those canthariphilous insects which vigorously feed on cantharidin sources, peculiar behaviours may be additionally observed. Within or in the near vicinity of cantharidin traps *Atrichopogon trifasciatus* (but not other canthariphilous ceratopogonids) shows an increased frequency of copulations (Fig. 8.2H). Before a copulation, which lasts several hours, females are suddenly attacked by males (M. Frenzel, unpubl.). Moreover, male anthicid beetles of *N. monoceros* show an increased courtship behaviour after having fed on natural or synthetic cantharidin. After ingestion of the drug the frequency of everting and presenting their anal vesicles towards females is significantly increased (C. Hemp, S. Thießen, unpubl.; Fig. 8.2I). Finally, predatory activities of canthariphilous insects such as *A. trifasciatus* females may be increased because these gnats preferably attack conspecifics and other ceratopogonid species (Fig. 8.2F,G).

Fig. 8.2. Interactions of cantharidin and insects. **A–D, F, G** Attractancy and ingestion of synthetic and natural cantharidin. Ceratopogonid flies of *Atrichopogon oedemerarum* (**A**) and anthicid beetles of *Notoxus monoceros* (**D**) ingesting traces of cantharidin from a water droplet (**A**) or from impregnated filter paper (**D**). *Arrows* in **A** indicate that faecal water droplets from other specimens are still attractive. Pyrochroid beetle *Schizotus pectinicornis* feeding on oedemerid beetle *Oedemera virescens* (**B**). Remnants of one *Oedemera* specimen after 24-h feeding activity of canthariphilous anthicids of *N. monoceros* (**C**). Female of *A. trifasciatus* feeding on *A. oedemerarum* female (**F**). One female of *A. trifasciatus* (*arrow*) after feeding on 12 specimens of other canthariphilous *Atrichopogon* species (**G**). **E** Feeding of a non-cantharidin-tolerant empidid fly on an *A. trifasciatus* female. **H, I** Behaviour at cantharidin sources. Copulation of *A. trifasciatus* at cantharidin traps (**H**); courtship of *Notoxus monoceros*, indicating male everting (*arrow*) its abdominal vesicles (*left* female; **I**). **J–L** SEM views of elytral notch (*arrows*) of a male *N. monoceros* at different magnifications (**J, K**); KOH-macerated elytral apex of same species showing cross section through notch and internal pore plate supplied with numerous tubules from glandular cells (**L**)

8.4 Transfer of Cantharidin Through Developmental Stages and Between Individuals in Canthariphilous Species

Depending on the season, phenological appearance, temperature, moisture of air, day time or weather conditions, the sex ratio of attracted canthariphilous insects may considerably fluctuate. However, during several years a characteristic sex ratio was observed at cantharidin sources for many canthariphilous taxa which often significantly deviates from the sex ratio in the field and may therefore reflect the different roles of cantharidin in both sexes.

As documented by fossil analysis, attractancy results with yellow-coloured traps, emergency studies or laboratory observations (e.g. K. Dettner, unpubl.; Havelka and Caspers 1981; Szadziewski and Krzywinski 1988), especially females are attracted in most canthariphilous ceratopogonid flies (sex ratio of males to females: *Atrichopogon* species, 3:5.2; Frenzel et al. 1992; Mayer 1962), *Anthomyia pluvialis* populations (Anthomyiidae) from southern France (sex ratio: 1:7; K. Dettner, unpubl.) and some canthariphilous mirids (Young 1984d). On the other hand the sex ratio of canthariphilous *Atrichopogon trifasciatus* (1:4), sciarid species or certain Anthicidae (e.g. *Formicomus* species: 1.5:2) may be more balanced (K. Dettner, C. Hemp, unpubl.), whereas most coleopteran and heteropteran canthariphiles can show a distinct male dominance at cantharidin baits (sex ratio of various genera of Pyrochroidae and Anthicidae: 2:0.01; Young 1984a,b).

As in many meloid beetles also females of different canthariphilous taxa may transmit cantharidin to their eggs, as was recorded in *Schizotus pectinocornis* with deuterated cantharidin (Holz et al. 1994). Labelled cantharidin is subsequently transferred into following developmental stages which exhibit decreasing titres per individual or per dry weight (Holz et al. 1994). Since developmental stages of canthariphiles are usually not attracted by the toxin, their cantharidin load is derived from their mothers.

It was also fascinating to recognize that male canthariphilous pyrochroid beetles (Eisner 1988; Holz et al. 1994) or male anthicid beetles (K. Dettner, C. Hemp and S. Thießen, in prep.) may transfer significant quantities of cantharidin to their female partners during copulation (Fig. 8.1B). This means that females which may not be capable of sequestering exogenous cantharidin are only chemically protected after a copulation. Since females of these insects only copulate a few times during their lives, there should be the tendency to maximize the nuptial gift, i.e. the cantharidin titre, by sexual selection by females. Actually, females of these canthariphilous insects show a peculiar courtship behaviour and hereby gustatorily test the cantharidin titre of the potential male copulatory partners by biting into a male exocrine gland (Eisner 1988; Eisner et al. 1996a; K. Dettner, C. Hemp and S. Thießen, in prep.). These glands are either located on head surfaces (Pyrochroidae: Eisner 1988; Eisner et al. 1996a; C. Holz and K. Dettner, in prep.) or elytral apices of male anthicid beetles (Schütz and Dettner 1992), and their glandular tissues are able to excrete variable amounts of cantharidin depending on the internal cantharidin titre of the male (Fig. 8.2J–L). Modified apices of anthicid elytra originally were interpreted as sense organs for cantharidin (Young 1984c). Behavioural tests where preference of females was recorded when they had to select between male partners previously fed with and without cantharidin showed that females of *Schizotus* and *Notoxus* significantly selected mates with higher cantharidin loads (Eisner 1988; K. Dettner, C. Holz, Thießen and C. Hemp, in prep.).

Within *Anthomyia pluvialis* females apparently are not able to transfer the toxin to their offspring. However, in *Atrichopogon trifasciatus* also eggs, larvae and adults of a generation have been shown to contain cantharidin with considerably increased titres when females were previously fed with cantharidin (M. Frenzel, unpubl.; Fig. 8.1B).

8.5 Interspecific Transfer of Cantharidin

8.5.1 From Producers and Canthariphiles to Cantharidin-Tolerant and Nontolerant Predators

In contrast to many exocrine reactive defensive compounds of insects which are unspecifically targeted against and generally damage living organisms, cantharidin may be tolerated by a distinct fraction of animals that dispose mechanisms in order to detoxify cantharidin (see Sect. 8.6.1).

Records on cantharidin-tolerant organisms are mainly based on predation of meloid beetles (Bologna 1991), oedemerid beetles (K. Dettner and G. Streil, unpubl.), anthicid beetles (K. Dettner and S. Thießen, unpubl.) and experiments with authentic cantharidin (Görnitz 1937). Due to the high attractivity of even cantharidin traces, carcasses, cantharidin-containing faeces and other remnants of these cantharidin-tolerant animals may be again used as attractants and as cantharidin sources by canthariphiles if the anhydride has not been destroyed after gut passage in the predator.

Diverse mites (Görnitz 1937) and especially spiders (*Aculepeira*, *Araneus*, *Argiope*, *Coryna*, *Latrodectus*, *Pardosa*, *Theridion*), after attacking the abovementioned beetles, are not injured at all after contact with cantharidin-containing prey. Moreover, they even contained cantharidin after ingestion of cantharidin-containing prey (K. Dettner, G. Streil and S. Thießen, unpubl.) which is now available for further transfer into canthariphilous insects.

Remarkably, many stored-product insects (e.g. *Anthrenus*, *Dermestes*, *Ptinus*, *Sitotrepa*) feed on cantharidin-containing drugs of Spanish flies. It was suggested that these insects avoid incorporation of cantharidin during the passage of food through the gut (Görnitz 1937). Several cantharidin-tolerant arthropods must be found among predators (e.g. Asilidae, Ephippigeridae, Formicidae, Reduviidae) and parasitoids (e.g. Bombyliidae, Calliphoridae,

Eulophidae/Chalcidoidea: *Melittobia*, Phoridae: *Phora*) of meloid beetles (Bologna 1991). Larvae of Myrmeleonidae (K. Dettner and S. Thießen, unpubl.) represent further cantharidin-tolerant predators.

Among vertebrates numerous records exist of species which take up cantharidin-containing prey and are not injured (Eisner et al. 1990) after ingestion of meloids. According to Bologna (1991), these species include frogs and toads (Eisner et al. 1990), Iguanidae and many birds (e.g. Coraciidae, Emberizidae, Gruës, Hirundinidae, Laniidae, Meropidae, Muscicapidae, Paridae, Pici, Tyranni). Among mammals only primitive insectivorous species such as hedgehogs or to some extent bats, armadillos and rabbits may tolerate cantharidin-containing food (Kelling et al. 1990; Bologna 1991). It has been reported that avain faeces may contain meloids or oedemerids which seems to be the reason for the faecal atractiveness for canthariphilous insects.

Due to its extraordinary toxicity, cantharidin-containing food cannot be consumed by most mammals. Further nontolerant species especially include predacious insects such as ground beetles (Carrel and Eisner 1974), Heteroptera and ants (K. Dettner and S. Thießen, unpubl.) but also herbivores such as Orthoptera, beetles (e.g. *Phyllopertha*), lepidopteran larvae (e.g. *Bombyx*, *Lymantria*, *Malacosoma*, *Vanessa*), aphids or stick insects (Görnitz 1937).

8.5.2 From Producers to Canthariphiles and Within Canthariphiles

Data on the habitats and the biology of canthariphilous insects are of special interest because it is desirable to know how these insects may have access to the precious, patchily distributed cantharidin sources. Is the capacity to sequester cantharidin exclusively restricted to winged adult stages after they have perceived the terpene from a great distance? In the case of canthariphilous males it would then be only necessary to return to the habitat in order to copulate and to transfer the drug into their copulation partner and offspring. If both sexes are canthariphilous, copulation and cantharidin transfer could take place in the vicinity of natural cantharidin sources. Then only females have to return to their habitats for deposition of eggs. In principle, it should be also possible that moderately vagile larval stages may sequester cantharidin in their habitat in order to increase their fitness. In the latter case, the prerequisite exists that canthariphilous insects and cantharidin producers at least temporarily share common habitats. Many canthariphilous insects are associated with rotting wood which usually contains fungi. This could be a suitable habitat for both canthariphilous insects and certain cantharidin producers from the Oedemeridae. In order to gain cantharidin there may also be a possibility that canthariphilous species purposively search for typical and temporary domiciles of meloids and oedemerids such as umbellifer flowers.

In several cases canthariphiles are associated exclusively with fungi (e.g. Handsome fungus beetles, certain Sciaridae). Therefore it would be of interest if cantharidin-like compounds may also occur as fungal metabolites. Otherwise cantharidin should be gathered from other co-occurring canthariphilous species.

The following compilation includes biological and taxonomic notes on all hitherto known canthariphilous taxa which are important for an evolutionary interpretation of this phenomenon (Fig. 8.1). Certainly, it is often not possible to decide with certainty whether insects from cantharidin traps are virtually canthariphilous. This may depend on the design, the position and the sticky material on the bottom of the traps, the geographical area, the season, the biocoenosis in each particular site and the biology of the trapped arthropods. Principally other volatiles emitted from trapped canthariphiles may be sufficient to deceive the presence of a canthariphilous taxon. On the other hand, repeated catches of specific taxa (maybe positively correlated with the presence of true canthariphiles) and their absence in control traps without cantharidin may indicate canthariphilous relationships. However, the active space of sublimated cantharidin from individual traps is not known and it may be possible that small numbers of canthariphilous insects also appear in control traps which are placed in the neighbourhood of cantharidin-baited traps. This seems especially important for those canthariphiles (e.g. Sciaridae, Diapriidae) which are found especially in the vicinity of the traps and where cantharidin shows no extreme close-range attractancy as in Pyrochroidae, Anthicidae, Anthomyiidae or Ceratopogonidae. Moreover, it was usually not possible to analyse these species by GC-MS in order to determine the brought or ingested cantharidin.

There are several observations from North and South America on attractancy and feeding of Miridae on meloid beetles (Pinto 1978; Young 1984c,d; Mafra-Neto and Jolivet 1994). The nearly 30 known canthariphilous mirid species are found within the two subfamilies Orthotylinae (*Hadronema*) and Bryocorinae (*Caulotops*, *Cryptocapsus*, *Eurychilella*, *Halticotoma*, *Neuleucon*, *Pycnoderes*, *Sixeonotus*, *Sysinas*, *Thentecoris*). Most canthariphilous mirids are observed inserting their mouthparts into membranous areas of such meloids as *Epicauta*, *Lytta* or *Meloe*. They feed also on meloid cadavers, haemolymph and authentic cantharidin (Pinto 1978). The meloids are constantly surrounded by these bugs and are even pursued by flying or walking canthariphilous species. The beetles try to defend against these aggressive ectoparasites by scraping with their legs. Since the defensive capacity of meloids is reduced during mating they are then especially molested by the plant bugs (Pinto 1978) just as observed in ceratopogonid flies which attack copulating meloids. It seems remarkable that Bryocorinae with most representatives of canthariphilous species often feed on ferns. Within Orthotylinae there are both generalist predators and phloem sap feeders. According to Young (1984d), canthariphilous mirids possess no mouthparts which would be adapted for arthropod prey. It is not clear whether canthariphilous mirids were primarily phytophagous (Pinto 1978) and the ingestion of cantharidin would represent an ancillary food or vice versa (Young 1984d).

Small numbers of an undetermined species of Tingidae were regularly found associated with *Epicauta* beetles (Mafra-Neto and Jolivet 1994). Due to the lack of data, it is not possible to evaluate the significance of canthariphily within this phytophagous family which is closely allied with mirids.

Most records dealing with canthariphilic associations are from Anthicidae (Görnitz 1937; Young 1984c,d; Schütz and Dettner 1992) which amply reflects that canthariphily is distributed in many anthicid beetle taxa including Notoxinae (*Mecynotarsus, Notoxus, Pseudonotoxus*), Tomoderinae/Tomoderini (*Thomoderus*), Anthicinae/Anthicini (*Anthicus, Formicilla, Hirticomus, Pseudoleptaleus, Sapintus*), Anthicinae/Microhorini (*Acanthinus, Aulacoderus, Microhoria, Tenuicomus, Vacusus*), Anthicinae/Formicomini (*Formicomus*) and Anthicinae/Endomiini (*Endomia, Anisotria, Pedilus*). Anthicid beetles are reported to attack all kinds of dead meloids such as *Meloe* or *Lytta* (Görnitz 1937; Havelka 1980) but also may feed on dead conspecifics. They also attack living triungulid larve of *Meloe* and feed on them (Görnitz 1937). Canthariphilous anthicid beetles are capable of sequestering considerable amounts of cantharidin which is transferred to testes and accessory glands in males and to ovaries in females (Schütz and Dettner 1992). Concerning attractancy and interindividual transfer of cantharidin there are obviously two strategies (a,b) realized in Anthicidae. Within canthariphilous species of type a, long-range attractancy for cantharidin is mainly restricted to males (e.g. *Notoxus monoceros*; Fig. 8.2D). Females, although positively orientating to the anhydride, are only found in minor amounts at cantharidin baits, which may correspond to a reduced perceptive capacity for cantharidin or with a reduced manouevrability in female anthicids (Schütz and Dettner 1992). Males of these anthicids possess peculiar elytral notches which in fact represent glandular structures in order to excrete minor amounts of cantharidin depending on the internal titre of the toxin (Fig. 8.2J–L). During close-range courtship behaviour, females bite into these elytral glands in order to measure the cantharidin titre of the potential mate prior to copulation (Schütz and Dettner 1992). This kind of sexual selection ensures that females only accept those copulation partners that contain significant amounts of cantharidin (Schütz and Dettner 1992). After copulation males have transferred significant amounts of cantharidin to females which transfer this precious wedding gift to their eggs and larvae.

In anthicid beetles of type b (such as *Formicomus* species), both sexes are attracted to cantharidin and obviously can sequester the terpene. During copulation, males can transfer cantharidin to females, but obviously both sexes are able to sequester cantharidin for their offspring. Males of type b possess no elytral notches (Schütz and Dettner 1992).

Anthicid beetles are relatively common in leaf litter but may be also found on flowers; some species are egg predators and others are probably scavengers.

Guts of adult anthicids very often contain mycelia and spores of fungi and it has been suggested that all anthicids primarily feed on fungi in addition to various arthropods (Görnitz 1937). Larvae live in decaying vegetation.

Young (1984a) and Eisner (1988) independently discovered that *Neopyrochroa flabellata* (Pyrochroidae) could be attracted by cantharidin. Later, canthariphily could be found especially in males of other representatives of subfamilies Pyrochroinae (*Neopyrochroa*, *Pyrochroa*, *Schizotus*) and Pedilinae (here treated as a subfamily of Pyrochroidae: *Anisotria*, *Pedilus*), although both sexes show equal abundance (see Sect. 8.4). The species *Anisotria shooki* could even be described for the first time by using cantharidin-baited filter papers (Young 1984b). However, members of the family have also been observed chewing elytra of *Meloe* and *Epicauta* meloid beetles (LeSage and Bousquet 1983; Young 1984d) and feeding on oedemerid beetles (Fig. 8.2B). Male representatives of *Neopyrochroa* and *Schizotus* possess peculiar glands on the dorsal side of their heads where a fraction of internal cantharidin is excreted depending on the titre of internal cantharidin (Eisner 1988; Eisner et al. 1996a; C. Holz and K. Dettner, in prep.). Both males and females may incorporate 37 pg cantharidin/μg dry weight from exogenous sources. Apart from the head glands, cantharidin is especially enriched in male accessory glands (Holz 1995; Eisner et al. 1996b). During close-range courtship behaviour, females may effectively test cantharidin titres of potential males and accept only those males which were previously successful in gathering and sequestering increased amounts of cantharidin. By copulation males may transfer significant amounts of cantharidin from their accessory glands to the females which subsequently incorporate this nuptial gift into their eggs (Eisner 1988; Holz et al. 1994). Deuterium-labelled cantharidin from males can even be detected in following larval stages (Holz et al. 1994).

Short-lived adults of Pyrochroidae may be collected at light, in bait traps, on shrubs and flowers (e.g. umbellifer) and under bark. The five larval stages of Pyrochroinae are also found under bark, where they feed on rotting cambial tissue and fungal hyphae (Parker 1982). Larvae of Pedilinae live in decaying plant debris.

Endomychidae include five canthariphilous species within subfamilies Eumorphinae (*Aphorista*, *Lycoperdina*) Xenomycetinae (*Xenomycetes*) and Stenotarsinae (*Danae*) which feed and mate on cantharidin (Young 1984d, 1989). These beetles and their larvae are found beneath bark in fungus and rotton wood, in dung, and in decaying mouldy fruit. Also, adults feed and mate on fungi and moulds. The canthariphilous, brachypterous *Lycoperdina* species are usually found within puffballs; sometimes they can also be collected from forest litter. One possible reason for the peculiar leaving of fruiting bodies and wandering by the *Lycoperdina* larvae (Pakaluk 1984) might be their canthariphilous behaviour.

Canthariphily could be recently recorded from two genera of galerucine chrysomelid beetles (Chrysomelidae). Out of 196 genera of subtribe Luperina (Seeno and Wilcox 1982) several specimens of an African *Bonesioides* species were attracted to cantharidin traps (C. Hemp, unpubl.). Moreover, one South American *Diabrotica* (*Aristobrotica*) species (out of 40 genera of subtribe Diabroticina; Seeno and Wilcox 1982) was observed to be probably attracted to and feeding on dead or thanatosis-displaying meloid beetles (Mafra-Neto and Jolivet 1994). The *Diabrotica* beetles live on the ground and perforate the abdominal sternites of the meloids in order to feed on the internal organs. It is not clear if meloid faeces together with the solanaceous secondary plant chemicals or cantharidin act as attractant for this unusual entomophagous leaf beetle (Mafra-Neto and Jolivet 1994). Just as other galerucines, *Diabrotica* beetles usually represent polyphagous herbivores which use chemical cues for host plant selection. Because there is also a strong tendency in Galerucinae to sequester toxic or bitter secondary plant constituents (Dettner 1987), it seems highly probable that entomophagy was a secondary event in the evolution of *Diabrotica angullicollis*.

Single specimens of different species of Staphylinidae are usually found in cantharidin traps. Among pollen and nectar-feeding species of adults of the gneus *Eusphalerum* (Omaliinae: Eusphalerini), one species (*E. minutum*) was found to be canthariphilous and all analysed specimens contained considerable amounts of the terpene anhydride (about 400 pg/μg; K. Dettner and C. Holz, unpubl.). However, *E. minutum* feeds mainly on pollen (Klinger 1983) and incorporation of cantharidin must therefore be interpreted as pharmacophagy. Since predacious feeding by adult Omaliinae has evolved multiple times from saprophagy or mycophagy (Newton and Thayer 1995), canthariphily might represent a secondary event in *Eusphalerum* evolution. On the other hand, ancestors of *Eusphalerum* were probably zoophagous (Klinger 1983). Therefore, the pollinivorous lifestyle of all existing representatives of the genus has been derived and receptors for cantharidin could be an historical heritage, at least in European *Eusphalerum* species. It is highly interesting that females of the canthariphilous ceratopogonid *Atrichopogon lucorum* were observed more or less successfully attacking living *E. minutum* by piercing interesegmental membranes, elytra and prothorax of these rove beetles (Klinger 1979).

Cleridae as a family has been reported to be canthariphilous (Bologna and Havelka 1984). Both larvae and adults of most clerid beetles are predacious and are usually found on and under bark. Moreover, they may also feed on pollen and may be attracted to carrion. However, cantharidin could not be identified from a Chinese *Trichodes* species (Juanjie et al. 1995).

Apart from heteromeran canthariphilous beetles, about 20 species from the genera *Atrichopogon* (Fig. 8.2A), *Culicoides* and *Forcipomyia* of the ceratopogonid subfamily Forcipomyiinae represent the most important faunal elements on all kinds of cantharidin sources (Ceratopogonidae). Canthariphilous *Atrichopogon* species (subgenus *Meloehelea*) were often observed feeding on living and especially dead meloids (*Meloe*, *Epicauta*) and oedemerids (*Oedemera*, *Chrysanthia*; Mayer 1962; Wirth 1980). For example, females of *A. lucorum* attacked both sexes of living oedemerid *Oedemera femorata*. The flies were also observed successfully piercing intersegmental membranes of tarsal joints and abdominal sclerites of living beetles (Havelka 1979). Favourable attacks may occur during copulation of meloids, which may be attacked by mixed swarms of hovering *Meloehelea* species (Havelka 1980; Wirth 1980). The high sensitivity of ceratopogonids to cantharidin is illustrated by their behaviour in sucking even on meloid faeces (Havelka 1979). Also, water droplets on a canthardin-impregnated filter are voraciously devoured by the flies although cantharidin is practically insoluble within water (Fig. 8.2A). The feeding flies deplete water droplets from their hind gut which are attractive and subsequently fed on by other ceratopogonid specimens. By using their strong salivary proteinases (Downes 1978), canthariphilous ceratopogonids may also feed on dry cantharidin-containing prey from spider webs.

It is conceivable that the aggregated occurrence of meloid and oedemerid cantharidin sources represents a possibility for predatory canthariphilous species to gain further access to the already sequestered toxin of other canthariphilous species. In this case obviously dead canthariphiles are significantly preferred. For example, females of *Atrichopogon trifasciatus* fed on dead (but also on living: Fey 1954) specimens of *Anthomyia pluvialis* and tried to attack living anthicid beetles (Mayer 1962). The female flies of *A. trifasciatus* hold an exceptional position since they are very aggressive and attack not only larger prey but also other canthariphilous *Atrichopogon* species including even conspecific males and females (Fig. 8.2F,G). Unfed or cantharidin-fed *A. oedemerarum* or *A. lucorum* (but not *A. brunnipes*) were observed to be grasped by the predator's midlegs and sucked during 5–10 min by female *A. trifasciatus* (M. Frenzel, unpubl.). The prey does not survive this attack. Even noninsectivorous, normally nectar-feeding males of *A. trifasciatus* were observed in grasping and handling an *A. oedemerarum* female with their mouthparts. In contrast, predacious activities were never observed in males of the other three ceratopogonid species (M. Frenzel, unpubl.). Within the subfamily Forcipomyiinae females of most species attack larger insects and arachnids in order to take up protein-rich food, without doing serious injury (Downes 1955). At the same time both sexes of these insectivorous species visit flowers such as umbellifers to gain sugar-containing nectar (Downes 1958). According to Downes (1958), a polyphagous insectivorous food represents a primitive nutritional type in Forcipomyiinae. This could indicate that also canthariphily could be a primitive character which indeed was retained only within few subgenera of *Atrichopogon*.

Known immature stages of canthariphilous species are found on the underside of wet rotting logs (Wirth 1980), under decaying foliage or on wet stones within or at the border of small brooks (Bangerter 1933).

Among Anthomyiidae, only *Anthomyia pluvialis* and *Delia trispinosa* (*Hylemyia*) are attracted to cantharidin. Larvae of anthomyid flies live in plant stems, decaying plant material or are parasitic in insect larvae; some adults are predacious. Whereas nothing is known about the life history of *Delia trispinosa*, adults of *A. pluvialis* are found on plants infested by aphids, various flowers, plant resins, faeces of birds, dung and dead fishes (Görnitz 1937; Hennig 1976). *Anthomyia* larvae are saprophagous and phytophagous and are found within similar substrates as adults (additionally in fungi and nests of birds). In the laboratory, *Atrichopogon brunnipes* was observed feeding on *A. pluvialis* (Fey 1954).

Adults of four Sciaridae species (Lycoriidae) and undetermined species of Cecidomyidae (Lestremiinae, Cecidomyiinae) have been attracted by cantharidin (K. Dettner, nnpubl.; Görnitz 1937; Young 1984d). Cecidomyids live in decaying organic matter (fungi feeders), may be predacious or parasitic and feed on fresh plants. Larvae of sciarids generally feed on fungi (often under bark or within decaying wood) but also take up decaying or fresh plant materials. Adults are further recorded from animal excrements and many species can be reared from old nests of birds and mammals (Freeman 1983). According to Lengersdorf (1930), the four known canthariphilous species belong to two subfamiles Megalosphyinae (*Scythropochroa*) and Lycoriinae (*Heterosciara*, *Schwenckfeldina*, Gen. sp.). The larvae of *Heterosciara* and *Scythropochroa* probably feed on decaying wood (Freeman 1983).

Within Chloropidae one species of *Goniopsita* was regularly attracted to cantharidin and palasonin traps together with anthomyid flies (K. Dettner, unpubl.). Most species of frit flies live in plants or flowers; some species are also found in decaying animal material or are predacious on insects or insect eggs. Interestingly, a pharmacophagous African species of this family was also recorded, which is attracted by and feeds on pyrrolizidin alkaloids (Boppré, 1986).

Among four genera of canthariphilous Braconidae from Blacinae (*Blacus*, *Syrrhizus*) and Euphorinae (*Microtonus*, *Streblocera*), females of two canthariphilous braconid species of genera *Microtonus* and *Syrrhizus* have also been reported to parasitize on adult *Notoxus* anthicids (Görnitz 1937; Young 1984c,d). Obviously, cantharidin represents a kairomone for these braconids. The euphorines and blacines are koinobiont endoparasitoids, especially of beetles. It remains of interest than certain *Microtonus* species may also successfully attack developmental stages of their hosts and two other Euphorini genera are reported to parasitize on mirid bugs (Shaw and Huddleston 1991). Further, Bologna (1991) reports braconids of *Melittobia* as parasitoids of meloid larvae of *Zonitoschema*.

Traps supplied with cantharidin or several analogues in the area of northern Bavaria and southern France repeatedly contained representatives of proctotrupoid family Diapriidae (especially Diapriinae). These parasitoids are both solitary and gregarious and primarily attack larvae and pupae of Diptera. Various host records are from Mycetophilidae, Sciaridae, Ceratopogonidae and cycloraphan families such as Chloropidae, Anthomyiidae, Muscidae, Tachinidae, Calliphoridae, Sarcophagidae and Tephritidae (Hennig 1976; Gauld and Bolton 1988). It is suggested that cantharidin acts as a kairomone for these parasitoids.

Together with true canthariphilous taxa, the traps regularly contained several specimens from other taxa. Whether these species are attracted by cantharidin is questionable, because few records are from control traps. These taxa include representatives of Chironomidae, Psychodidae, Tipulidae, Mycetophilidae, Scatopsidae, Phoridae, Sphaeroceridae, Empididae, various Muscidae, Thysanoptera, Psocoptera, Lagriidae, Chalcidoidea and Inchneumonidae. Concerning phorid flies it seems remarkable that representatives of *Phora* may parasitoid on meloid beetles (Bologna 1991). These humpbacked flies infest carrion of vertebrates and insects; some species may also parasitize living arthropods or develop within fungi. Adults often visit flowers, particularly Umbelliferae (Schmitz 1981). Because phorids were regularly caught in control traps, it is not clear whether they represent true canthariphiles. Among caught lagriid beetles, *Lagria* larvae were exclusively captured. It would be astonishing if larval stages of these heteromeran beetles were canthariphilous. Very often a few thysanopteran species from Thripidae (abundant: *Limothrips denticornis*) were also caught in the cantharidin traps. Most species of thrips are phytophagous but also predators especially of soft-bodied arthropods. Other species may ingest fungal spores or pollen grains (Schliephake and Klimt 1979). Further studies must show if these typically show any preference for cantharidin-containing arthropods.

True representatives of "background fauna" seem to be represented by Collembola, aphids, cicadas, other heteropteran families, psyllids, ants and various beetles such as phalacrids, nitidulid, curculionid or rove beetles.

Recently, the presence of cantharidin has also been reported gas chromatographically from nymphal stages (about 0.15%) and adults (about 0.06%) of the Chinese homopteran species *Lycorma delicatula* (Fulgoridae) (Feng et al. 1988). It is not known from what kind of cantharidin sources this herbivorous planthopper species feeding on *Ailanthus* and *Melia* trees may gain its toxic terpene.

8.6 Evolution of Canthariphily

8.6.1 Cantharidin May Increase Individual Fitness

Due to its overall toxicity, production, sequestration and transfer of cantharidin may represent a significant selective advantage for cantharidin-producing or adapted canthariphilous insects. These taxa are therefore chemically protected against a wide array of noncantharidin-tolerant predators and parasitoids and may thereby increase their individual fitness.

Remarkably, vagile adults of most canthariphilous taxa which have to search actively for this precious compound just represent those stages which are already chemically well protected by various exocrine defensive glands or haemolymph toxins. Canthariphilous beetle species may secrete methylbutanoic acids, esters, E-2-hexenal (abdominal defensive gland of *Eusphalerum*: Dettner and Reissenweber 1991), iridoid compouds (thoracic defensive gland of anthicids: Hemp and Dettner 1997), anthraquinones (haemolymph compounds of galerucine chrysomelids: Dettner 1987) or may possess defensive glands (pyrochroids: Dettner 1984). In the same way, mirid and tingid heteropteran species are effectively protected by defensive secretions (Aldrich 1988).

As a consequence, chemical protection is already guaranteed by defensive secretions before adult canthariphiles have gained access to natural cantharidin sources or even in those geographical areas where natural cantharidin sources are rare and cantharidin titres of canthariphiles remain very low. This indicates that cantharidin seems of special advantage for nonmobile stages eggs and pupae but also for larvae which are usually devoid of defensive glands. Female pyrochroids and anthicid larvae, both with low cantharidin titres, were partly fed with cantharidin in the laboratory. Subsequently, developmental stages with (pyrochroid eggs, anthicid larvae) and without cantharidin were offered to ants of the genus *Myrmica*. In both cases only cantharidin-containing samples exhibited repellencies and deterrencies against the ants (C. Holz, U. Hofmann, A. Geisen, unpubl.). In the same way, predatory larvae of the coccinellid beetle *Coleomegilla maculata* rejected cantharidin-containing eggs of pyrochroid *Neopyrochroa flabellata* (Eisner et al. 1996b).

A deterrent activity of cantharidin in adult canthariphiles is finally illustrated by the following experiment. The small empidid fly *Platypalpus* (and probably other empidid species) was repeatedly observed within cantharidin traps where it was attacking and successfully sucking (up to 10 min) on attracted ceratopogonids, which did not survive in spite of their thanatosis (Fig. 8.2E). If this empidid fly was attracted to cantharidin, this compound could represent a kairomone for this predator. However, when cantharidin-fed living ceratopogonids were presented as prey only short *Platypalpus* attacks were observed. Afterwards the prey survived and the predator showed cleaning behaviour of its mouthparts (Frenzel and Dettner 1994).

However, cantharidin does not only protect from attacks by predators or parasitoids. This remarkable anhydrid further possesses potent antifungal activities against *Trichophyton* and *Microsporum* species and may therefore protect developing embryos and all developmental stages from entomopathogenic fungi (Bologna 1991; Blum 1996).

There remains the question of whether cantharidin-containing arthropods are endangered by attack by canthariphilous species. In fact, canthariphiles primarily attack faeces, dead, moribund or copulating cantharidin-containing organisms (see Sect. 8.5.2). As a rule, living attacked species survive the encounter with canthariphiles such as ceratopogonids. Moreover, violent aggression may be mainly restricted to the laboratory situation. Therefore, interactions between both types of arthropods do not necessarily increase the mortality of attacked species, but may be rather interpreted as a kind of molestation. Whether immobile cantharidin-containing stages (eggs, pupae) or larvae significantly suffer from attacks by canthariphiles has to be elucidated.

Since cantharidin-producing beetles geographically dominate in subtropical or tropical areas, the cantharidin supply for canthariphiles may drastically vary depending upon the geographical area. This was amply illustrated by studying cantharidin titres of pyrochoid and anthicid beetles and also from ceratopogonid flies collected in several areas of central Europe (Schütz and Dettner 1992; Frenzel and Dettner 1994). Increased amounts of the anhydride were only detectable if there were sympatric meloid and/or oedemerid beetles. Sometimes the selective value of canthariphily seems even limited, because the geographical areas of canthariphilous *Atrichopogon* species and their potential cantharidin-containing hosts seem only partly overlapping (Görnitz 1937; Downes 1955; Frenzel et al. 1992). On the other hand, one has to keep in mind that not only meloid beetles but also oedemerids are suitable cantharidin sources.

8.6.2 Detoxication of Cantharidin in Producers, Canthariphilous Insects and Cantharidin-Tolerant Animals

Since cytosolic PP2As are universally present in organisms, this raises the question of how not only cantharidin or even palasonin producers, but also canthariphiles and other cantharidin-tolerant animals protect themselves from being damaged by their toxic metabolites or food. Since organic solvents only extract a small portion of cantharidin from a biomatrix ("free" cantharidin) and a higher fraction may be only extracted after strong pretreatment with a base or an acid ("bound" cantharidin), it seems probable that detoxication of cantharidin is achieved by a tight binding to an unknown compound. It must be assumed that all cantharidin-tolerant organisms (producers, consumers) should possess such unknown compounds which should inhibit penetration of this terpene anhydride through cell membranes and should therefore avoid its contact with cytosolic PP2As.

If both cantharidin fractions are analysed in anthicid or pyrochroid specimens from the field or beetles previously fed with authentic cantharidin, the titre of free cantharidin does not seem to surpass certain levels, depending on species and the compartment analysed. In contrast, the titre of bound cantharidin may reach elevated values which depend upon the compartment investigated and the amounts of anhydride previously ingested. Within cantharidin-producing oedemerids, meloids and canthariphilous anthicids, pyrochroids and ceratopogonids, the titre of free cantharidin always lies between 6 and 50% of the total cantharidin (Frenzel and Dettner 1994; K. Dettner, C. Holz and S. Thießen, unpubl.). Since the conservative PP2As of cantharidin-tolerant organisms seemingly are not adapted to tolerate the toxic anhydride, the titre of free cantharidin is probably not allowed to surpass certain toxic levels which may depend upon species and compartment. Therefore it may be suggested that cantharidin shows a different toxicity for producers and cantharidin-tolerant organisms.

This assumption is amply corroborated if equimolar cantharidin and analogue solutions are injected into *Oedemera* producers (K. Dettner, unpubl.) and if appropriate oral toxicities are registered for canthariphilous *Atrichopogon oedemerarum* specimens (M. Frenzel, unpubl.). Compared with analogues **2–6** (see Fig. 8.3), cantharidin exhibits the least toxicity for the oedemerid producers in spite of its pronounced mouse toxicity (Fig. 8.3), which indicates that cantharidin producers are adapted to their toxin. In general, toxicity of injected analogues for *Oedemera* beetles does not correlate with mouse toxicity. In contrast, cantharidin shows an increased but medium toxicity for ceratopogonids if it is compared with the same analogues (**2–6**). It is astonishing that endothallactone (**4**), in spite of its moderate attractancy for the flies, shows an extraordinary high oral toxicity for *Atrichopogon*.

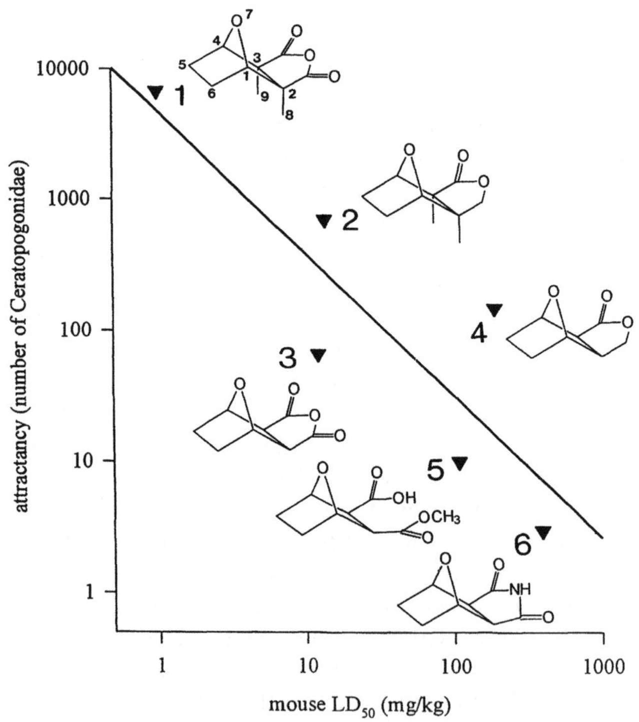

Fig. 8.3. Attractancy of canthariphilous *Atrichopogon* ceratopogonid flies (1990–1994; northern Bavaria) to sticky traps baited with equimolar amounts of authentic cantharidin (**1**) and cantharidin analogues **2–6** against mouse toxicity which was expressed as LD_{50} values (mg/kg body weight; intraperitoneal application; Matsuzawa et al. 1987; Wang 1989). Regression: $\log y = 3.68 - 1.08 \times \log x$ ($r = -0.86$; $p = 0.029$)

Therefore it may be expected that cantharidin producers tolerate high titres of cantharidin, whereas canthariphiles and tolerant organisms may only survive medium titres of free cantharidin which contrasts to nontolerant organisms where the anhydride displays a high toxicity (Fig. 8.3). Moreover, it may be expected that cantharidin tolerance is especially found within those fractions of heteromeran beetles where a biosynthesis of this unusual compound evolved long ago (Meloidae, Oedemeridae) and where canthariphilous taxa were identified (Anthicidae, Pyrochroidae).

8.6.3 Evolution of Attractancy of Toxic Cantharidin in Canthariphilous Insects

The evolution of canthariphily is nearly equivalent with the question concerning the evolution of chemoreceptors for cantharidin which are responsible for this unusual attractancy towards an extremely toxic compound. It may be suggested that most worldwide known canthariphilous insects possess cantharidin receptors and originally preyed upon meloid (Young 1984d) and/or oedemerid beetles, which synthesize this chemically unusual but biochemically potent and toxic compound (hypothesis 1). Therefore, canthariphiles could represent a poor remainder of a formerly more extensive group of arthropods associated with meloid and oedemerid beetles. Later on, when the nutritional biology of these species changed secondarily, the original chemoreceptors for cantharidin could persist, because cantharidin may increase fitness and its patchy distribution may additionally ensure meeting of copulation partners. Apart from canthariphilous heteromeran anthicid and pyrochroid beetles and canthariphilous braconid species which parasitize on these heteromeran beetles, this assumption may be true for those canthariphilous insects which originally preyed upon living and dead arthropods (e.g. Miridae, Ceratopogonidae, Anthomyiidae) and which are characterized by a small but significant percentage of canthariphilous genera and species (Fig. 8.4).

In general, the percentage fraction of canthariphiles within the worldwide known inventory of total numbers of genera and species of canthariphilous taxa favours the following interpretation (Fig. 8.4). Cantharidin biosynthesis and internal concentration could only evolve in conjunction with cantharidin detoxication which should have happened within a fraction of more or less closely allied heteromeran beetles such as Oedemeridae, Meloidae, Pyrochroidae, Anthicidae (Dettner 1987), perhaps including a few other families (Lagriidae?). Later on, the capacity to synthesize this unusual compound would have been lost secondarily in Pyrochroidae and Anthicidae; however, detoxication of cantharidin would have been retained and would represent a predisposition. This is underlined by the relatively high fraction of canthariphilous species and genera in nearly all subfamilies and tribes of Pyrochroidae and Anthicidae. Whether evolution of chemoreceptors and

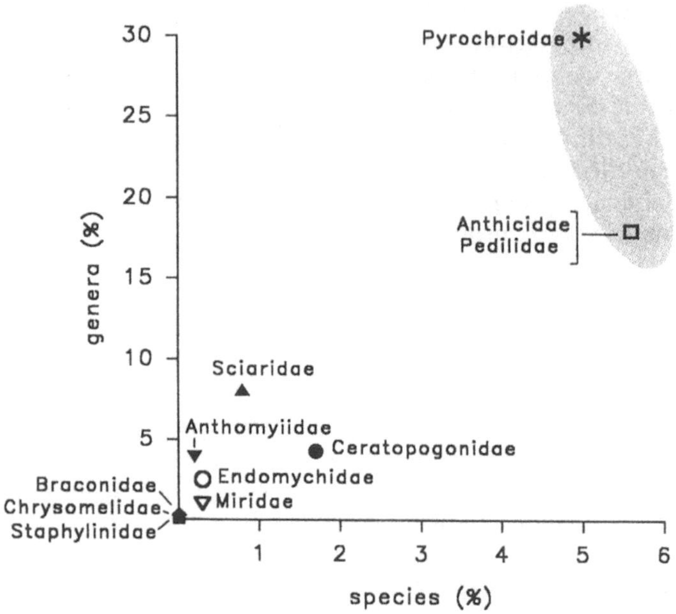

Fig. 8.4. Percentage fraction of canthariphilous genera and species as compared with the known worldwide total numbers of genera and species per taxon (according to Parker 1982). In order to compare numbers of all taxa at a given time, data for Anthicidae and Pedilidae were combined according to Parker (1982)

attractancy for cantharidin were secondary events in Pyrochroidae and Anthicidae or whether receptors were already present originally in cantharidin producers remains questionable. The latter alternative could be supported by the presence of typical anthicid elytral notches in primitive meloids of *Protomeloe* (Abdullah 1964). However, it is not known whether these small beetles synthesize cantharidin (as all other meloids) or whether they possess cantharidin receptors. Furthermore, the exact taxonomic position of these living fossils is not clear.

Finally, canthariphilous taxa remain whose life history is correlated with fungi (Endomychidae, Sciaridae) or other kinds of various substrates (Chrysomelidae, Staphylinidae, Cleridae). Since cantharidin-like natural compounds are not known from fungi, there may be the vague possibility that these fungi might always co-occur with decaying wood which would be a possibility to establish contact with cantharidin-containing organisms and could be a path to gain cantharidin from those oedemerid larvae which live in decaying wood materials. Moreover, it seems possible that cantharidin receptors and detoxication of the anhydride evolved convergently later several times. This may be true for those huge taxa such as Staphylinidae and Chrysomelidae which only possess one or few canthariphilous species (Fig. 8.4) and originally were phytophagous. The huge fraction or cantharidin-tolerant organisms seems to indicate that detoxication of cantharidin is not so difficult.

A second hypothesis for the evolution of attractancy for cantharidin within nonheteromeran arthropods could be that other sources for cantharidin or cantharidin mimics from living or dead arthropods, higher plants of fungi might exist (hypothesis 2; Young 1984d). Due to its pronounced physiological skin-irritating activities at low concentrations which allows us to identify traces of the compound, certainly no other cantharidin sources exist in nature. This might be also underlined by the following experiment (K. Dettner, unpubl.). During June 1993, all arthropods from several specimens of a *Ranunculus* and an umbellifer species were quantitatively collected together with canthariphilous specimens of rove beetle *Eusphalerum minutum*, which was found in blossoms of both plant species. A subsequent hydrolysis and trace analysis of cantharidin by GC-MS exclusively showed the presence of the anhydride within all specimens of the canthariphilous *Eusphalerum* rove beetle species from both plants. Following co-occurring organisms contained no cantharidin at all: blossoms, leaves and stems of umbellifer and of *Ranunculus*, noncanthariphilous Omaliinae, Mordellidae, Melyridae, Nitidulidae, Elateridae, longhorn beetles, tenthredinids, ants, various lepidopteran larvae and adults, Syrphidae, Tipulidae, Thysanoptera, aphids, larvae of earwigs and of chrysopids. However, it is possible that other cantharidin mimics or completely unrelated compounds might exist with similar structural elements as cantharidin which may attract canthariphiles. It is astonishing that certain faunal elements from cantharidin traps are also found in the trap flowers of *Aristolochia clematitis* (Aristolochiaceae, Ceratopogonidae with *A. lucorum*, Cecidomyidae, Chironomidae, Scatopsidae, Phoridae; Havelka 1982), in umbellifer flowers (Szadziewaki and Krzywinski 1988) and to some extent in *Arum* species (Araceae, Psychodidae; Dobat, pers comm.; Sphaeroceridae; Rohacek et al. 1990). Since these faunal lists only roughly correspond, certain cantharidin mimics might exist which are also produced by certain plants in order to attract pollinators. These kinds of plants simulate odours of microbially decaying protein or faeces which are usually released by carrion or dung.

Additional information may be obtained if attractiveness of natural and synthetic cantharidin analogues for canthariphilous insects is observed. Since several unrelated canthariphilous insects are transiently associated with flowers of Apiaceae (e.g. ceratopogonids, *Eusphalerum* and cardinal beetles), it was suggested that other secondary compounds from plants might also be attractive for canthariphiles (Frenzel et al. 1992). Moreover, many secondary plant constituents exist (e.g. pyrano- and furanocoumarines from Apiaceae; protanemonin from Ranunculaceae; tulipalin from Liliaceae) which externally exhibit a similar dermatitis in mammals as cantharidin. Therefore attractancy of plant-derived compounds and cantharidin analogues was registered for ceratopogonid flies (Frenzel et al. 1992; Fig. 8.3). Whereas none of the plant-derived furano- or pyranocoumarines were attractive to the flies, several cantharidin analogues to a minor extent may attract ceratopogonids (especially females of *Atrichopogon oedemerarum*). The attractancy of cantharidin analogues was not correlated with the vapour pressure of the tested compounds. Apart from cantharidin, only a few closely related compounds which have substituents as a five-ring or carboxylic groups in the exo-position (Fig. 8.3, 2–6) are active. Attractancy for *A. oedemerarum* is significantly reduced if the anhydride ring of cantharidin (Fig. 8.3, 1) or endothalanhydride (Fig. 8.3, 3) is replaced by a γ-lactone (Fig. 8.3, 2, 4) or if 2,3-dimethyl groups of cantharidin are eliminated (Fig. 8.3, 3). Opening of the anhydride ring of endothalanhydride (Fig. 8.3, 5), replacing of bridgehead ether oxygen by a methylene group, change of ring geometry by introducing a 5,6-double bond or replacement of the central oxygen in the endothalanhydride moiety by an NH group (Fig. 8.3, 6) result in a drastic or complete loss of attractancy. In order to have effective interaction with an unknown chemoreceptor, a 7-oxabicyclo[2.2.1]heptane moiety in conjunction with an exo-annulated γ-lactone or an anhydride ring is the minimum structural requirement.

In the meantime, attractancy of palasonin (3-demethylcantharidin) and hexachloroethane (Hennig 1976) for European canthariphilous species was also registered (K. Dettner, unpubl.). Compared with cantharidin, the demethylcantharidin was less attractive for ceratopogonids (ratio of individuals cantharidin/palasonin: 6.5:1) and *Anthomyia pluvialis* (28:1). Contrary to Hennig (1976), the chlorinated compound exhibited no attractancy at all for the anthomyid species. Remarkably, a moderate attractancy of palasonin has also been observed for the anthicid beetle *Notoxus monoceros* (ratio of individuals cantharidin/palasonin: 3.7:1; K. Dettner and U. Hofmann, unpubl.). Altogether these experiments with analogues and their partial attractancy for canthariphilous insects indicate that the second hypothesis cannot be rejected. Indeed, other more volatile flower constituents and compounds unrelated to cantharidin should be tested for their attractancy to these insects. Moreover, differences in the perception of male and female flies and between the different species for different analogues were observed. Therefore it is necessary to gather more information on the chemoreceptors and especially the transport and receptor proteins of these canthariphilous insects.

Considering both hypotheses, additional speculation concerning the highly sensitive and selective perception and ingestion of a toxic compound may be based on recent biochemical data on perception and transport of odour molecules in the olfactory system.

It is at first remarkable that cantharidin inhibits the target molecule PP2A, which plays a regulatory role in governing the responsiveness of at least mammalian olfactory neurons. In order to activate the phosphorylated and thus inactivated odorant receptor proteins, the phosphate groups must be rapidly removed by PP2A (Kroner et al. 1996). This means that the odorant cantharidin not only exhibits its toxicity after reaching certain organs in the interior of target organisms, but also even acts peripherally in the olfactory system.

Moreover, terrestrial insects possess olfactory receptor neurons which exist in an aqueous extracellular environment. Within this extracellular fluid surrounding olfactory neurons and the membrane-bound odorant receptors, small, water-soluble odorant binding proteins (OBP) in various species were identified which solubilize hydrophobic odour molecules and facilitate their transport through the aqueous neural environment up to the olfactory receptors (Vogt et al. 1990; Krieger et al. 1996). Cantharidin with its hydrophobic properties should represent such a ligand to be bound on OBPs of canthariphilous insects. Among these OBPs, three subfamilies with different odorant binding specificities were identified. One interspecifically highly divergent protein subfamily I is sex-specific and more specialized and probably more fine-tuned to its particular pheromone ligands. Two other more conserved OBP subfamilies II and III (called general odorant binding proteins: GOBPs) are present in both sexes and appear to be associated with general odorant-sensitive neurons responding to food and host odours (Vogt et al. 1990; Pelosi and Maida 1995; Krieger et al. 1996). Further OBPs have been recently identified from the *Drosophila* olfactory system and most of these proteins were recognized as belonging to the lipocalin family (McKenna et al. 1994). These lipocalins share well-conserved, three-dimensional structures and are often found outside the olfactory system. It is remarkable that two members of this lipocalin family, which were identified from the heteromeran beetle *Tenebrio molitor*, are not expressed in the antennae but in the male accessory glands (McKenna et al. 1994).

In view of these biochemical data, the evolution of cantharidin receptors and binding proteins might be speculated:

1. Since canthariphily is taxonomically so widely distributed within both sexes of various unrelated insect orders outside of heteromeran beetles (e.g. Staphylinidae, Chrysomelidae, Diptera, Hymenoptera, Heteroptera), protein candidates for antennal binding of cantharidin could be the widely distributed extracellular GOBPs of the lipocalin family. These proteins bind and transport small hydrophobic ligands, share a characteristic and conserved three-dimensional structure, are taxonomically widely distributed and are expressed in both sexes.
2. Because attractancy of cantharidin and certain analogues to ceratopogonid flies is roughly correlated with their mouse toxicity (Fig. 8.3), which obviously corresponds with the degree of interaction of these ligands with the extremely conservative PP2As, the cantharidin binding proteins in the antennal lymph should also represent rather conservative proteins, just as GOBPs of the lipocalin family. Since possible binding partners of GOBPs have been ascribed as generally binding terpenoid odorants and plant and host odours (Vogt et al. 1990; Pelosi 1994), it seems probable that also specific animal odours or volatiles from a decaying animal food or host could bind to GOBPs. Preliminary comparisons of amino acid sequences of PP2A with known OBPBs and GOBPs indeed did not indicate significant similarities (H. Breer, pers. comm.). However, it is necessary to identify the cantharidin-binding proteins in the olfactory system and subsequently

prove the exact binding site of cantharidin at PP2A and GOBPs in order to compare the appropriate tertiary structures of these proteins which could fold into similar structures. Recently, the three-dimensional structure of the protein phosphatase calcineurin has been elucidated (Griffith et al. 1995). Moreover, it has been proved that change in single amino acids in the β12/β13 loop of PP2A results in a dramatic loss of potency of various toxins such as okadaic acid, or microcystine (Barford 1996). Therefore it may be suggested that cantharidin also interacts with the β12/β13 loop of the metalloenzyme PP2A. Although few sequences of GOBPs have been reported, the possible binding site of cantharidin with these proteins remains unknown (Pelosi and Maida 1995). However, it was speculated that GOBPs bind odorants that could be the same for various insect species (Pelosi and Maida 1995).

3. Apart from detoxication, selective transfer and specific distribution of toxic cantharidin within the body of cantharidin producers and canthariphiles have to be achieved by various transport molecules. Since male accessory glands of heteromeran meloid beetles, but also of canthariphilous heteromeran pyrochroids and anthicids, act as cantharidin kidneys, male accessory glands, in particular, but also female ovaries and other body compartments should contain specialized transport molecules for cantharidin. Therefore lipocalin proteins, which do not only carry odours in the insect olfactory system, but are also expressed in male accessory organs of heteromeran tenebrionid beetles (McKenna et al. 1994), could be candidates for both perception and selective transfer of cantharidin through the membranes and body compartments of these insects.

8.6.4 Evolution of Cantharidin Transfer Through Trophic Levels

With respect to attractancy, sequestration, intra- and interspecific transfer of adaptively significant chemicals, canthariphilous insects resemble other pharmacophagous arthropods which, for example, selectively take up pyrrolizidine alkaloids (Boppré 1986; Dettner 1987; Dettner et al., this Vol.; Dussourd et al. 1988) or inorganic compounds such as sodium which are precious to herbivores (Smedley and Eisner 1996). The paternal contribution of inorganic and organic chemicals to eggs is remarkable and may be more widespread than hitherto suspected. Whereas the male gift may be smaller than that of the female in those insects transferring pyrrolizidine alkaloids (Dussourd et al. 1988), cantharidin or sodium are primarily sequestered and transferred by males. Therefore it is conceivable that an optimal supply of offspring is mainly achieved by female sexual selection for those copulation partners which are especially enriched with the essential compounds. This evaluation of male qualities occurs prior to copulation (Eisner et al. 1996a). Transfer of cantharidin from males to females may not only increase egg qualities, but also offspring survivorship (see Sect. 8.6.1). If the ability of females to collect cantharidin is reduced, the survivorship of the female may also be increased after copulation. The judgement of females, of course, cannot be based on an examination of the male's genotype but depends on an assessment of his phenotype. It makes sense that increased cantharidin titres in individual males may be indicators of resource gathering skill, vagility, physiological soundness and optimal health (Thornhill and Alcock 1983).

Remarkably, several natural serine/threonine phosphatase and especially PP2A inhibitors exist apart from cantharidin (MacKintosh and MacKintosh 1994). These chemically unrelated toxins include the polyketide calyculin A (from the marine sponge *Discodermia*), the toxic cyclic peptides microcystines (from bacteria, blue-green algae), nodularins and the polyether carboxylic acid okadaic acid (including related derivatives such as dinophysistoxin-1, acanthifolicin), which was first isolated from marine sponges *Halichondria okadaii* (Cohen et al. 1990; Shenolikar 1994). It is highly interesting that okadaic acid and related dinophysistoxin-1 may also circulate within trophic levels and now in aquatic ecosystems (Dettner et al., this Vol.). These toxic compounds (okadaic acid: mouse intraperitoneal LD_{50}: 0.2 mg/kg) are produced by marine dinoflagellates and may be concentrated by filter-feeding either in marine sponges or in midgut glands of mussels (Cohen et al. 1990). Although detailed ecological and detoxication data on tolerant organisms are missing, consumption of these mussels causes severe gastroenteritis and the illness, designated diarrhetic shellfish poisoning, was shown to be caused by okadaic acid and dinophysistoxin-1. Therefore the driving force for the evolution and circulation of such chemically diverse and unusual toxins as okadaic acid in aquatic and cantharicin in terrestrial systems might be their inhibition of an identical, conservative and centrally important cytosolic molecular target PP2A in eukaryotic organisms. These cytotoxicities and known molecular targets (such as proteins, nucleic acids, biomembranes and electron chains) of various natural compounds such as alkaloids might be the key to understanding their ecological significance even in multitrophic systems (Wink 1993; Brown and Trigo 1995). The biochemical studies on detoxication of receptor and transfer molecules for such toxins in adapted organisms amply illustrate that the science of chemical ecology partially has to develop from a micromolecular natural product chemistry towards biochemistry, in order to understand complex chemically mediated interactions within ecosystems.

Acknowledgements. This chapter is dedicated to my honoured Colleague Prof. Dr. H. Zwölfer. I thank my Ph.D. students Dr. M. Frenzel, Dr. C. Hemp, U. Hofmann, Dr. C. Holz, B. Rath, G. Streil and S. Thießen for their extraordinary engagement. For cooperation and providing of samples of cantharidin analogues the help of Prof. Dr. W. Boland (Bonn) is gratefully acknowledged. Prof. Dr. H. Breer (Stuttgart) and Prof. Dr. F. X. Schmidt (Bayreuth) provided important information on protein biochemistry. Furthermore, the help of Dr. G. Havelka (Karlsruhe) and G. Uhmann (Pressath) was greatly appreciated. The comments of G. Bauer, C. Holz and W. Völkl on earlier drafts were highly important to improve the manuscript. Finally, I thank the German Research Organization (DFG) for financial support.

References

Abdullah M (1964) *Protomeloe crowsoni*, a new species of a new tribe (Protomeloini) of the blister beetles (Coleoptera, Meloidae), with remarks on a postulated new pheromone (cantharidin). Entomol Ts Arg 86:43–48

Aldrich JR (1988) Chemical ecology of the Heteroptera. Annu Rev Entomol 33:211–238

Bangerter H (1933) Mücken-Metamorphosen V. Konowia 12:248–259

Barford D (1996) Molecular mechanisms of the protein serine/threonine phosphatases. TIBS 21:407–412

Blum MS (1996) Semiochemical parsimony in the Arthropoda. Annu Rev Entomol 41:353–374

Bochis RJ, Fisher MH (1968) The structure of palasonin. Tetrahedron Lett 16:1971–1974

Bologna MA (1991) Coleoptera Meloidae, Fauna d'Italia, vol XXVIII, Calderini, Bologna

Bologna MA, Havelka P (1984) Nuove segnalazioni di attrazione della cantharidina dei Meloidae su Coleotteri e Ditteri. Boll Assoc Rom Entomol 39:77–82

Boppré M (1986) Insects pharmacophagously utilizing defensive plant chemicals (pyrrolizidine alkaloids). Naturwissenschaften 73:17–26

Bowers MD (1990) Recycling plant natural compounds for insect defense. In: Evans DL, Schmidt JO (eds) Insect defences. State University of New York Press, Albany

Brown KS, Trigo JR (1995) The ecological activities of alkaloids. In: Cordell GA (ed) The alkaloids, vol 47. Academic Press, San Diego, pp 227–354

Carrel JE, Eisner T (1974) Cantharidin: potent feeding deterrent to insects. Science 183:755–757

Carrel JE, McCairel MH, Slagle AJ, Doom JP, Brill J, McCormick JP (1993) Cantharidin production in a blister beetle. Experientia 49:171–174

Cohen P, Holmes FB, Tsukitani Y (1990) Okadaic acid: a new probe for the study of cellular regulation. TIBS 15:98–102

Dettner K (1984) Description of defensive glands from cardinal beetles (Coleoptera, Pyrochroidae) – their phylogenetic significance as compared with other heteromeran defensive glands. Entomol Basil 9:204–215

Dettner K (1987) Chemosystematics and evolution of beetle chemical defenses. Annu Rev Entomol 32:17–48

Dettner K, Reissenweber F (1991) The defensive secretion of Omaliinae and Proteininae (Coleoptera: Staphylinidae): its chemistry, biological and taxonomic significance. Biochem Syst Ecol 19:291–303

Downes JA (1955) The food habits and descriptions of *Atrichopogon pollinivorus* sp. n. (Diptera: Ceratopogonidae). Trans R Entomol Soc Lond 106:439–453

Downes JA (1958) The feeding habits of biting flies and their significance in classification. Annu Rev Entomol 3:249–266

Downes JA (1978) Feeding and mating in the insectivorous Ceratopogoninae (Diptera). Mem Entomol Soc Can 105:1–61

Dussourd DE, Ubik K, Harvis C, Resch J, Meinwald J, Eisner T (1988) Biparental defensive endowment of eggs with acquired plant alkaloid in the moth *Utetheisa ornatrix*. Proc Natl Acad Sci USA 85:5992–5996

Eisner T (1988) Insekten als fürsorgliche Eltern. Verh Dtsch Zool Ges 81:9–17

Eisner T, Conner J, Carrel JE, McCormick JP, Slagle AJ, Gans C, O'Reilly JC (1990) Systemic retention of ingested cantharidin by frogs. Chemoecology 1:57–62

Eisner T, Smedley SR, Young DK, Eisner M, Roach B, Meinwald J (1996a) Chemical basis of courtship in a beetle (*Neopyrochroa flabellata*): cantharidin as precopulatory "enticing" agent. Proc Natl Acad Sci USA 93:6494–6498

Eisner T, Smedley SR, Young DK, Eisner M, Roach B, Meinwald J (1996b) Chemical basis of courtship in a beetle (*Neopyrochroa flabellata*): cantharidin as "nuptial gift". Proc Natl Acad Sci USA 93:6499–6503

Feng Y, Jianqi M, Zhongren L, Tianpeng G (1988) A preliminary investigation of the cantharidin resources of Shaanxi province. Acta Univ Septentrion Occident Agric 16:28

Feuell AJ (1965) Insecticides. In: von Wiesner J (ed) Die Rohstoffe des Pflanzenreiches, Lief 4. J Cramer, Weinheim 244 pp
Fey F (1954) Beiträge zur Biologie der canthariphilen Insekten. Beitr Entomol 4:180–187
Freeman P (1983) Sciarid flies, Diptera, Sciaridae. Handbooks for the identification of British insects 9/6. Royal Entomological Society of London, London
Frenzel M, Dettner K (1994) Quantification of cantharidin in canthariphilous Ceratopogonidae (Diptera), Anthomyiidae (Diptera) and cantharidin producing Oedemeridae (Coleoptera). J Chem Ecol 20:1795–1812
Frenzel M, Dettner K, Wirth D, Waibel J, Boland W (1992) Cantharidin analogues and their attractancy for ceratopogonid flies (Diptera: Ceratopogonidae). Experientia 48:106–111
Gauld I, Bolton B (1988) The Hymenoptera. British Museum, Oxford University Press, Oxford, 332 pp
Görnitz K (1937) Cantharidin als Gift und Anlockungsmittel für Insekten. Arb Phys Angew Entomol Berlin-Dahlem 4:116–157
Graziano MJ, Pessah IN, Matsuzawa M, Casida JE (1988) Partial characterization of specific cantharidin binding sites in mouse tissues. Mol Pharmacol 33:706–712
Griffith JP, Kim JL, Kim EE, Sintchak MD, Thomson JA, Fitzgibbon MJ, Fleming MA, Caron PR, Hsiao K, Navia MA (1995) X-ray structure of calcineurin inhibited by the immunophilin-immunosuppressant FKBP12-FK506 complex. Cell 82:507–522
Havelka P (1979) *Atrichopogon lucorum* (Meigen, 1818) [Diptera, Ceratopogonidae] - ein neuer, temporärer, canthariphiler Ektoparasit am Ölkäfer *Meloe violaceus* Mrsh., 1802 [Coleoptera, Meloinae]. Arbeitsgem Österr Entomologen 30:117–119
Havelka P (1980) *Meloe violaceus* MARSH, 1802 (Coleoptera, Meloinae) und seine cantariphilen Begleiter an einem Standort nördlich Karlsruhe. Beitr Naturkd Forsch Südwestdschl 39:153–159
Havelka P (1982) Die Ceratopogonidenfauna der Osterluzei (*Aristolochia clematitis*). Mosq News 42:524
Havelka P, Caspers N (1981) Die Gnitzen (Diptera, Nematocera, Ceratopogonidae) eines kleinen Waldbaches bei Bonn. Decheniana Beih 25:1–100
Hemp C, Dettner K (1997) Mesothoracic glands in anthicid beetles (Coleoptera: Anthicidae): morphology and chemical analyses of the gland secretion in *Formicomus pedestris* Rossi, 1790, *F. gestroi* Pic, 1894, *F. rubricollis* LaFerté, 1848, and *Microhoria terminata* Schmidt, 1848. Entomol Gen (in press)
Hennig W (1976) Anthomyiidae. In: Lindner E (Hrsg) Die Fliegen der Paläarktischen Region, VII/1, 1.–3. Teilband. Schweizerbart'sche Verlagsbuchhandlung, Stuttgart
Holz C (1995) Die Bedeutung des Naturstoffs Cantharidin bei dem Feuerkäfer *Schizotus pectinicornis* (Pyrochroidae). Verh Westdtsch Entomol Tag 1994:73–78
Holz C, Streil G, Dettner K, Dütemeyer J, Boland W (1994) Intersexual transfer of a toxic terpenoid during copulation and its paternal allocation to developmental stages: quantification of cantharidin in cantharidin-producing oedemerids (Coleoptera: Oedemeridae) and canthariphilous pyrochroids (Coleoptera: Pyrochroidae). Z Naturforsch 49c:856–864
Juanjie T, Youwei Z, Shuyong W, Zhengji D, Chuanxian Z (1995) Investigation on the natural resources and utilization of the Chinese medicinal beetles. Acta Entomol Sin 38:324–331
Kelling ST, Halpern BP, Eisner T (1990) Gustatory sensitivity of an anuran to cantharidin. Experientia 46:763–764
Klinger R (1979) Eine Sternaldrüse bei Kurzflügelkäfern. Systematische Verbreitung sowie Bau, Inhaltsstoffe und Funktion bei *Eusphalerum minutum* (L.) (Coleoptera: Staphylinidae). Dissertation Universitat Frankfurt, Frankfurt
Klinger R (1983) Eusphaleren, blütenbesuchende Staphyliniden, 1) zur Biologie der Käfer. Dtsch Entomol Z N F 30:37–44
Krieger J, von Nickisch-Rosenegk E, Mameli M, Pelosi P, Breer H (1996) Binding proteins from antennae of *Bombyx mori*. Insect Biochem Mol Biol 26:297–307
Kroner C, Boekhoff I, Breer H (1996) Phosphatase 2A regulates the responsiveness of olfactory cilia. Biochim Biophys Acta 1312:169–175

Lengersdorf F (1930) Lycoriidae (Sciaridae). In: Lindner E (Hrsg) Die Fliegen der Paläarktischen Region, II/1. Schweizerbart'sche Verlagsbuchhandlung, Stuttgart.
LeSage L, Bousquet Y (1983) A new record of attacks by *Pedilus* (Pedilidae) on *Meloe* (Meloidae: Coleoptera). Entomol News 94:95–96
Li YM, Casida JE (1992) Cantharidin – binding protein, identification as protein phosphatase 2A. Proc Natl Acad Sci USA 89:11867–11870
MacKintosh C, MacKintosh RW (1994) Inhibitors of protein kinases and phosphates. TIBS 19:444–448
Mafra-Neto A, Jolivet P (1994) Entomophagy in Chrysomelidae: adult *Aristobrotica angulicollis* (Erichson) feeding on adult meloids (Coleoptera). In: Jolivet PH, Cox ML, E Petitpierre (eds) Novel aspects of the biology of Chrysomelidae. Kluwer, Dordrecht, pp 171–178
Matsuzawa M, Graziano MJ, Casida JE (1987) Endothal and cantharidin analogues: relation of structure to herbicidal activity and mammalian toxicity. J Agric Food Chem 35:823–829
Mayer MK (1962) Untersuchungen mit Cantharidin-Fallen über die Flugaktivität von *Atrichopogon (Meloehelea) oedemerarum* Storå, einer an Insekten ektoparasitisch lebenden Ceratopogonidae (Diptera). Z Parasitenkd 21:257–272
McCormick JP, Carrel JE (1987) Cantharidin biosynthesis and function in meloid beetles. In: Blomquist GD, Blomquist GJ (eds) Pheromone biochemistry, Prestwich. Academic Press, Orlando, pp 307–350
McKenna MP, Hekmat-Scafe DS, Gaines P, Carlson JR (1994) Putative *Drosophila* pheromone-binding proteins expressed in a subregion of olfactory system. J Biol Chem 269:16340–16347
Meyer D, Schlatter C, Schlatter-Lanz I, Schmid H, Bovey P (1968) Die Zucht von *Lytta vesicatoria* im Laboratorium und Nachweis der Cantharidinsynthese in Larven. Experientia 24:995–998
Newton AF, Thayer MK (1995) Protopselaphinae new subfamily for *Protopselaphus* new genus from Malaysia, with a phylogenetic analysis and review of the Omaliine group of Staphylinidae including Pselaphidae (Coleoptera). In: Pakaluk J, Slipinski SA (eds) Biology, phylogeny and classification of Coleoptera Museum i Institut Zoologii PAN, Warszawa, pp 219–320
Pakaluk J (1984) Natural history and evolution of *Lycoperdina ferruginea* (Coleoptera: Endomychidae) with descriptions of immature stages. Proc Entomol Soc Wash 86:312–325
Parker SP (1982) Synopsis and classification of living organisms, vol 2. McGraw Hill, New York
Pelosi P (1994) Odorant-binding proteins. Crit Rev Biochem Mol Biol 29:199–228
Pelosi P, Maida R (1995) Odorant-binding proteins in insects. Comp Biochem Physiol 111B:503–514
Pinto JD (1978) The parasitization of blister beetles by species of Miridae. Pan-Pac Entomol 54:57–60
Purevsuren G, Koblicova Z, Trojanek J (1987) Cantharidinimide, a novel substance from *Mylabris mongolica*. Dokth. Cesk Farm 36:32–34
Rohacek J, Beck-Hang I, Dobat K (1990) Sphaeroceridae associated with flowering *Arum maculatum* (Araceae) in the vicinity of Tübingen, SW Germany (Insecta: Diptera). Senckenb Biol 71:259–268
Schliephake G, Klimt K (1979) Thysanoptera, Fransenflügler, die Tierwelt Deutschlands, Teil 66. Fischer, Jena
Schmitz H (1981) 33. Phoridae. In: Lindner E (Hrsg) Die Fliegen der Paläarktischern Region, IV7. Schweizerbart'sche Verlagsbuchhandlung, Stuttgart
Schütz C, Dettner K (1992) Cantharidin secretion by elytral notches of male anthicid species (Coleoptera: Anthicidae). Z Naturforsch 47c:290–299
Seeno TN, Wilcox JA (1982) Leaf beetle genera (Coleoptera: Chrysomelidae). Entomography 1:1–222
Shaw MR, Huddleston T (1991) Classification and biology of braconid wasps (Hymenoptera: Braconidae). Handbooks for the identification of British insects 7/11:1–126. Royal Entomological Society of London, London
Shenolikar S (1994) Protein serin/threonine phosphatases – new avenues for cell regulation. Annu Rev Cell Biol 10:55–86

Smedley SR, Eisner T (1996) Sodium: a male moth's gift to its offspring. Proc Natl Acad Sci USA 93:809–813

Szadziewski R, Krzywinski R (1988) Biting midges of the genus *Culicoides* visiting umbelliferous flowers in Poland. In: Olejnicek J (ed) Medical and veterinary dipterology, pp 155–158

Thornhill R, Alcock J (1983) The evolution of insect mating systems. Harvard University Press, Cambridge, 547 pp

Vogt RG, Prestwich GD, Lerner MR (1990) Odorant-binding-protein subfamiles associated with distinct classes of olfactory neurons in insects. J Neurobiol 22:74–84

Wang GS (1989) Medical uses of *Mylabris* in ancient China and recent studies. J Ethnopharmacol 26:147–162

Wink M (1993) Allelochemical properties or the raison d'être of alkaloids, 1–118. In: Cordell GA (ed) The alkaloids, vol 43. Academic Press, San Diego.

Wirth WW (1980) A new species and corrections in the *Atrichopogon* midges of the subgenus *Meloehelea* attacking blister beetles (Diptera: Ceratopogonidae). Proc Entomol Soc Wash 82:124–139

Young DK (1984a) Field studies of cantharidin orientation by *Neopyrochroa flabellata* (Coleoptera: Pyrochroidae). Great Lakes Entomol 17:133–135

Young DK (1984b) *Anisotria shooki*, a new genus and species of Pedilinae (Coleoptera: Pyrochroidae) with a note on the systematic position of *Lithomacratarìa* Wickham and a key to the genera. Coleopterist's Bull 38:201–208

Young DK (1984c) Cantharidin and insects: an historical review. Great Lakes Entomol 17:187–194

Young DK (1984d) Field records and observations of insects associated with cantharidin. Great Lakes Entomol 17:195–199

Young DK (1989) Notes on the bionomics of *Xenomycetes morrisoni* Horn (Coleoptera: Endomychidae) another cantharidin-orienting fungus beetle. Pan-Pac Entomol 65:447–448

9 Survival in a Hostile Environment. Evaluation of the Developmental Success of the Oligophagous Leaf Beetle *Chrysomela vigintipunctata* (Scop)

W. Topp

9.1 Introduction

Many herbivorous insects are restricted to a limited number of host plant species (Strong et al. 1984). Restriction mainly is explained by the "feeding-specialization hypothesis" which predicts that polyphagous herbivores handle the sophisticated chemical defences of any single food-plant less efficiently than monophagous or oligophagous species. Diet specialization should mediate more efficient utilization of food resources and allow easier degradation of secondary metabolites in foods (Howe and Westley 1988). The willow leaf beetle *Chrysomela vigintipunctata* (Scop.) is such an oligophagous insect which successfully develops on some but not on all willow species.

Among willows, secondary metabolites differ considerably and fall into at least two groups on the basis of their secondary compounds. There are some species (e.g. *Salix fragilis*) which are characterized by phenolglycosides in their leaves which act as qualitative toxins, unless an herbivore has a defence against their special activity. Others (e.g. *Salix alba*) do not contain phenolglycosides within their leaves but are rich in proanthocyanidins, a form of condensed tannins, which act as quantitative toxins and generally reduce or inhibit ingestion. Flavonoids such as flavonols, flavanones, flavones, chalcones and dihydrochalcones are widely spread within both groups (Hegnauer 1973; Shao 1991).

Earlier studies have shown the remarkable influence of secondary compounds of willows on the distributional pattern of insects. Rowell-Rahier (1984) showed when using published food plants listed for weevels, sawflies and the caterpillars of the British moth that the *Salix* species with phenolglycosides tend to be the food of the specialized herbivores and are avoided by generalists. Conversely, *Salix* species without phenolglycosides tend to be eaten by more generalist insects and are avoided by the more specialists. Moreover, within the latter group, the food selection patterns of herbivores followed closely the phenolglycoside spectra of the willow species (Tahvanainen et al. 1985).

These observations were partly confirmed by our studies on the leaf beetle *C. vigintipunctata*. In the study area this beetle was found feeding on four different species, namely *S. fragilis*, *S. alba*, *S. viminalis* and *S. caprea*. A fifth species, *S. triandra*, which grows nearby, was strictly avoided. The avoidance

Ecological Studies, Vol. 130
Dettner et al. (eds.) Vertical Food Web Interactions

pattern of *C. vigintipunctata* can be explained by the phenolglycoside salidroside which is characteristic in *S. triandra* but absent in all other willows mentioned above. Nevertheless, the preference for the *Salix* species, one (*S. fragilis*) characterized by phenolglycosides and the others (*S. alba*, *S. viminalis*, *S. caprea*) known for their high concentrations of condensed tannins but without phenoglycosides within their leaves, was not expected.

Until the 1980s the willow beetle *C. vigintipunctata* was a rare species throughout most of central Europe with only a few records during the last decades. Since the mid-1980s several outbreaks of the beetle have been observed. *C. vigintipunctata* became so abundant as to partly or completely defoliate single trees of *S. fragilis*, *S. alba* and its hybrid *S.* × *rubens* or of ample willow stands growing in the tributaries of the River Rhine and in areas situated further east (Topp and Beracz 1989; Erbeling and Terlutter 1995).

In the following study we evaluate the developmental success of the chrysomelid *C. vigintipunctata* when feeding on either of the preferred host plants, *S. alba* and *S. fragilis*, which deviate tremendously regarding their chemicals. *S. alba* is assumed as a host plant favouring polyphagous species because it is characterized by digestibility reducers. *S. fragilis* is assumed to be preferred by monophagous or oligophagous herbivores which are adapted to the specific phenolglycosides dominant within leaves of this host plant.

Comparing the beetle populations, we estimated the performance on each of the two host plants. To do this we measured the food preference of the adults, the feeding efficiencies, growth and survival of larvae and adults as well, calculated the mortality caused by predators or parasitoids, measured the mortality of adults during overwintering and calculated the life tables for the populations feeding on both host plants. In addition, we analysed some of the secondary compounds of the individual trees the beetles fed on.

9.2 Chemical Defence in Adults and Larvae

Beetles show an impressive array of defences to avoid being eaten. One widespread method of advertising unpalatable qualities is by conspicuous coloration. Warning colours include red, orange, yellow, black and white (Bowers 1993). The adults of *C. vigintipunctata* look bright with a typically aposematic coloration on their elytra consisting of 20 black dots of different size on a yellow, orange or red background. When disturbed, the beetles ooze secretions from their defence glands which open into pores on the surface of the pronotum and the elytra. The defence secretions of *C. vigintipunctata* are characterized by the presence of isoxazolinone glycosides and a large quantity of lipids. The deterrent effect and the toxicity of isoxazolin-5-one derivatives were demonstrated towards the ant *Myrmica rubra* (Pasteels et al. 1988). The adults of the genus *Chrysomela* apparently do not sequester salicin for their own defence, but females are able to sequester salicin in their eggs for the benefit of both eggs and neonate larvae (Pasteels et al. 1986; Rowell-Rahier and Pasteels 1989).

Larvae of *C. vigintipunctata* are mostly black. They secrete salicylaldehyde and benzaldehyde by exocrine glands (Pasteels et al. 1982). These white glands give a conspicuous contrast to the black segments of the thorax and abdomen when extruded. Pasteels et al. (1983) were able to demonstrate that salicylaldehyde is derived from salicin, a phenolglycoside present in the leaves of some host plants of this beetle. The transformation of salicin to salicylaldehyde takes place in the exocrine glands of the beetles which are highly suitable as a site for high β-glucosidase enzyme activity. Salicylaldehyde as a deterrent is advantageous for the larvae. Because of its higher volatility in comparison to salicin it is a more suitable compound with which to repel enemies (Pasteels et al. 1983). The same strategy in producing defensive secretions is evolved in several congeneric species (Smiley et al. 1985; Pasteels et al. 1988). The transformation from salicin to salicylaldehyde is also beneficial in another way; the studies of Pasteels et al. (1983) also showed that the concentrations of glucose and salicylaldehyde in the secretions were far from being equimolar, which indicates that the glucose formed by salicin hydrolysis is largely recovered.

From these studies it seems plausible that willow species containing salicin within their leaves will be more advantageous for larval development of *C. vigintipunctata* than willow species not containing salicin. With respect to the host plants investigated this means that *S. fragilis* should be better for *C. vigintipuntata* development than *S. alba*, a species lacking phenolglycosides.

9.3 Feeding Preference and Oviposition. Not All Mothers Know Best

From the defence strategies of larvae described above the following hypothesis is drawn: *S. fragilis* is the preferred host plant in comparison with *S. alba*. However, field observations contradict this hypothesis.

In *C. vigintipunctata*, aggregations for overwintering are often formed at conspicuous features of the landscape. As with Coccinellidae, these may be peaks of hills or posts on the slopes. When dispersing from the overwintering sites to the valley floor where the host plants grew in the years of observation, bud burst and leaf flush had already occurred. However, length of leaves varied between species. In 1987, for example, at the time of dispersions *S. alba* leaves were totally expanded whereas leaves from *S. fragilis* had only reached about one-quarter of their maximum size (Topp et al. 1989).

During the time interval of dispersion (12–14 April, Fig. 9.1) in a choice test carried out in the laboratory, the adults consumed nearly as much *S. alba* foliage as *S. fragilis* foliage. In the field, however, during the same time interval about 70% of the beetles settled and fed on *S. alba* leaves whereas *S. fragilis* was selected by about 20% of the beetles observed in the river valleys. During similar field observations, a conspicuous preference for *S. alba* trees was found by Erbeling and Terlutter (1995).

At the end of April the adults were more selective in feeding. Choice tests carried out from 24 April until 26 April (Fig. 9.1) indicated a preference for *S. fragilis* foliage. Preference may be caused by salicylates which apparently characterize leaf chemistry within this time interval but probably do not during the time interval of early leaf flush (Topp and Bell 1992). In the field several beetles which initially settled on *S. alba* trees may have dispersed to *S. fragilis* thereafter. Nevertheless, testing the distributional pattern of the beetles at the end of April and beginning of May we still found 60–70% on *S. alba* trees.

Using choice tests in the laboratory which were carried out in July with adults of the following generation, we were able to exclude conditioning of adult preference by previous adult or larval experience. Conditioning is also irrelevant in a further congeneric willow leaf beetle, *C. aenicollis*, with host plant preference based on salicylate chemistry (Rank 1992).

The deviations between laboratory tests and field observations may rely on differences in preference and selection measurements for both studies. In the laboratory, using choice tests for single individuals in separated plastic containers, the preference for feeding and ovipositing on *S. fragilis* may be mediated exclusively by the salicylates salicin and/or tremulacin. Salicin is known to act in conjunction with populin and luteolin-7-glucoside to attract *C. vigintipunctata costella* (Matsuda and Matsuo 1985). Moreover, Denno et al. (1990) showed that mature leaves of *S. fragilis* are richer in phenolglycosides than young ones.

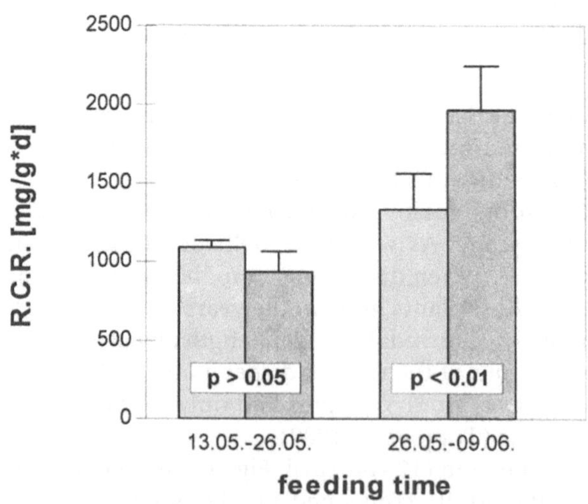

Fig. 9.1. Feeding preference ($\bar{x} \pm$ SE) of *C. vigintipunctata* adults on *S. fragilis* (*light columns*) and *S. alba* (*dark columns*) at different times of the season ($38 < n < 40$)

In the field, however, host plant chemistry cannot be seen as the only parameter influencing food plant choice of the dispersing and migrating individuals. Additionally colour, shape, size and phenology of host plants are known to affect host selection in a wide variety of insect herbivores (Courtney and Kibota 1990). In our study area the high *S. alba* trees with their totally expanded leaves may be more attractive to the dispersing and migrating beetles than the smaller *S. fragilis* trees with hardly expanded leaves. Finally, when settled, the selection by herbivorous insects is more akin to a series of take-it-or-leave-it situation than "comparison shopping" (Thorsteinson 1960).

The probability that a plant will be found is also important. When the most suitable hosts are less abundant, the herbivore's fitness may be higher if cues from lower ranked plants became acceptable and if individuals expand consumption until the most suitable plant is found (Miller and Strickler 1984). Optimal diet models have been developed to explain trade-offs between findability and suitability. For animals that search randomly, optimal foraging models predict that the time or energy expenditure at which a resource becomes acceptable depends not on the abundance of that plant, but on the abundance of more suitable plants (Pyke et al. 1977; Jaenike 1978; Roitberg et al. 1982). Suitability differs between species and will change with season. Within the study area only single *S. fragilis* trees occurred, but *S. alba* trees were in groups, and there were many hybrids between both species (= *S.* × *rubens*).

In early spring *S. alba* trees were settled not only by *C. vigintipunctata* but also by the alder leaf beetle *Agelastica alni*. The latter, in contrast to *C. vigintipunctata*, dispersed to the alder as soon as leaf flush of this species occurred. Not a single individual remained feeding on *S. alba* foliage. The behaviour of *C. vigintipuncta* feeding and ovipositing on *S. alba* trees may be explained because individuals of this species are less choosy than *A. alni* individuals with respect to the host plants in question. Moreover, when testing the food choice of single individuals we even found a polymodal reaction pattern among willow leaf beetles. Most of the individuals tested clearly preferred *S. fragilis* foliage. However, there were some individuals which found *S. alba* foliage most appropriate (Topp et al. 1989). From these observations we concluded that there is a polymorphism in respect to plant preference. Most individuals will prefer *S. fragilis*, some will show a neutral reaction pattern, but a few will prefer *S. alba*.

9.4 Fecundity

Plant secondary chemicals and nutritional properties of host plants may affect reproduction (Ohmart et al. 1985). In no-choice trials, we counted the number of eggs laid per female when either *S. fragilis* foliage or *S. alba* foliage was offered. This technique measures a combination of acceptance and fecundity

Table 9.1. A comparison of the fecundity of *C. vigintipunctata* adults fed leaves from *S. alba*, *S. fragilis* and leaves from *S. fragilis* when the salicin content was increased from 2 to 6% dry wt.

Host plant	Number	Eggs/female	SE
S. alba	12	98[a]	20
S. fragilis	24	195[b]	27
S. fragilis (+ salicin)	24	325[c]	37

Significances (Duncan-test, $p < 0.05$) are indicated by different letters.

and characterizes food quality in terms of herbivore fitness. The results confirmed the data above. When feeding on *S. fragilis* foliage the females produced more eggs than when feeding on *S. alba* foliage. However, there were some females which were obviously more fecund when feeding on *S. alba* foliage than "average" females feeding on *S. fragilis* foliage (Topp and Bell 1992).

Fecundity of females is increased by salicin (Table 9.1). When the amount of salicin in the *S. fragilis* foliage was increased from 20 to 60 mg/g dry wt. the beetles exhibited higher fecundity and higher longevity. The proportion of variance (ANOVA) explaining the effect of salicin for both parameters was in the range 7.5–11%.

9.5 Feeding Performance of Larvae and Developmental Rates

Feeding is one of the most fundamental behaviours. The adaptive strategies of consumers and the broad consequences for consumer fitness have likely lead to its evolution as a highly regulated behaviour (Slansky 1993). Foraging strategies are therefore rigorously shaped by natural selection to maximize the net nutrient gain from feeding and to minimize the risks to survival (Hassel and Southwood 1978). Many well-documented examples with gravimetric measurements based on dry weight to assess food allocation budgets gave evidence of the adaptive behaviour of insects (Waldbauer 1968; Slansky and Scriber 1985).

We compared two different treatments of larval feeding performance on both the host plants mentioned. One of the treatments was carried out 14 days earlier in the season than the other. This procedure was adopted because egg laying of *C. vigintipunctata* took place over a period of several weeks. The first treatment considered eggs which were collected in the middle of May in the field and which were afterwards immediately incubated at 20 °C and a light-dark cycle of LD 16/8. Larval development occurred from 13 May until 26 May. For the second treatment, we considered egg batches which were collected at the same time but which were left in the field at lower temperatures and incubated into the breeding chambers 14 days later. Larval development occurred from 26 May until 9 June. Following this procedure, the quality of food was different by seasonal variation (Schultz et al. 1982) but any effect of maternal age upon fitness of progeny could be excluded (Wassermann and Asami 1985).

As for food choice, stimulants and deterrents may influence food consumption. In this study the feeding rate was calculated as commonly used in quantitative food utilization studies as the amount of food consumed per larva over its whole life expressed on a "per day per mg body mass" basis (relative consumption rate, RCR).

Consumption rates of insect herbivores even under the same abiotic environmental factors is far from being constant. Slansky and Scriber (1985) found a general increase in RCR as foliage water (% fresh weight) and nitrogen (% dry weight) increased. In our studies foliage water was almost the same in all treatments (Topp et al. 1989). Although foliage N decreased slightly with season, there was no relationship between RCR and N concentration of foliage (Topp et al. 1989).

Within the first treatment larvae of both host plants exihibited almost the same RCR values (Fig. 9.2). Larvae consumed about there own body weight per day. Within the second treatment, later in the season, consumption rates of larvae increased. This increase was only slight when larvae fed on *S. fragilis* foliage but was more pronounced when *S. alba* foliage was offered. Larvae consumed almost twice as much as their own body weight. The increase in food consumption which was shown for RCR (dry wt.) as well as for RCR (fresh

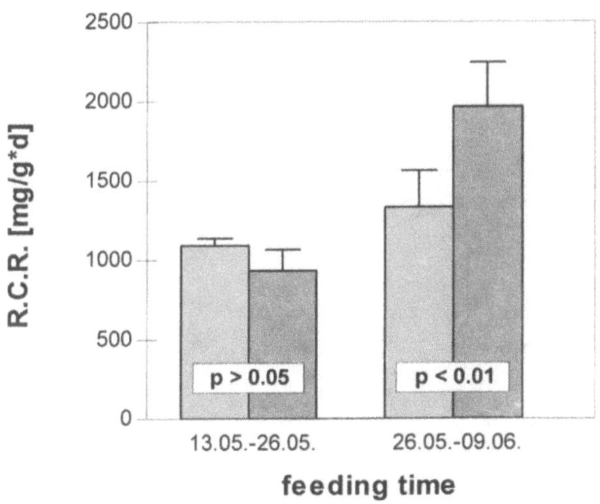

Fig. 9.2. Relative consumption rates (RCR, $\bar{x} \pm SD$) of *C. vigintipunctata* larvae when feeding on *S. fragilis* (*light columns*) and on *S. alba* (*dark columns*) foliage at different times of the season ($30 < n < 70$)

wt.) is interpreted as a compensatory response to lowered nutrient levels. Similar effects were demonstrated for several lepidopteran caterpillars feeding on different host plants or on artificial diets differing in nutrient levels (Slansky 1993). Such responses prevent large reductions in larval growth which otherwise could occur when feeding on food of suboptimal quality.

Not all herbivorous insects appear to exhibit compensatory feeding with respect to season. For example, winter moth caterpillars (*Operophtera brumata*), which fed on the same two willow trees as *C. vigintipunctata*, exhibited constant feeding rates or even decreased their consumption with decreasing water and nutrient levels of host plants, which change with season (Kirsten and Topp 1991). The lack of any compensatory responses in winter moth caterpillars may be explained by a stronger sensitivity to allelochemicals which additionally affect food intake. Insects may even avoid nutrient-rich leaves of their food plants, because these have higher concentrations of potentially deleterious allelochemicals (Williams et al. 1983; Johnson et al. 1985).

Compensatory responses as exemplified by *C. vigintipunctata* will cause more per capita damage on host plants and finally may result in an increasing biosynthesis of inductive allelochemicals (Slansky 1993). In our studies a higher mortality of larvae and adults feeding on the deteriorating *S. alba* foliage instead of deteriorating *S. fragilis* foliage was obvious. Mortality may be provoked by the combined effect of constitutive and inductive allelochemicals (see Sect. 9.8).

Survival rates during pre-imaginal development are indicated in parentheses in Fig. 9.3 and confirm our assumptions. When feeding early in the season, survival of larvae is highest and only marginally different between host plants.

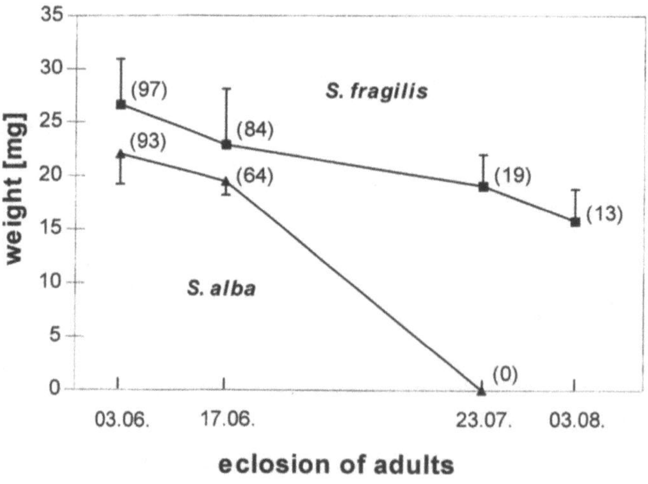

Fig. 9.3. Weight ($\bar{x} \pm$ SD, mg fresh wt.) of *C. vigintipunctata* adults ($38 < n < 41$) at the data of eclosion when feeding on either *S. fragilis* or *S. alba* foliage at different times of the year. *Numbers* in *parentheses* indicate the survival rates (%) during larval development

However, mortality increased with the season. Larvae exhibiting a seasonal delay always had a far better chance of surviving when they got the opportunity to feed on *S. fragilis* foliage (Fig. 9.3). Larvae which developed extremely late in the season survived only when feeding on *S. fragilis* foliage.

The weight (mg fresh wt.) of eclosing adults was found to be highest when seasonal larval development took place as early as possible. Feeding later in the season resulted in a lower adult weight (Fig. 9.3). The weight of adults feeding on either host plant was only slightly different, as long as eclosion of adults did not occur later than June. These findings are in accordance with the results obtained by Rank (1994). Larval growth of *C. aenicollis* feeding either on salicylate-rich or on salicylate-poor willow leaves did not differ among the willow species, but varied among the individual plants in respect of the water content.

Slowing growth and lengthening developmental time while maintaining a constant consumption rate is another means of increasing overall food consumption in accordance with nutrient-poor leaves. Larvae of *C. vigintipunctata* did not follow this strategy. The disadvantage of this strategy could be foreseen because larvae will suffer higher mortality and lower pupation weight if development is lengthened (Fig. 9.3).

9.6 Impact of Predators and Parasitoids

C. vigintipunctata produces deterrents to avoid predation and, indeed, generalist invertebrate predators are repelled by the volatile larval defence secretions which are produced in the dorsal glands of the thorax and abdomen. One of these generalist predators is the assassin bug *Rhinocoris iracundis* (Heteroptera, Nabidae). Individuals of this species were repelled by larval defence secretions even when starved. In the laboratory the assassin bugs died instead of taking advantage of one of the chrysomelid larvae offered.

When checking the distributional pattern of beetles within trees we found some predatory insects that apparently were unaffected by the chemical defence of larvae and successfully fed on *C. vigintipunctata* larvae. One was the specialist predator *Parasyrphus nigritarsis* (Diptera, Syrphidae). This species layed its eggs on the *Chrysomela* clutch. When the syrphid larvae hatched earlier than the beetle larvae they preyed on *C. vigintipunctata* eggs. Finally, when the syrphids attacked *C. vigintipunctata* third-instar larvae they were able to circumvent the defence strategies of their prey by grasping them on the underside of the thorax or abdomen. Studies of other chrysomelid-eating *Parasyrphus* species obtained similar results (Rank and Smiley 1994).

Our observations on the life cycles further indicated that *P. nigritarsis* is a specialist for chrysomelids which feed early in the season. The life histories of both species were synchronized by diapause. Diapause in the beetle was manifested in the adult stage (see Sect. 9.7) whereas it occurred in the last larval stage in the predator. The beetles dispersed as early as March or April. Larval diapause of syrphids was terminated in March. After a short period of pupation hover flies dispersed in April.

P. nigritarsis is not the only specialist predator that feeds on *C. vigintipunctata*. The shieldbug *Troilus luridus* (Heteroptera, Pentatomidae) is another. The bug was observed consuming half of the egg clutches or an even greater proportion of the hatchlings.

Arthropods are potentially an ideal food for insectivorous birds. In many insects the patterns of feeding may be related to strategies of predator avoidance (Heinrich 1979). Larvae and adults of *C. vigintipunctata*, similarly to aposematic caterpillars, did not restrict foraging to leaf undersides or to nighttime, did not move away from damage, and did not snip damaged leaves. Moreover, third-instar larvae often climb to the end of leafless twigs to pupate. At these exposed places they are visible to predators from a considerable distance. Regarding this behaviour, the chemical defence mechanisms evolved by *C. vigintipunctata* (see Sect. 9.2) seem to be suitable to repel insectivorous birds. In several species chemical protection against predation by birds has been demonstrated (Schuler and Hesse 1985; Evans and Schmidt 1990).

In order to investigate the palatability of *C. vigintipunctata* larvae to birds, we studied the searching behaviour of tits and nuthatch in the field when feeding their nestlings. Suggesting that the willow leaf beetles' salicylaldehyde and benzaldehyde might serve as protection against birds, we carried out laboratory experiments by testing blackcap.

The following questions were addressed:

1. Do insectivorous birds search for *C. vigintipunctata* larvae to feed their nestlings?
2. Has any avoidance pattern evolved due to the chemical compounds sequestered from the dorsal glands of the larvae?
3. If there is an avoidance pattern, what is the minimum concentration of repellents to initiate one?

For field studies we placed bird nests in the willow trees where outbreaks of the willow leaf beetles were observed in the previous year. During the breeding periods of the tits and nuthatch in the following year we observed the searching flights of the birds and the localities where the prey were captured. Finally, we determined the prey the adults had selected for their nestlings (Floren 1989; Topp and Bell 1992).

The results confirmed the repellent effect of the chemicals. During the year of investigaton, the breeding success of great tit, blue tit and nuthatch were all about 80%. However, none of the birds searched for prey in the willow trees where the nests were put up and where outbreaks of *C. vigintipunctata* occurred. Instead, all adults flew over long distances out of the willow trees and preyed mostly on Lepidoptera collected from the trees growing nearby. None of the nestlings checked in one of 24 different nests was fed with a chrysomelid larva or even a larva of *C. vigintipunctata*.

In a first laboratory experiment, several individual blackcaps were offered individually different prey; a choice of mealworms, mealworms which were in contact with *C. vigintipunctata* larvae and thus had adopted the beetle's smell, and larvae of *C. vigintipunctata.* In a second experiment, an artificial diet with different concentrations of salicylaldehyde and benzaldehyde was offered. All these studies clearly showed the rejection of *C. vigintipunctata* larvae and the repellent effect of both chemicals. A concentration of $0.2\,\mu M/\mu l$ of salicylaldehyde added to the artificial diet resulted in a reduced consumption. An artificial diet with a concentration of $0.5\,\mu M/\mu l$ in addition was totally rejected. When the artificial diet without any allelochemicals was offered to the blackcap, the body weight on average increased by 9%. The addition of $0.5\,\mu M/\mu l$ salicylaldehyde resulted to a weight loss of 9% within 8 h.

Two parasitoids, *Schizonotus sieboldi* (Hymenoptera, Chalcicoidea) and *Cleonice (Steiniella) callida* (Diptera, Tachinidae) were the most common species feeding on larvae and pupae of *C. vigintipunctata.*

9.7 Pre-diapause Feeding of Adults and Induction of Diapause

In the years of study adult eclosion occurred in June. In the field, weights of hatching adults were slightly different with respect to their food plants before they started pre-diapause feeding. Largest adults were collected when larvae consumed *S. fragilis* foliage. These differences in weight of newly hatched adults also were obtained under laboratory conditions when specimens were reared under the same abiotic conditions (20 °C, LD 16/8, RH = 100%, cf. Fig. 9.3), and it is assumed that the deviations measured result from differences in food quality during larval development (Wassermann and Mitter 1978; Werner 1979).

Individuals preparing to enter diapause show an increased food consumption and an accumulation of greater lipid reserves in comparison with non-diapause individuals (Schopf et al. 1995). However, weight gain may be different with respect to host plant and also to lipid metabolism in insects, which is affected by a number of neuroendocrinological, physiological and environmental influences (Downer 1984), and could be affected by secondary chemicals different in either host plant. For example, a caterpillar experiencing increased energy costs may accumulate less high-energy lipid, but it may gain the same dry mass as one experiencing lower energy costs (Slansky 1993).

Adult size during eclosion, subsequent weight gain during pre-diapause feeding and accumulation of lipid reserves may be adaptive. The assumption is made that larger females with the greatest energy reserves will have the best chance of surviving the dormant phase during winter which lasts 9–10 months. Furthermore, larger females may lay more eggs, which might be heavier and result in a higher percentage of larvae surviving to pupation than eggs from smaller females (Palmer 1985).

For the data presented here, weight gain of adults and the amount of lipid reserves stored after pre-diapause feeding was almost the same with respect to both *Salix* species. When adults flew to the overwintering sites the weight has increased by about the factor 2.5 in comparison with the eclosion weight, whereas increase in lipid reserves was fourfold (Fig. 9.4). In a study of compensatory feeding by a caterpillar, Slansky and Wheeler (1991) found that the lipid content (% dry wt.) declined as feeding rate increased with diet dilution.

At the beginning of diapause, lipid reserves of adults feeding on either host plant did not differ (Fig. 9.5). Lipid content of the beetles was in the range 19–24% dry wt. When diapause lasted 1 month, up to 20% of the diapausing adults which formerly fed on *S. alba* foliage died. Because lipid reserves of these beetles were still relatively high (Fig. 9.5) starvation was not seen to be responsible for the mortality. Individuals which fed exclusively on *S. fragilis* foliage survived this first period of diapause.

Seven months later, at the end of February, mortality of diapausing adults continued irrespective of host plant. At this time interval, weight of moribund adults had decreased sharply and lipids were mobilized tremendously. Because lipid reserves within both beetle groups were as low as at the date of eclosion starvation as a further mortality factor is assumed.

In *C. vigintipunctata* mortality due to the depletion of available reserves should be minimized if accumulation of lipids is enhanced, if diapause is shortened or if winter temperatures are lower than in the year of investigation (cf. Grigo and Topp 1980). A fourfold increase in lipids was also found in the bark beetle *Ips typographus*, which successfully overwinters for a period of about 7 months (Schopf et al. 1995).

C. vigintipunctata is an early season feeder and exhibits behaviour patterns which allow development as early as possible. In this respect the development resembles that of winter moth larvae (*Operophtera brumata*) which co-occur

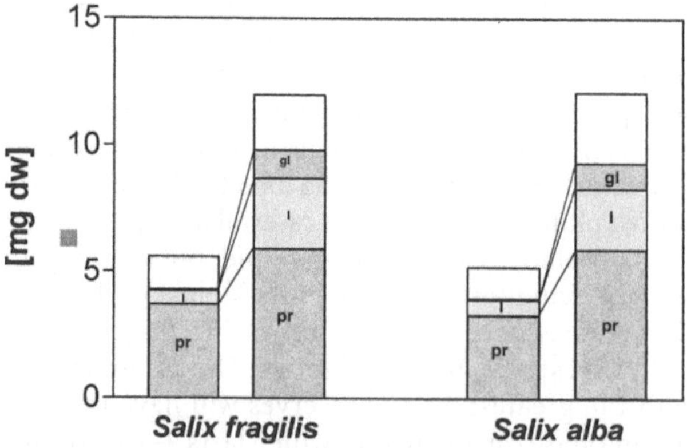

Fig. 9.4. Weight gain (mg dry wt.) during pre-diapause feeding period of *C. vigintipunctata* adults when feeding on either *S. fragilis* or *S. alba* foliage and proportions of metabolites per beetle (mg dry wt.) ($12 < n < 15$). *pr* protein; *l* lipid; *gl* glucose

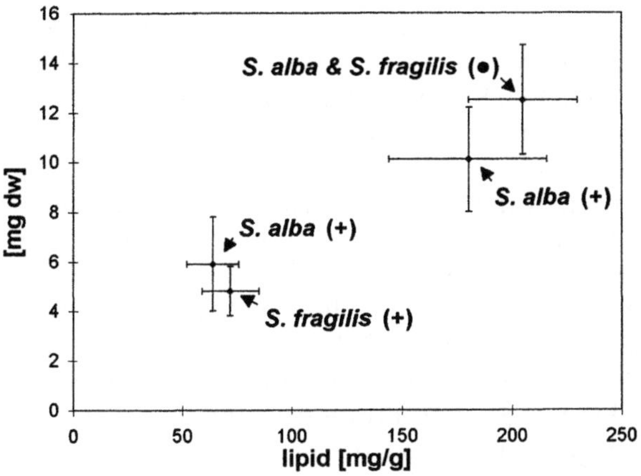

Fig. 9.5. Decline in weight (mg dry wt.) and lipid content (mg/g) of moribund *C. vigintipunctata* (+) adults during diapause in comparison with adults (●) when entering diapause ($\bar{x} \pm$ SD, $7 < n < 12$)

on the willow trees investigated. However, the length of diapause in both species is different. Winter moth larvae start feeding at bud burst; the larval feeding period is followed by a pupal diapause which is induced as early as the end of May and lasts about 6 months (Topp and Kirsten 1991). Larval development of the beetles starts some weeks later and is preceded by the post-diapause feeding period of adults. Adult diapause of *C. vigintipunctata* was induced at the beginning of July and lasted 9–10 months. Although temperature requirements for *C. vigintipunctata* were sufficient to develop a second generation per year, adult diapause was obligatory (Topp et al. 1989). This is adaptive because of the deteriorating plant quality with season. Because of this, *C. vigintipunctata* does not have the chance to be successful in producing a second generation on either host plant.

C. vigintipunctata is not the only willow leaf beetle with an obligatory univoltine development. However, other willow leaf beetle species either diapause later in the season or they are polymorphic in respect of reproductive diapause. For example, the willow leaf beetle *Plagiodera versicolora* produces one or two generations per year and minimizes the risk of being starved in this way.

9.8 Secondary Compounds of *S. alba* and *S. fragilis* Leaves

C. vigintipunctata individuals which fed either on *S. alba* or on *S. fragilis* foliage had to cope with extremely different sets of secondary compounds. Feeding early in the season was advantageous on both host plants; feeding later in the season resulted in a lower survival rate, in a lower efficiency to convert food into body mass, and in lower growth rates. The willow leaf beetle performance was depressed still further when *S. alba* foliage was consumed instead of *S. fragilis* foliage (see Sect. 9.5). Seasonal changes in the composition and/or in the concentration of secondary compounds specific to each host plant may cause these alterations and be responsible for the reducing fitness of the individuals with season. Especially the situation during adult pre-diapause feeding period at the beginning of June was crucial. During outbreaks, within this time interval, trees were almost defoliated. Since the Salicaceae produce new leaves during most of the growing season the photosynthetic area after a severe attack in June was largely restored by late July.

In *S. alba* the foliage available for the beetles during the pre-diapause feeding period was characterized by a high concentration of tannins (Fig. 9.6). Tannins present in *S. alba* foliage, which affect the nutritional value of the protein present in the leaf, may have similar effects on *C. vigintipunctata* as on the winter moth *Operophtera brumata* feeding on oak (Feeny 1970). *S. alba* trees, which were almost defoliated by the beetles, additionally exhibited high concentrations of some flavonoids. The seasonal change of apigenin-7-0-(4-p-coumaroylglucoside) is indicated in Fig. 9.6.

Large increases in phenolic compounds also were observed rapidly following damage to silver birch (*Betula pendula*) foliage artificially or by insect grazing (Hartley 1988). Neither of two Lepidoptera species tested showed any significant preferences to phenolic levels and appeared indifferent to damage. Preferences did not differ even when phenolic levels of birch leaves were manipulated by inhibition of PAL, which is a key enzyme in the biosynthesis of phenolic compounds (cf. Fig. 9.7). Hartley and Lawton (1990) have suggested that phenolic induction by herbivorous insects on birch may be directed pri-

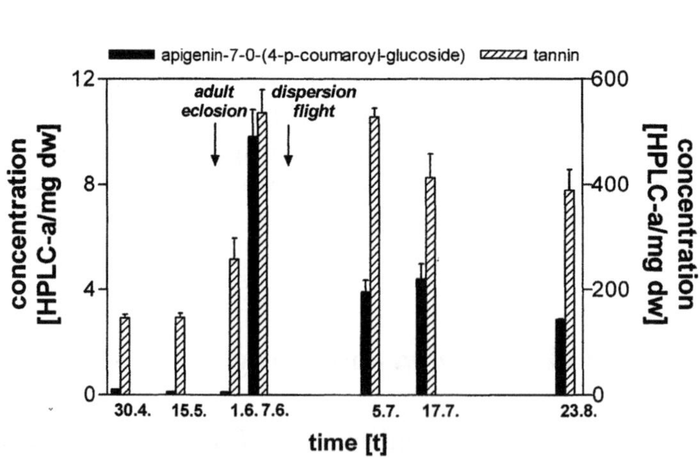

Fig. 9.6. Seasonal variation in the concentration of condensed tannins and apigenin-7-0-(4-p-coumaroylglucoside) in *S. alba* leaves ($\bar{x} \pm SD$, n = 4–5 for each value). The pre-diapause feeding period of *C. vigintipunctata* preceding the dispersion flight is indicated

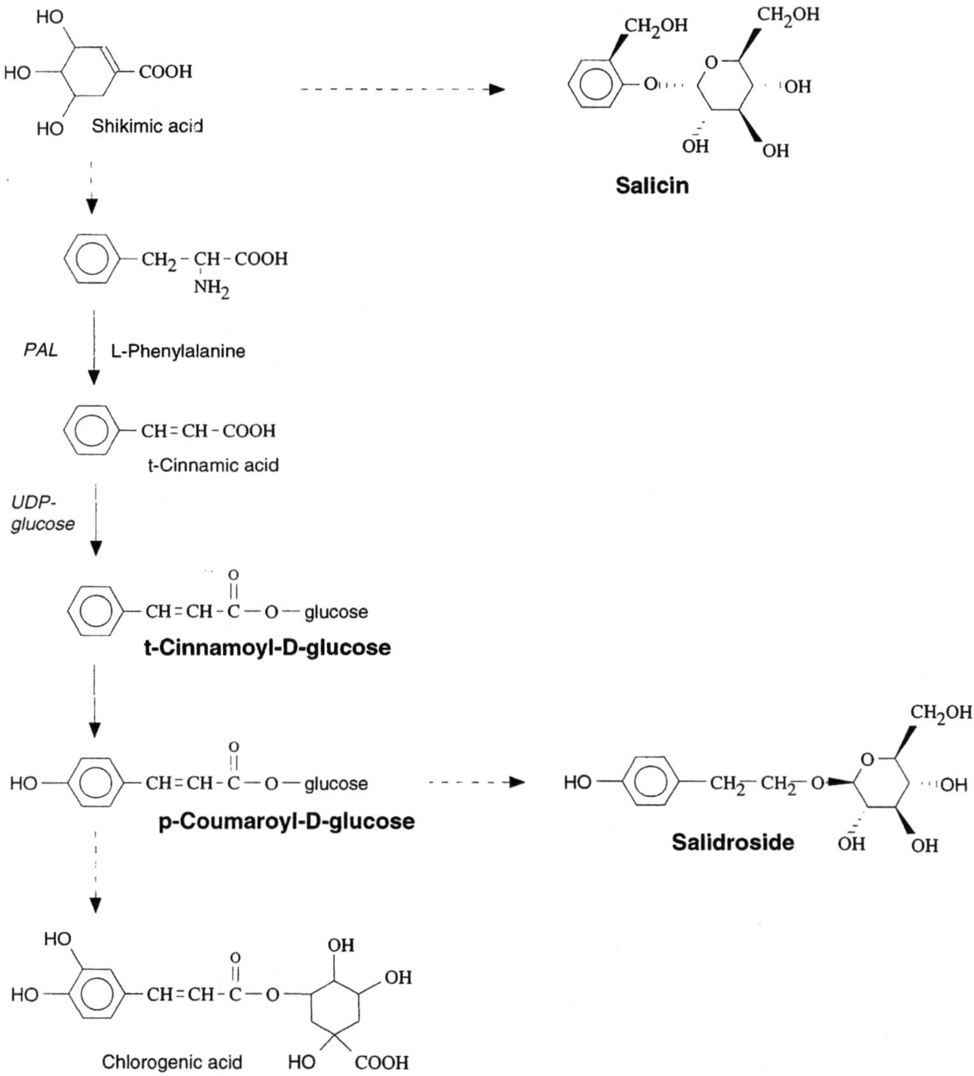

Fig. 9.7. Chlorogenic acid biosynthesis pathway and derived phenolglycosides characteristic of *S. fragilis* and *S. triandra* leaves. (Modified from Villegas and Kojima 1986)

marily at invading plant pathogens, rather than the herbivores responsible for the defoliation. Furthermore, Hunter and Schultz (1994) fed gypsy moth larvae with leaves of red oak trees and demonstrated that phytochemical induction can inhibit a pathogen of the herbivore responsible for the defoliation.

In our study a total of 23 different phenolic compounds were isolated from *S. alba* foliage (Baak et al. 1989). To get some idea which of these compounds is responsible for the lower fitness and increased mortality of *C.*

vigintipunctata, as a first step we used correlation analyses to test the concentrations of the secondary compounds with respect to the feeding activity of *C. vigintipunctata* and to the mortality of the caterpillars of *Phalera bucephala*. We tested *P. bucephala* because this bivoltine species is obviously adapted to the increasing content of tannins in *S. alba* with season. Tolerance to ingested tannins is known from several species out of different groups of insects (Bernays et al. 1981; Manuwoto and Scriber 1986). In these species detergency of gut fluids counteracts the potential of tannins to precipitate dietary protein (Martin and Martin 1984).

By testing *P. bucephala* for the effect of flavonoids, the effect of tannins otherwise relevant in *C. vigintipunctata* was excluded. Our studies were based on nine feeding intervals using five individual trees different with respect to their predation by herbivores (Baak et al. 1989). From the 23 chemicals isolated, only three secondary metabolites exhibited correlations to the herbivore attack and to the mortality of *P. bucephala* caterpillars (Fig. 9.8). To date, we have not been able to classify the effect of one single compound or even of the additive effect of the compounds in question.

One of the three flavonoids, rutin (quercetin-3-0-rutinoside) is known as a major trichome component in tomato. Larval growth of the fruitworm is significantly inhibited by the addition of rutin to an artificial diet (Isman and Duffey 1982).

The results mentioned above do not indicate the response pattern of *C. vigintipunctata* beetles to secondary compounds. However, the studies revealed further complexity in the response of plants to physical injury. Herbivory by *C. vigintipunctata* declined the quality of *S. alba* foliage for the caterpillar *P. bucephala*, which occurs later in the season. The same chemicals which probably inhibited the caterpillars already exhibited high concentrations during the time interval of pre-diapause feeding in *C. vigintipunctata*. Therefore, we assume that these flavonoids, singly or in concert, may cause mortality of the willow leaf beetles during diapause.

Wound-induced responses in plants have been the subject of much research over the past 20 years (Schultz and Baldwin 1982; Karban and Myers 1989). Some studies such as those of Fowler and MacGarvin (1986) indicated that induced wounding affected herbivore population dynamics, whereas others such as Wratten et al. (1984) or Hartley (1988) found little evidence for population effects. Harrison and Karban (1986) demonstrated that lupine previously damaged by the feeding of one herbivore was a less suitable food for a second herbivore later in the season. A reduction in fecundity in excess of 20% was observed when *Plagiodera versicolora* females consumed naturally and artificially damaged willow foliage (Raupp and Denno 1984). Benz and Abivardi (1991) demonstrated a temperature-dependent wound-induced response.

When feeding on *S. fragilis* instead of on *S. alba* foliage *C. vigintipunctata* benefits from salicin (see Sect. 9.3). Further phenoglycosides found in *S. fragilis* leaves were tremulacin, 2′-0-acetylsalicortin and salicortin. A further

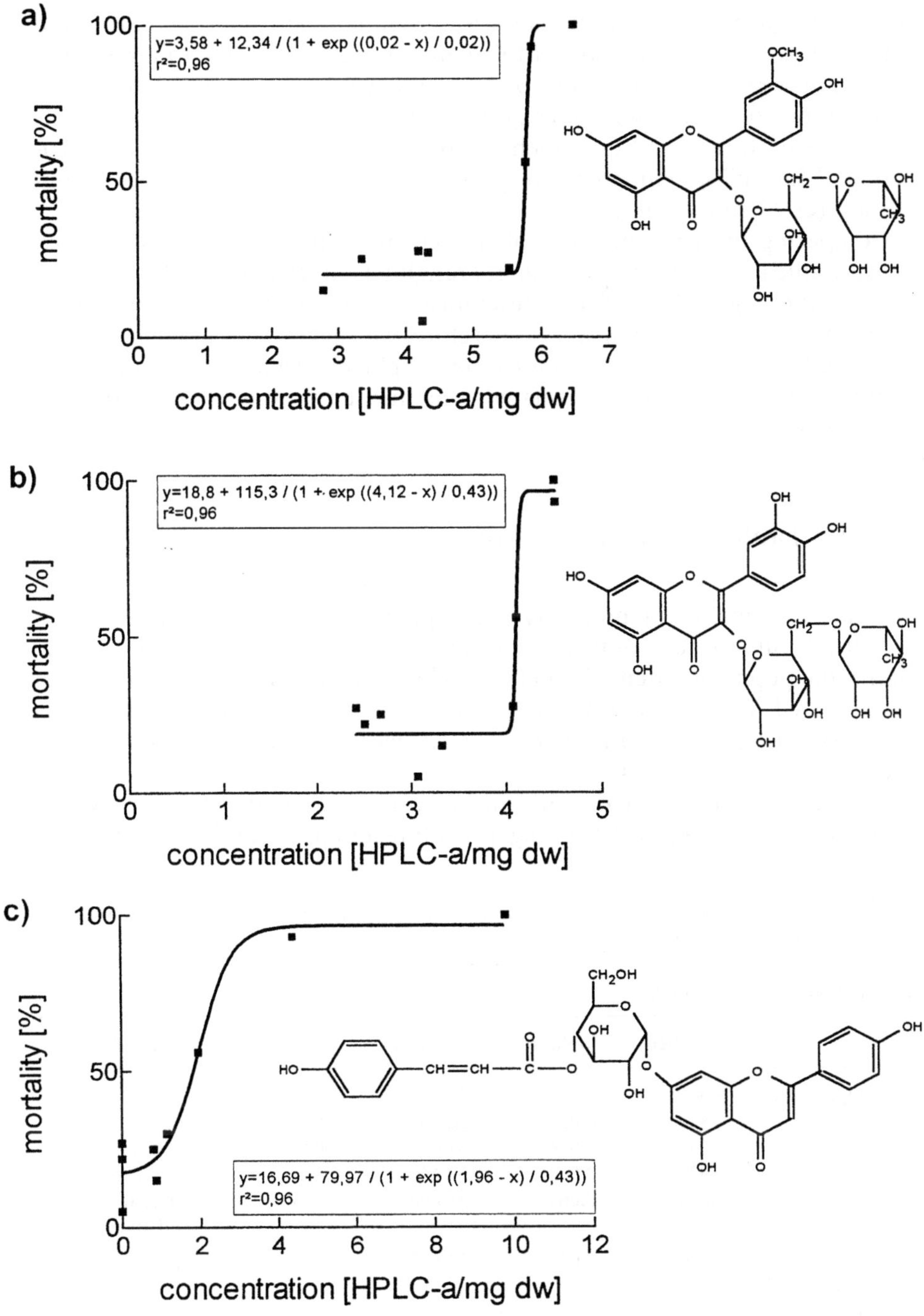

Fig. 9.8a–c. Correlations between the concentration of flavonoids of *S. alba* leaves and the mortality of *P. bucephala* caterpillars. **a** Isorhamnetin-3-0-rutinosid. **b** Quercetin-3-0-rutinoside (rutin). **c** Apigenin-7-0-(4-p-coumaroylglucoside)

main constituent of *S. fragilis* leaves was t-cinnamoyl-D-glucose (Groß and Topp 1997). The accumulation of this constituent may be due to a relatively low turnover rate to p-coumaroyl-D-glucose (Fig. 9.7). T-cinnamoyl-D-glucose is not known from *S. fragilis* trees growing in its centre of distribution. At the tributaries of the River Rhine which compose the western border of the distribution of *S. fragilis*, t-cinnamoyl-D-glucose may characterize a distinct chemical race (cf. Mizuno et al. 1989).

Comparing the concentration of this chemical between individual trees, we were able to reveal large differences ranging from 19.6 to 54.1 mg/g dry wt. at the beginning of May. These differences paralleled the impact of beetles occurring on the individual trees. Therefore, we suggest that grazing damage increases the concentration of t-cinnamoyl-D-glucose, which in turn affects the herbivore. Bioassays with the isolated constituent are necessary to prove our hypothesis.

9.9 Life Tables and *k*-factor Analysis

In Table 9.2 sources of mortality are expressed in terms of killing power (k-values) numerically equivalent to the differences between log(n + 1) and log(n), the population densities before and after a designated mortality factor has operated. The situation during a *C. vigintipunctata* outbreak with respect to either host plant, *S. fragilis* or *S. alba*, were evaluated on the basis of two individual host plants.

Sources of population decline during pre-imaginal development which arose from higher trophic levels, e.g. predators and parasitoids, accounted for

Table 9.2. Life table data for *C. vigintipunctata* when development occurred on either *S. fragilis* or *S. alba* trees

Age interval	*S. fragilis*		*S. alba*		
	Number	k-Value	Number	k-Value	Mortality factor
Egg–pupa	8.300		38.200		
	7.850	0.024	36.330	0.022	Predators
	7.070	0.046	33.080	0.041	Parasitoids
	6.970	0.006	32.660	0.005	Rainfall, storm
	6.000	0.065	25.000	0.116	Unknown effects and starvation
Summer adults	5.680	0.024	23.675	0.024	Antagonistic effects
Overwintering adults	5.570	0.009	20.100	0.071	Antagonistic effects
Spring adults	1.360	0.612	–	4.303	Antagonistic effects and starvation

about 15% with respect to *S. fragilis* and about 13% to *S. alba*. In regards to the patchy distribution of predators and parasitoids within trees, especially that of one of the most common, *Schizonotus sieboldi*, differences were not significant. Even if attractants to specialist parasitoids and predators are mediated by defence secretions of larvae and metabolites of injured plants as well, they should be much more concentrated and more effective over long distances when *C. vigintipunctata* feeds on *S. fragilis* foliage.

Pronounced differences in mortality during pre-imaginal development were caused by several factors which are summarized in Table 9.2 under the heading "unknown effects and starvation". The observations in the field refer to a considerable number of dead third-instar larvae and pre-pupae mostly attached to the leaves of *S. alba*. The causes of mortality in these individuals are unknown and may rely on different effects. Mortality by intraspecific competition and consequently starvation is one, mortality by toxic plant secondary compounds is another. Thirdly, individuals feeding on host plants may suffer from viral and fungal infections.

From a phytocentric point of view, chemical defences that are effective against herbivores and microbial pathogens are needed. Willow leaf beetles which are adapted to phenolglycosides present in *S. fragilis* foliage may simultaneously benefit from infectional defences against microorganisms. Several phenolics that are constitutively present in plants and that are believed to confer disease resistance include simple phenolic compounds (Waterman and Mole 1994). When adapted to these compounds which are otherwise toxic to microbes and generalist herbivores, specialists may experience a lower mortality by viral and fungal infections.

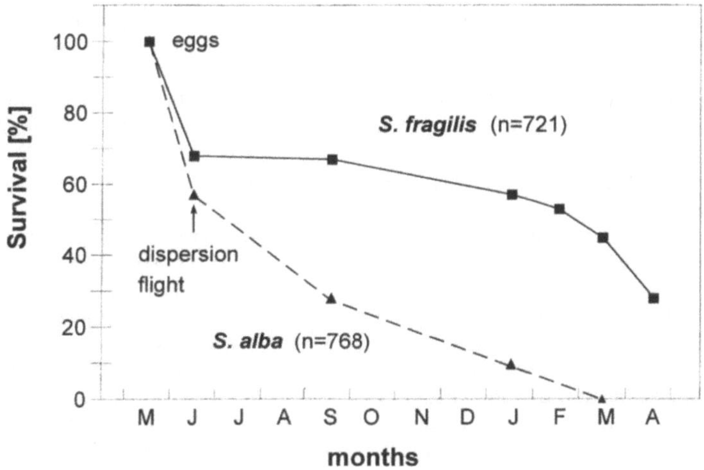

Fig. 9.9. Survival rate of *C. vigintipunctata* in the course of the year when feeding either on *S. fragilis* or on *S. alba* foliage. Mortality after dispersal flight of adults to the overwintering sites revealed maximum effects (cf. Table 9.2)

The most pronounced *k*-values were evaluated for the adult stage. With respect to *S. alba*, mortality during the pre-diapause feeding period and at the beginning of adult diapause may be mainly caused by plant secondary compounds. Thus, sources of dramatic population change did not arise from a higher trophic level but from a lower one.

Individuals which formerly fed on *S. fragilis* foliage mostly survived until the spring of the next year and then experienced a high rate of mortality (Fig. 9.9). Shortage of fat reserves, not sufficient for the 9–10 month long phase of overwintering, may be responsible for population decline when feeding on this host plant.

The situation in the field is even more complex than the one described here. During outbreaks, *C. vigintipunctata* did not exclusively settle on *S. fragilis* and *S. alba* trees; *S. viminalis* and *S. caprea* were two further host plants growing in the study area. Moreover, in contrast to all other host plants mentioned, plant quality of *S. viminalis* with respect to the fitness of *C. vigintipunctata* did not decline, but rather improved with season (Topp and Beracz 1989).

Acknowledgements. I wish to thank K. Dettner (University of Bayreuth) and G. Bauer (University of Freiburg) for providing valuable information. I am also grateful to Fred Bartlett, Philadelphia for improving the English text and to Hans-Jörg Groß, Regina Häusler and Anna Herzog for their help.

References

Baak S, Kirsten K, Topp W, Weissenböck G, Wray V (1989) Flavonoid glycosides and condensed tannins in *Salix alba* leaves: inhibitory effects on the development of *Phalera bucephala* (Notodontidae) larvae. 37th Annu Congr Med Plant Res, Braunschweig, 1989. Planta Med:99–100

Benz G, Abivardi C (1991) Preliminary studies on wound- and PIIF-induced resistance in some solanaceous plants against *Spodoptera littoralis* (Boisd.) (Lep., Noctuidae). J Appl Entomol 111:349–357

Bernays EA, Chamberlain DJ, Leather EM (1981) Tolerance of acridids to ingested condensed tannin. J Chem Ecol 7:247–256

Bowers MD (1993) Aposematic caterpillars: life-style of the warningly colored and unpalatable. In: Stamp NE, Casey TM (eds) Caterpillars. Ecological and evolutionary constraints on foraging. Chapman & Hall, London, pp 331–371

Courtney SP, Kibota TT (1989) Mother doesn't know best: selection of hosts by ovipositing insects. In: Bernays EA (ed) Insect plant interactions, vol 2. CRC Press, Boca Raton, pp 161–187

Denno RF, Larsson S, Olmstead KL (1990) Role of enemy-free space and plant quality in host-plant selection by willow beetles. Ecology 71:124–137

Downer RGH (1984) Lipid metabolism. In: Kerkut GA, Gilbert LI (eds) Comprehensive insect physiology biochemistry and pharmacology, vol 10. Biochemistry. Pergamon Press, Oxford, pp 77–113

Erbeling L, Terlutter H (1995) Ein Massenauftreten von *Chrysomela* (*Melasoma*) *vigintipunctata* (Scop.) (Coleoptera: Chrysomelidae) im Sauerland 1994. Natur Heimat 55:17–22

Evans DL, Schmidt JO (1990) Insect defenses. Adaptive mechanisms and strategies of prey and predators. State University of New York Press, Albany

Feeny PP (1970) Seasonal changes in oak leaf tannins and nutrients as a cause of spring feeding by winter caterpillars. Ecology 51:565–581

Floren A (1989) Nahrungsökologische Untersuchungen bei einem Massenauftreten des Weidenblattkäfers *Melasoma vigintipunctata* (Scop.). Diplom-Arbeit, Universität Köln, Köln

Fowler SV, MacGarvin M (1986) The effects of leaf damage on the performance of insect herbivores on birch, *Betula pubescens*. J Anim Ecol 55:565–573

Grigo F, Topp W (1980) Einfluß der Adaptationstemperatur und Photoperiode auf den Sauerstoffverbrauch bei Staphyliniden (Col.) in Diapause und Non-Diapause. Zool Anz 204:19–26

Groß HJ, Topp W (1997) β-D-glucopyranose-1-0-trans-cinnamate: a new constituent from foliage of *Salix fragilis*. Biologia (Bratisl) (in press)

Harrison S, Karban R (1986) Effects of an early season folivorous moth on the success of a later season species, mediated by a change in the quality of the shared host, *Lupinus arboreus* Sims. Oecologia 69:354–359

Hartley SE (1988) The inhibition of phenolic biosynthesis in damaged and undamaged birch foliage and its effect on insect herbivores. Oecologia 76:65–70

Hartley SE, Lawton JH (1990) Damage-induced changes in birch foliage: mechanisms and effects on insect herbivores. In: Watt AD, Leather SR, Hunter MD, Kidd NAC (eds) Population dynamics of forest insects. Intercept, Andover, pp 147–155

Hassell MP, Southwood TRE (1978) Foraging strategies of insects. Annu Rev Ecol Syst 9:75–98

Hegnauer R (1973) Chemotaxonomie der Pflanzen, Bd 6. Birkhäuser, Basel

Heinrich B (1979) Foraging strategies of caterpillars. Oecologia 42:325–337

Howe HF, Westley LC (1988) Ecological relationshops of plants and animals. Oxford University Press, New York

Hunter MD, Schultz JC (1994) Induced plant defenses breached? Phytochemical induction protects an herbivore from disease. Oecologia 94:195–203

Isman MB, Duffey SS (1982) Toxicity of tomato phenolic compounds to the fruitworm, *Heliothis zea*. Entomol Exp Appl 31:370–376

Jaenike J (1978) On optimal oviposition behaviour in phytophagous insects. Theor Popul Biol 14:350–356

Johnson ND, Brian SA, Ehrlich PR (1985) The role of leaf resin in the interaction between *Eriodictypon californicum* (Hydrophyllaceae) an its herbivore, *Trirhabda diducta* (Chrysomelidae). Oecologia 66:106–110

Karban R, Myers JH (1989) Induced plant responses to herbivory. Annu Rev Ecol Syst 20:331–348

Kirsten K, Topp W (1991) Acceptance of willow species for the development of the winter moth *Operophtera brumata* (Lep., Geometridae). J Appl Entomol 111:457–468

Manuwoto S, Scriber JM (1986) Effects of hydrolyzable and condensed tannin on growth and development of two species of polyphagous Lepidoptera: *Spodoptera eridania* and *Callosamia promethea*. Oecologia 69:225–230

Martin MM, Martin JS (1984) Surfactants: their role in preventing the precipitation of proteins by tannins in insect guts. Oecologia 61:342–345

Matsuda K, Matsuo H (1985) A flavonoid, luteolin-7-glucoside, as well as salicin and populin, stimulating the feeding of leaf beetles attacking saliceous plants. Appl Entomol Zool 20:305–313

Miller JR, Strickler KL (1984) Finding and accepting host plants. In: Bell W, Cardé R (eds) Chemical ecology. Sinauer, Sunderland, pp 127–157

Mizuno M, Kato M, Iinuma M, Tanaka T, Kimura A, Ohashi H, Sakai H, Kajita T (1989) Two chemical races in *Salix sachalinensis* Fr. Schmidt (Salicaceae). Bot Mag Tokyo 102:403–411

Ohmart CP, Stewart LG, Thomas JR (1985) Effects of nitrogen concentration of *Eucalyptus blakelyi* foliage on the fecundity of *Paropsis atomaria* (Coleoptera: Chrysomelidae). Oecologia 68:41–44

Palmer OJ (1985) Life-history consequences of body-size variaton in the milkweed leaf beetle, *Labidomera clivicollis* (Coleoptera: Chrysomelidae). Ann Entomol Soc Am 78:603–608

Pasteels JM, Braekman JC, Daloze D (1982) Chemical defence in chrysomelid larvae and adults. Tetrahedron 38:1891–1897

Pasteels JM, Rowell-Rahier M, Braekman JC, Dupont A (1983) Salicin from host plant as precursor of salicylaldehyde in defensive secretion of chrysomeline larvae. Physiol Entomol 8:307–314

Pasteels JM, Daloze D, Rowell-Rahier M (1986) Chemical defense in chrysomelid eggs and neonate larvae. Physiol Entomol 11:29–37

Pasteels JM, Braekman JC, Daloze D (1988) Chemical defense in the Chrysomelidae. In: Jolivet P, Petitpierre E, Hsiao TH (eds) Biology of Chrysomelidae. Kluwer, Dordrecht, pp 233–252

Pyke GH, Pulliam HR, Charnov EL (1977) Optimal foraging: a selective review of theory and tests. Q Rev Biol 52:137–154

Rank NE (1992) Host plant preference based on salicylate chemistry in a willow leaf beetle (*Chrysomela aeneicollis*). Oecologia 90:95–101

Rank NE (1994) Host-plant effects on larval survival of a salicin-using leaf beetle *Chrysomela aeneicollis* Schaeffer (Coeloptera: Chrysomelidae). Oecologia 97:342–353

Rank NE, Smiley JT (1994) Host-plant effects on *Parasyrphus melanderi* (Diptera: Syrphidae) feeding on a willow leaf beetle *Chrysomela aeneicollis* (Coeloptera: Chrysomelidae). Ecol Entomol 19:31–38

Raupp MJ, Denno RF (1984) The suitability of damaged willow leaves as food for the leaf beetle, *Plagiodera versicolora*. Ecol Entomol 9:443–448

Roitberg BD, van Lenteren JC, van Alphen JJM, Galis F, Prokopy RJ (1982) Foraging behavior of *Rhagoletis pomonella*, a parasite of hawthorn (*Crataegus viridis*) in nature. J Anim Ecol 51:307–325

Rowell-Rahier M (1984) The presence or absence of phenolglycosides in *Salix* (Salicaceae) leaves and the level of dietary specialisation of some of their herbivorous insects. Oecologia 62:26–30

Rowell-Rahier M, Pasteels JM (1989) Phenolglycosides and interactions at three trophic levels: Salicaceae-herbivores-predators. In: Bernays EA (ed) Insect plant interactions, vol 2. CRC Press, Boca Raton, pp 75–94

Schopf R, Krauße-Opatz B, Köhler U (1995) Saisonale Veränderungen von Lipiden und Aminosäuren beim Buchdrucker *Ips typographus* (Col., Scolytidae). Mitt Dtsch Ges Allg Angew Entomol 10:47–50

Schuler W, Hesse E (1985) On the function of warning coloration: a black and yellow pattern inhibits prey-attack by naive domestic chicks. Behav Ecol Sociobiol 16:249–255

Schultz JC, Baldwin IT (1982) Oak leaf quality declines in response to defoliation by gypsy moth larvae. Science 217:149–151

Schultz JC, Nothnagle PJ, Baldwin IT (1982) Seasonal and individual variation in leaf quality of two northern hardwood species. Am J Bot 69:753–759

Shao Y (1991) Phytochemischer Atlas der Schweizer Weiden. Dissertation ETH Zürich Nr 9532, Zürich

Slansky F (1993) Nutritional ecology: the fundamental quest for nutrients. In: Stamp NE, Casey TM (eds) Caterpillars. Ecological and evolutionary constraints on foraging. Chapman & Hall, London, pp 29–91

Slansky F, Scriber JM (1985) Food consumption and utilization. In: Kerkut GA, Gilbert LI (eds) Comprehensive insect physiology, biochemistry and pharmacology, vol 4. Pergamon, Oxford, pp 87–163

Slansky F, Wheeler GS (1991) Food consumption and utilization responses to dietary dilution with cellulose and water by velvetbean caterpillars, *Anticarsia gemmatalis*. Physiol Entomol 16:99–116

Smiley JT, Horn JM, Rank NE (1985) Ecological effects of salicin at three trophic levels: new problems from old adaptations. Science 229:649–651

Strong DR, Lawton JH, Southwood R (1984) Insects on plants. Blackwell, Oxford

Tahvanainen J, Julkunen-Tiitto R, Kettunen J (1985) Phenolic glycosides govern the food selection pattern of willow feeding leaf beetles. Oecologia 67:52–56
Thorsteinson AJ (1960) Host selection in phytophagous insects. Annu Rev Entomol 5:193–218
Topp W, Bell D (1992) *Melasoma vigintipunctata* (Scop.), ein Weidenblattkäfer mit Massenvermehrung. Faun-Ökol Mitt 6:267–286
Topp W, Beracz P (1989) Effect of host plant and changing seasonal development on consumption rates, utilization efficiencies and survival of *Melasoma 20-punctata* (Scop.) (Col., Chrysomelidae). J Appl Entomol 107:261–274
Topp W, Kirsten K (1991) Synchronisation of pre-imaginal development and reproductive success in the winter moth, *Operophtera brumata* L. J Appl Entomol 111:137–146
Topp W, Beracz P, Zimmermann K (1989) Distribution pattern, fecundity, development and survival of *Melasoma vigintipunctata* (Scop.) (Coleoptera: Chrysomelidae). Entomography 6:355–371
Villegas RJA, Kojima M (1986) Purification and characterization of hydroxycinnamoyl D-glucose. J Biol Chem 261:8729–8733
Waldbauer GP (1968) The consumption and utilization of food by insects. Adv Insect Physiol 5:229–288
Wassermann SS, Asami T (1985) The effect of maternal age upon fitness of progeny in the southern cowpea weevil, *Callosobruchus maculatus*. Oikos 45:191–196
Wassermann SS, Mitter C (1978) The relationship of body size to breadth of diet in some Lepidoptera. Ecol Entomol 3:155–160
Waterman PG, Mole SM (1994) Analysis of phenolic plant metabolites. Blackwell, Oxford
Werner RA (1979) Influence of host foliage on the development, survival, fecundity, and oviposition of the spear-marked black moth, *Rheumaptera hastata* (Lepidoptera: Geometridae). Can Entomol 111:317–322
Williams KS, Lincoln DE, Ehrlich PR (1983) The coevolution of *Euphydryas chalcedona* butterflies and their larval host plants. I. Larval feeding behavior and host plant chemistry. Oecologia 56:323–329
Wratten SD, Edwards PJ, Dunn I (1984) Wound-induced changes in the palatability of *Betula pubescens* and *B. pendula*. Oecologia 61:372–375

10 Ecdysteroids in Pycnogonids: Hormones and Interspecific Allelochemicals

K.-H. Tomaschko

10.1 Introduction

The search for biologically active substances from marine organisms during the last two or three decades has provided a vast variety of natural products. While in terrestrial systems plants are the dominating organisms with regard to the number of substances, in marine habitats the invertebrates are the most productive source of secondary metabolites (Faulkner 1994). Most of these compounds from marine invertebrates are assumed to have a defensive function. Consequently, the richest sources of secondary metabolites are those organisms which are sessile and/or soft-bodied, such as sponges, cnidarians, sea slugs, and tunicates (Pawlik 1993). Marine arthropods have been considered to lack chemical defences, as they are generally mobile and/or armoured.

In contrast to this assumption is the recent finding of chemical defence in *Pycnogonum litorale* (Tomaschko 1994a). This marine arthropod secretes ecdysteroids (ES) as a specific feeding deterrent against predatory crustaceans in its habitat. Being the moulting hormones of all arthropods, the ES fulfil a dual function in the pycnogonids: they act as hormones within the body and as interspecific semiochemicals. This duality in function raises a series of questions that will be discussed in this paper: What is the origin of the huge amounts of ES in the pycnogonids? Where are these compounds located in the body? Why do they not interfere with the moulting mechanism? How is the defensive secretion regulated? Another interesting aspect is the perception of ES in the crustaceans, where membrane-associated ES receptors are involved. Finally, evolutionary aspects of this unique defensive system are discussed.

10.2 Experimental Work with Pycnogonids

The pycnogonids are "living fossils". They have existed for at least 500 million years (Müller and Walossek 1986) and are considered an archaic group of arthropods (Bückmann and Tomaschko 1992). The class Pycnogonida is divided into two orders: the Palaeopantopoda to cover all fossil pycnogonids, and the order Pantopoda for all living species (Hedgpeth 1947). Due to their

Ecological Studies, Vol. 130
Dettner et al. (eds.) Vertical Food Web Interactions

spider-like bodies (Fig. 10.1a,c), the pycnogonids are often referred to as "sea spiders". They are exclusively marine and number to date about 1100 species in 84 genera (Müller 1993). Their leg spans reach from a few millimetres in littoral species to 60 cm in some deep-water species.

The majority of the species are epibenthic, a few are interstitial, and some are bathypelagic. Numerous commensal and parasitic pycnogonid species have been described, associated with cnidarians, poriferans, molluscan and echinoderm hosts (Arnaud and Bamber 1987). Although numerous reports on the interrelationship with other organisms exist, our knowledge about chemical interactions is restricted to the species *P. litorale*.

The experimental work on this species was made possible by a permanent culture that has existed for more than 25 years at the University of Ulm. The constant availability of large numbers of animals in all developmental stages rendered possible the investigation of the moulting physiology, the morphology and function of the defensive glands, and the defensive function of ES.

P. litorale (Fig. 10.1c) is approximately 1 cm in body length. It frequents exposed wave-washed shores in the northeast Atlantic littoral. Being an ectoparasite, it sucks the body fluid of cnidarians. In adaptation to this exposed way of living, it shows an extremely robust morphology, with stout, short legs, no auxiliary claws, but strong main claws for grasping the host tissue firmly.

The life cycle (Fig. 10.2) and the population dynamics of *P. litorale* have been reviewed in detail recently (Tomaschko et al. 1997; Wilhelm et al. 1997). Mating and egg production occur throughout the year. Preceded by a copulation that may last several weeks, deposition of 2000–9000 eggs takes a few hours. Embryonic development lasts about 6 weeks. Until hatching, the eggs are carried by the males with the ovigers, the third pair of the adult's appendages. The freshly hatched protonymphon larvae are 0.15 mm in body length and possess three pairs of larval appendages: two cheliphores – a characteristic which approaches them to the chelicerates – and two pairs of larval legs.

The larvae of *P. litorale* parasitize the hydroid polyp *Clava multicornis*. When a protonymphon larva is contacted by the tentacles of a polyp, it is seized and pulled to the mouth as it is the case with normal prey. However, some minutes later the larva is released unharmed. It grasps the hydranth with its legs and climbs down the hydrocaulus. Eventually, the larva is firmly attached to the stolons of the hydroid colony, beginning to ectoparasitize by sucking the hydroid's body fluid with its proboscis. During their larval period,

Fig. 10.1. **a** *Anoplodactylus lentus*, a dark-coloured pycnogonid from the Atlantic coast of North America, is unpalatable to *C. maenas*. **b** *C. maenas* seizes an adult female *P. litorale* with its chelum. **c** Adult female of *P. litroale*, dorsal aspect. **d** "ES print" of an adult female pycnogonid that was pressed towards a TLC plate with UV indicator, visualized at 254 nm. **e**, **f** Scanning electron micrographs of an adult female *P. litorale*: dorsal aspect (**e**) cuticula surface with opening of defensive gland and sensilla (*arrows*; **f**)

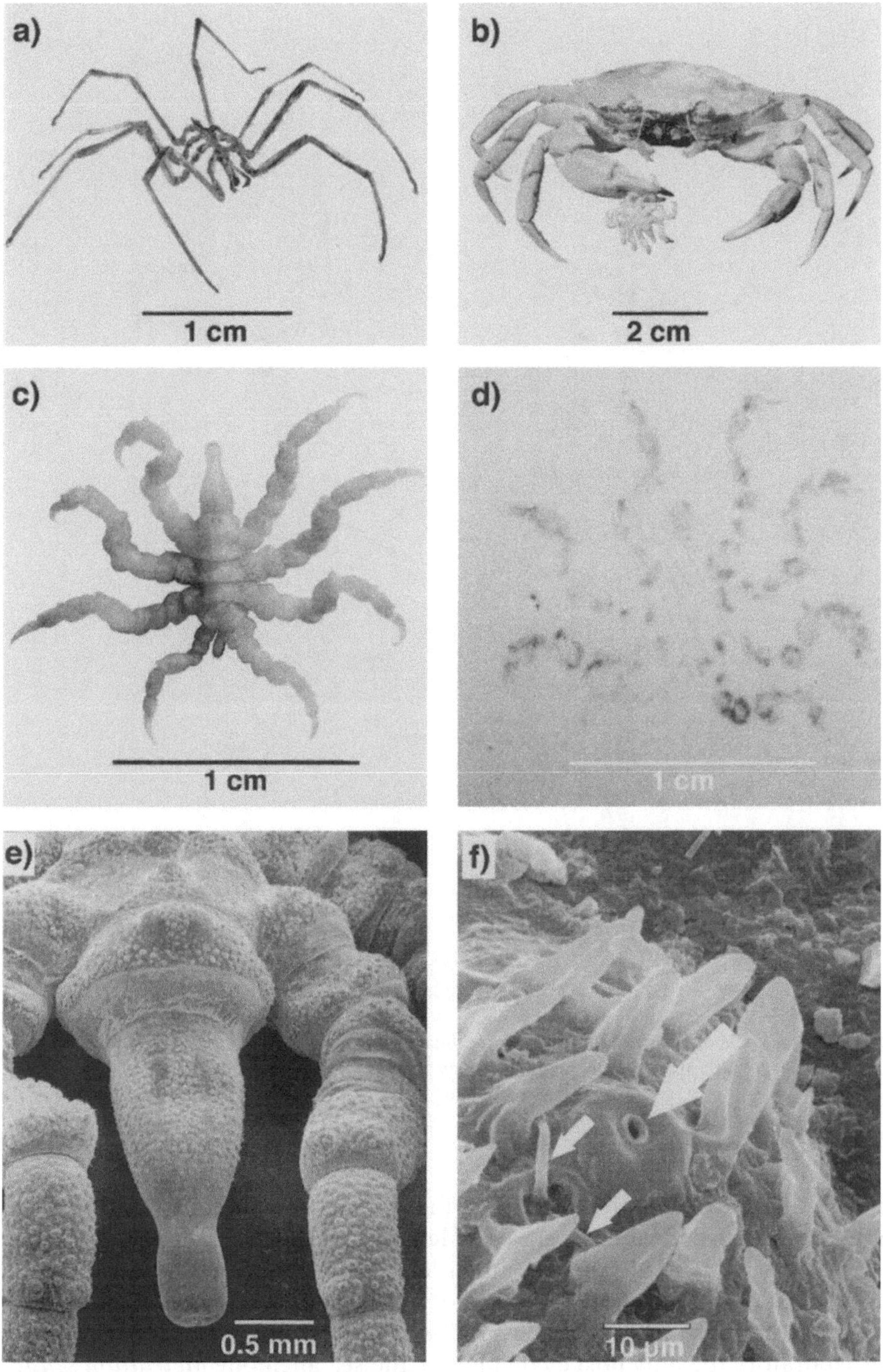

Fig. 10.1

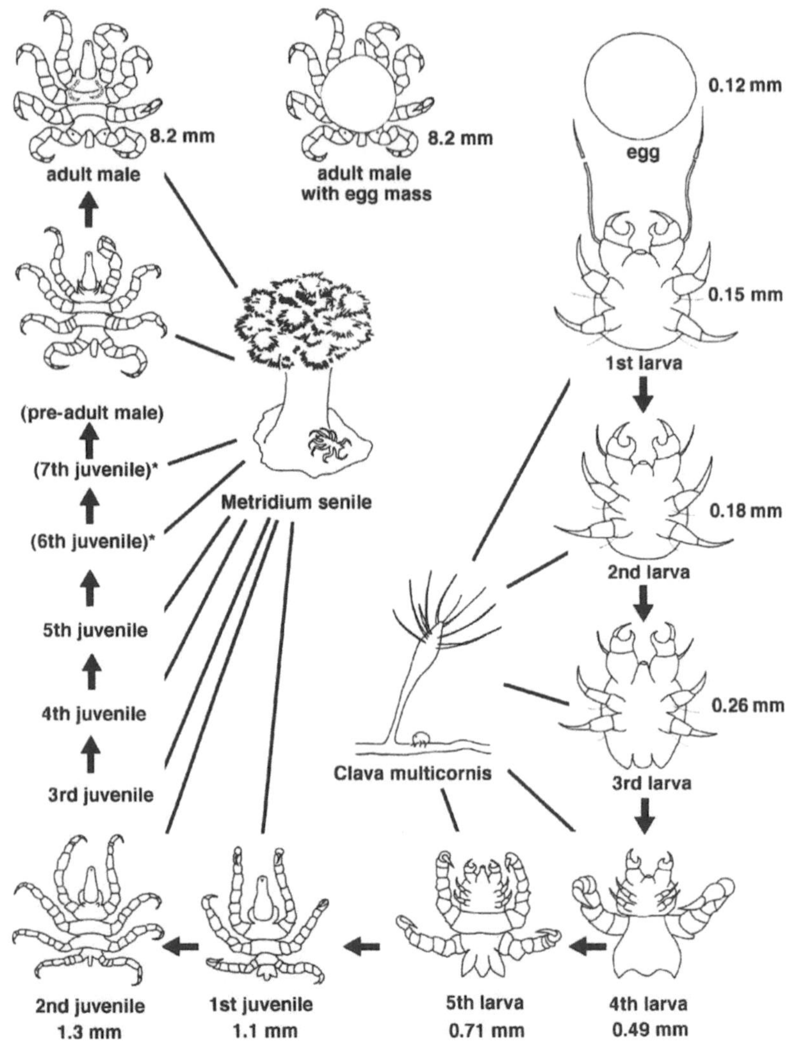

Fig. 10.2. Life cycle of *P. litorale*. Moults are indicate by *arrows*. *Lines* indicate the host-parasite relationships. *, sixth and seventh juvenile stages are facultative

the pycnogonids remain attached to the stolons for several months. They moult five times and reach 1 mm body length.

After their fifth moult, they lose their larval appendages and acquire a thick proboscis. From now on they are considered as juveniles. Juveniles and adults feed on actinians by inserting their proboscis into the host tissue for suction. In our laboratory the only host available was the sea anemone *Metridium senile*. Juveniles moult six to eight times until they become adult and show a terminal anecdysis (Lotz and Bückmann 1968). The ontogeny lasts between 6 and 18 months; the maximum age reached 9 years in the laboratory.

10.3 Ecdysteroids in *Pycnogonum litorale*

10.3.1 Quantitative and Qualitative Ecdysteroid Analyses

Behrens and Bückmann (1983) demonstrated for the first time the presence of ES in a pycnogonid. Their radioimmunological tests suggested that *P. litorale* contains unusually high levels of ES. The chromatographic investigation and subsequent analysis of several fractions in a bioassay and in a radioimmunoassay proved the presence of eight different ES (Tomaschko 1987). This led to a large-scale experiment in which 3000 adult animals (63 g dry weight) were extracted and eight ES (Fig. 10.3) isolated and identified by means of preparative high performance liquid chromatography (HPLC) and subsequent mass spectrometry and nuclear magnetic resonance (NMR) analysis (Bückmann et al. 1986).

The ES levels in *P. litorale* are the highest ever found in the animal kingdom. Except in freshly laid eggs and in young embryos, the eight ES occur in all developmental stages in similar amounts and ratios. In juveniles they reach a total of 5.9×10^{-4} M (almost 2000 nmol/g dry wt.) (Tomaschko and Bückmann 1993). This is equivalent to 0.1% of the body's dry wt. By comparison, the highest concentrations so far described in arthropods (locust embryos) were 10^{-4} M, i.e. about 0.025% of the dry weight. Common concentrations in arthropods are in the 10^{-6}–10^{-7} M range (Lafont and Horn 1989).

In freshly laid eggs all eight ES are already present, yet only in low amounts (17.2 nmol/g dry wt.). During the second half of the embryonic development, which lasts on average 50 days, total ES rise steeply and reach nearly 1000 nmol/g dry weight in the oldest embryos (Tomaschko and Bückmann 1993). Thus, the main increase in ES, reaching concentrations that are preserved in the same order of magnitude during all the rest of the life cycle, already occurs during embryonic development. The maintenance of these high concentrations during growth from the protonymphon larva to the adult affords the constant uptake and/or de novo synthesis of ES (see Sect. 10.3.2).

Not only are the high levels of ES unusual, but so are their chemical structures. The most abundant compound is the 20-hydroxyecdysone 22-acetate (20E22Ac) (Fig. 10.3), which in all developmental stages represents 66.2–85.8% of total ES. It has been found only during in vitro incubations in *Drosophila melanogaster* (Maroy et al. 1988) and in a snail (Garcia et al. 1986). The moulting hormone of all arthropods, 20-hydroxyecdysone (20E), makes up only a few percent of the total ES in all developmental stages (Tomaschko and Bückmann 1993). The other six ES are unique in the animal kingdom. The

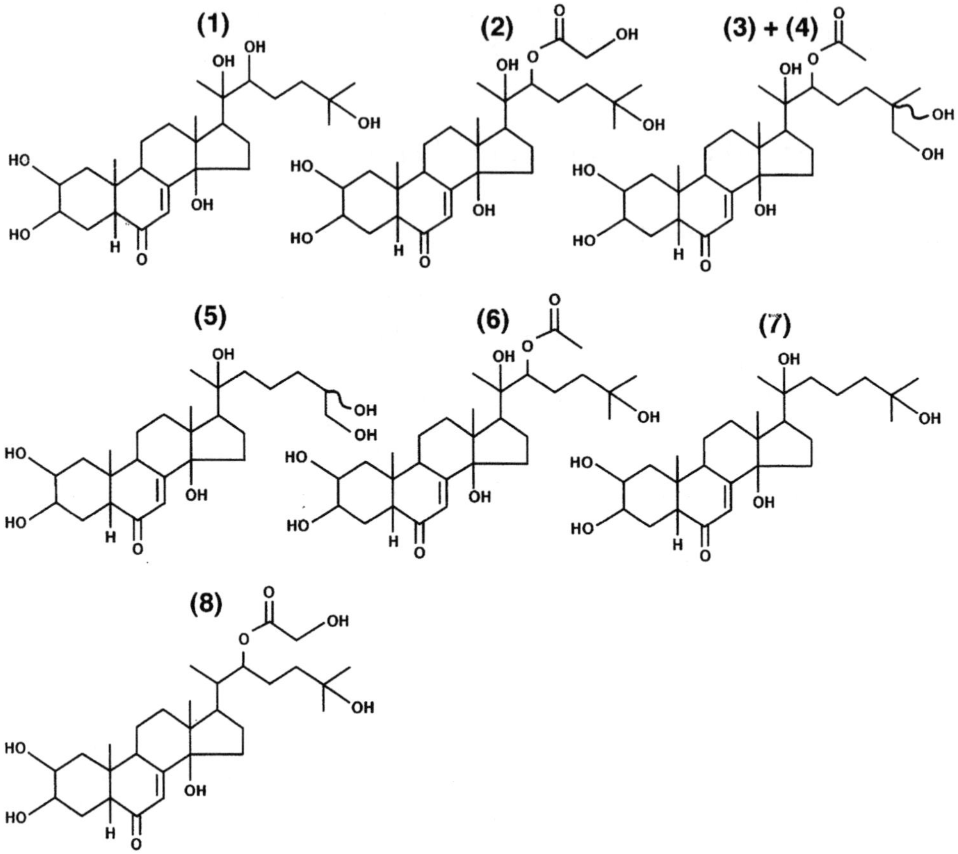

Fig. 10.3. ES in *P. litorale*. **1** 20-Hydroxyecdysone. **2** 20-Hydroxyecdysone 22-glycolate. **3** and **4** 25R and 25S isomers of 20,26-dihydroxyecdysone 22-acetate. **5** 22-Deoxy-20,26-dihydroxyecdysone. **6** 20-Hydroxyecdysone 22-acetate. **7** 22-Deoxy-20-hydroxyecdysone (taxisterone). **8** Ecdysone 22-glycolate

accumulation of such high amounts of ES appears, at the moment, as a feature unique among arthropods.

10.3.2 Origin of Ecdysteroids

There are two possible sources for the high ES levels in *P. litorale*:

1. *The Host Animals*. *P. litorale* is able to sequester ecdysone and 20E efficiently from its food, to metabolize these ES to the above described compounds and to store them in its epidermal glands (see Sect. 10.3.4.2) (Tomaschko and Bückmann 1990). *M. senile*, the host of juvenile and adult pycnogonids, contains no significant amounts of ecdysone or 20E (Habermehl

et al. 1976; Behrens 1981). Using an antiserum, we found 26 ng ecdysone equivalents/g dry weight in the sea anemone (Weigert 1995). In order to cover its ES requirements from its food, a pycnogonid would have to eat more than 10 kg of sea anemones. Thus, the ES in the food cannot account for the high ES levels in *P. litorale.*

2. *De novo synthesis.* Our knowledge about ES synthesis in arthropods derives mainly from insects and malacostracean crustaceans. Both groups synthesize ES from cholesterol. It is generally accepted that arthropods are not able to synthesize cholesterol from small molecules, e.g. acetate (Clayton 1964). Thus, cholesterol is a vitamin for arthropods which they must ingest with their food. *P. litorale* seems to have no lack of this vitamin, as cholesterol represents almost 1% of its dry weight (K.-H. Tomaschko and M. Feldlaufer, unpubl. data). This is not unusual, as cholesterol is the predominant sterol in almost all marine invertebrates. Its percentage of the total sterols in crustaceans is usually more than 90% (Goad 1978), and in horseshoe crabs about 80% (Bergmann et al. 1943). In *M. senile,* the host of *P. litorale,* cholesterol is by far the predominant sterol as well (Habermehl et al. 1976). Thus it seems likely that the pycnogonids sequester their cholesterol from their hosts.

No conclusive experimental evidence for a de novo synthesis of ecdysteroids from cholesterol in pycnogonids exist. Injections of radioactively labelled cholesterol and subsequent investigation of the labelled metabolites have not been successful as yet. This is probably due to the huge amounts of cholesterol in the pycnogonids, which compete with the labelled cholesterol for ES synthesis. More promising results were obtained by injections of (^{3}H)2,22-deoxyecdysone, an intermediate product in the synthesis of ES in insects and crustaceans. Twenty-four hours after injection of (^{3}H)2,22-deoxyecdysone into their haemolymph, adult females formed labelled 20E22Ac (Weigert 1995). This shows that pycnogonids are at least capable of performing the final hydroxylation steps in the ES synthesis pathway.

Another indication for de novo synthesis of ES from cholesterol might be the considerable increase in total ES during embryogenesis (Tomaschko and Bückmann 1993). However, in this case the increase in ES may result as well from maternal conjugates, as it is the case in the insects *Locusta migratoria* and *Schistocerca gregaria* (Hoffmann and Lagueux 1985).

The localization of a possible ES synthesis in pycnogonids is also unknown. In insects, ES synthesis takes place in prothoracic glands, ovaries, oenocytes, epidermis and testes (Delbeque et al. 1990), and in crustaceans in the Y-organ (Spindler et al. 1981). We have little information about ES-synthesizing organs or tissues in terrestrial chelicerates (review in Juberthie and Bonaric 1992), and none at all in horseshoe crabs. Incubations of isolated tissues of pycnogonids with (^{3}H)-2,22-deoxyecdysone suggest that the integument and the gut are involved in ES synthesis (Weigert 1995).

In conclusion, the high levels of ES in *P. litorale* derive most likely from both de novo synthesis and from the food. As both sources do not seem to be outstandingly productive, the unusually high ES levels can only be explained by an efficient sequestration and storage mechanism (see Sect. 10.3.4.2).

10.3.3 Hormonal Effects

Since Butenandt and Karlson (1954) isolated ecdysone from 500 kg of silkworms, ES has been identified in at least 8 major classes, 13 orders, and all major forms of arthropods (Jegla 1990). They appear in nearly all other protostomian phyla (Franke and Käuser 1989) and in many plants as well. To date we know about 260 different zoo- and phytoecdysteroids (Lafont and Wilson 1997). However, hormonal effects have so far been demonstrated convincingly only in arthropods, where the best-established function is the control of moulting.

In *P. litorale* the moult-inducing function of ES was proven in second stage larvae (Bückmann and Tomaschko 1992). Exogenous 20-hydroxyecdysone induces the same effects in the larvae as it does in other arthropods: according to its concentration and the time of application, it leads to premature or delayed moults, or to death (Bückmann and Tomaschko 1992).

There is also indirect evidence for 20E as a moulting hormone in *P. litorale*: in contrast to the total ES, it occurs in relatively low amounts in all developmental stages. Its maximum concentration is found in juveniles with 57 nmol/g dry wt. (3.5% of total ES). This concentration is comparable with the levels in other arthropods, where it acts as a moulting hormone. Among the pycnogonid ES, it is also the only one that exhibits a "conventional" moulting hormone titre course (Steel and Vafopoulou 1989), i.e. it shows a maximum at the time of apolysis of the cuticle (Tomaschko and Bückmann 1993). In contrast, all other ES reach their maxima at the time of ecdysis or right after a moult. Therefore, a moulting hormone function in the pycnogonids is not likely for these compounds.

This does not necessarily mean that these compounds do not have a potential as moulting hormones in arthropods. The major ES in pycnogonids, 20E22Ac, induced supermoulting in the tick *Ornithodoros moubata* at a 500-ppm level (Savolainen et al. 1991). However, such an effect could not occur in *P. litorale*, as the majority of the ES cannot circulate freely in the haemolymph, but is concentrated in epidermal glands (see Sect. 10.3.4.2).

10.3.4 Allelochemical Effects

10.3.4.1 Experimental Evidence for Defensive Functions of Ecdysteroids

Regarding the unusually high levels of ES in all developmental stages of *P. litorale*, even in those which do not moult, raises the question: are there non-hormonal functions of these compounds, and, if so, what are these functions? An important hint at a possible allelochemical function was contained in the results of Takahashi and Kittredge (1973). During feeding experiments with the lined shore crab, *Pachygrapsus crassipes*, the authors found that the normal feeding response of the crab was suppressed by seawater containing 10^{-6}–10^{-9} molar concentrations of 20E. The idea of looking for a possible defensive function of the ES in pycnogonids seemed tempting, as *P. litorale* is morphologically unprotected and has very limited locomotor ability. Living ectoparasitically in exposed positions on coelenterates or attached to hard substrates, it would be an easy prey for potential predators.

Carcinus maenas, the common shore crab, is a widespread generalist predator in the northeast Atlantic littoral (Crothers 1968), easily capable of coping with prey of the size of the pycnogonids (Fig. 10.1b). It reaches high abundance in the interstices of rocky shores and jetties, where it is often closely associated with *P. litorale* (K.-H. Tomaschko, pers. observ.). Laboratory experiments revealed that the pycnogonids are not eaten by the crabs. Fifty adult pycnogonids (25 males and 25 females) were offered to 50 different crabs that had not been fed for 24 h. Although all crabs were eagerly accepting parts of the mussel *Mytilus edulis*, they rejected the pycnogonids. Only five pycnogonids were slightly injured but survived this experiment. All others were either ignored or released unharmed after a few seconds (Tomaschko 1994a). Obviously, the pycnogonids are unpalatable to the crabs.

Similar results were obtained when lyophilized pycnogonid powder was added to semiartificial food pellets. Pellets containing 40% (w/w) of pycnogonid powder were completely rejected by all crabs. Even at a 5% level the food consumption was still significantly inhibited. This shows that the pycnogonids contain compounds that are responsible for the rejection behaviour of the crabs.

The effect of 20E and 20E22Ac on the feeding behaviour of the crabs was investigated in an experiment using semiartificial food pellets. Both control pellets and ES-treated pellets were seized by the crabs and carried to the mouth. While control pellets without ES were eaten readily within 10 s, the pellets containing ES were rejected dose-dependently by both sexes (Tomaschko 1994a). Artificial food consumption of pellets was completely inhibited by 20E at a 20×10^{-4} molar level, and 20E22Ac was less effective. It caused 95% inhibition at the same molar concentration. The minimum concentrations for significant reduction of food consumption were 1.25×10^{-4} M and 5.0×10^{-4} M, respectively. The other six ES that occur in *P. litorale* (Fig. 10.3) have not yet been investigated as to their feeding deterring effect. As these compounds are not commercially available, it is difficult to obtain sufficient amounts for profound experimental data. However, it is plausible to suspect that they have deterrent effects as well, as is the case with all other tested ES (see Sects. 10.3.4.2, 10.3.4.3).

The formation of a spectrum of unusual ES could be a special evolutionary adaptation to their function as feeding deterrents. It may represent a way to avoid counteradaptations by predators. Pasteels (1993) suggests that it is probably more difficult for predators to overcome a mixture of toxins with various physiochemical properties, and hence biochemical activities, than a single compound. In addition, possible synergistic effects of the ES among one another or with other (not yet identified) chemicals must be taken into account, as has been described for other defensive compounds in staphylinid beetles (Dettner 1993).

10.3.4.2 Storage and Secretion of Ecdysteroids

The majority of all ES in *P. litorale* are accumulated in epidermal glands and can be released in response to disturbance. This was demonstrated directly for 20E22Ac, the quantitatively predominant ES. Injected or ingested [^{3}H]20E is quickly metabolized to [^{3}H]20E22Ac (Tomaschko and Bückmann 1990) and then accumulated in epidermal glands (Tomaschko 1995). These glands occur on all parts of the body, except on the arthrodial membranes (Fig. 10.1e,f). They consist of epidermal cells containing large vacuoles. Each gland has a duct with a diameter of 1 μm which leads to the outer surface of the cuticle. According to this, the surface of the pycnogonids is perforated by numerous gland openings, numbering about 20000 in adults (Fig. 10.1e,f).

When mechanically irritated, *P. litorale* discharges a defensive effluent that contains all eight identified endogenous ES in concentrations reflecting almost exactly the proportions of the total ES within the body (Tomaschko 1994b). Hence, it follows that not only 20E22Ac, but also most of the other ES are stored in the defensive glands. Due to the uniform distribution of the defensive glands on the body surface, they provide an effective all-round defence against crustacean predators. This release of ES was demonstrated by a new autoradiographic technique for uneven surfaces (Tomaschko 1995). In addition, it could be visualized by a simple method: an adult female was pressed against a thin-layer chromatography (TLC) plate that was precoated with a fluorescent indicator. The resulting "ES print" becomes visible under UV light (Fig. 10.1c,d), due to its UV-absorbing properties.

Repeated intensive irritation of the entire body surface results in secretion of 98.6% of the total ES (Tomaschko 1994b). Indeed, pycnogonid specimens which had been deprived of their ES in this way were eaten by the crabs. However, under natural conditions such an intense molestion is unlikely to occur. During a predator attack, the pycnogonids economically administer their release of ES. Secretion is restricted to the irritated area of the body surface, e.g. exclusive artificial irritation of the left or the right body side results in exclusive secretion on the respective side. The same is true for single legs or for parts of legs. Thus, predators can be efficiently deterred without wasting too much ES. This precise secretion which is restricted to the area of contact requires specific sensory organs. They are represented by one or two forked setae associated with each pore (Fig. 10.1f). The ultrastructure of these sensilla clearly indicates a mechanoreceptive function (K.-H. Tomaschko and M. Heß, unpubl. data).

The secretion of defensive chemicals not only renders the pycnogonids unpalatable to the crabs, it also offers protection against serious injuries, as the crabs are already deterred by the secreted ES.

10.3.4.3 Perception of Ecdysteroids in Decapod Crustaceans

C. maenas recognizes potential food items by chemoreceptors that are located on the tips of the appendages and responds quickly to the touch of any food substance (Case and Gwilliam 1961; Gnatzy et al. 1984). Supposed food is seized and carried to the mouth by the legs and the chelae. When overcritical concentrations of ES are present, the crabs react within less than a second by removing the item with their mouth parts and chelae. Sometimes a curious behaviour can be observed: although an ES-containing food item has been rejected, it is seized again with the walking legs or with the chelae and carried to the mouth where it is removed again. This conflicting behaviour can be repeated several times. Eventually, the crab ignores the food item, e.g. a pycnogonid or a food pellet, although food items of a different kind, e.g. parts of *Mytilus edulis*, are eagerly accepted and eaten. The ability to "learn" that pycnogonids are unpalatable is not only beneficial for the encountered specimen, but also reduces predatory pressure on the whole pycnogonid population. This may be of vital significance for the pycnogonids, considering that regeneration of ES is slow and that repeated intensive irritation may result in a complete loss of almost all ES.

Another interesting point is the reception of ES in the crabs. From the feeding experiment it became clear that the recognition of the ES takes place solely in the mouth region. This is corroborated by the fact that application of ES solution to the walking legs, the carapax and the antennae did not evoke any visible reaction. As yet, two different areas of ES reception have been located: application of ES solution on the endopodites of the second maxilla results in intensive movements of the mouth appendages. Application to the dorso-anterior oesophagus leads to an immediate reflex-like dilatation of the entire oesophagus (Fig. 10.4; Tomaschko et al. 1995). Both reactions are part of the natural rejection behaviour and help to rinse the anterior oesophagus, i.e. to remove the ES.

The short latency of this response to ES can only be explained by interactions with membrane-associated receptors. In contrast, the classical mode of ES action is the specific activation of genes in the cell nucleus (Ashburner 1980). Little is known about extranuclear activities of ES. Membrane-associated ES receptors in *C. maenas* and other decapod crustaceans would provide a good possibility of increasing our knowledge about the basic principles of steroid action.

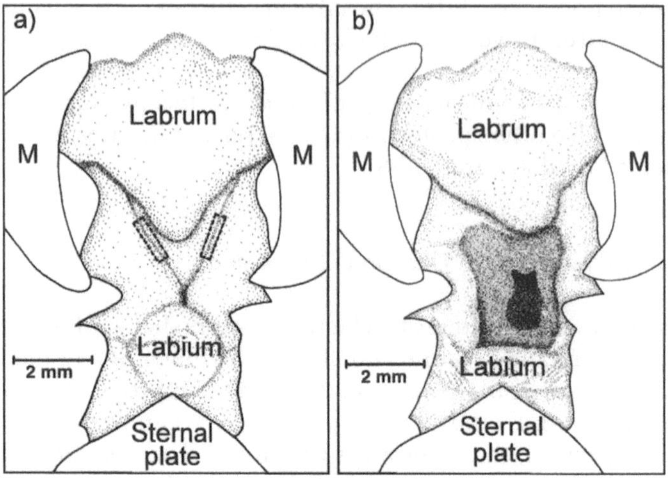

Fig. 10.4. Oesophagus of *C. maenas*, ventral aspect. **a** During the ODA (see Sect. 10.3.4.3), 1 μl of test solution is administered either to the right or to the left part of the antero-lateral oesophagus (areas marked with *dotted lines*). **b** Dilatated oesophagus (see Sect. 10.3.4.3)

In order to investigate the sensitivity and the specificity of these receptors, we have developed an easy and quick to perform bioassay (Tomaschko et al. 1995). The oesophagus dilatation assay (ODA) is based on the immediate dilatation of the oesophagus. This occurs when a solution containing overcritical amounts of ES is administered to the dorso-anterior oesophagus (Fig. 10.4).

From the preliminary results of the ODA (Table 10.1) we can conclude that the receptor is to a certain degree ES-specific. The low EC_{50} values (which indicate high sensitivity) are restricted to ES. The EC_{50} values for vertebrate-type steroids are one to three orders of magnitude higher. For progesterone we could not detect any effect at all. Water with different salinities (0–4%) and pH values (5–10) had no effect. As yet, we can only speculate about the structure specificity of the receptor and we are aware that the observed effects do not necessarily reflect the affinity of a single receptor protein to the different compounds, but may well be the result of complex nervous processing.

Ponasterone A, which has the highest biological activity in classical bioassays, is less effective than several other ES in the ODA. In contrast to other bioassays, our assay shows a stronger reaction towards ecdysone than to 20-hydroxyecdysone. It is also remarkable that the nonsteroidal ES agonists RH5849 and RH5992, which are able to mimic ES in several other bioassays, have no effect in the ODA. From these results, ES receptors in *C. maenas* differ significantly from the classical nuclear receptor with respect to their sensitivity and structure specificity. This is in agreement with the few reports available on membrane-associated ES receptors in other arthropods (e.g. Kaeuser et al. 1990). Further investigations are necessary in order to specify the biological

Table 10.1. Biological potency (given as EC_{50}) in *Carcinus* ODA. Compounds were dissolved in thistle oil containing 0.001–0.005% dimethylformamide (DMFA)

Substance	EC_{50}(μM)
Makisterone A[a]	2.2
Ecdysone	4.1
2-Deoxyecdysone[a]	7.4
Muristerone A[a]	7.4
2-Deoxy-20-hydroxyecdysone[a]	7.5
20-Hydroxyecdysone	9.6
Polypodine B[a]	61.0
Ponasterone A[a]	115.0
20-Hydroxyecdysone 22-acetate	144.0
3-Dehydro-20-hydroxyecdysone[a]	150.0
Testosterone	5190
Cholesterol	7050
Progesterone	No effect at 10000
ES agonist RH5849	No effect at 10000
ES agonist RH5992	No effect at 10000
Water (salinity 0–4%)	No effect
Water (pH 5–10)	No effect

[a] These compounds were a generous gift from Prof. René Lafont, Ecole Normale Supérieure, Paris.

and molecular properties of this type of steroid receptor. The easy accessibility of the ES receptors at the body surface of decapod crustaceans predestinates them for further experiments.

10.4 Conclusions

10.4.1 Defensive Secretion in Marine Arthropods

Secretion of defensive chemicals has been described in terrestrial and freshwater arthropods (Dettner and Schwinger 1980; Blum 1981; Lokensgard et al. 1993). In some insect taxa, chemical defence is such a common phenomenon that the morphology of the defensive glands and the chemical composition of their secretions contributes to the solution of systematic questions (Deroe and Pasteels 1982; Dettner 1993; Pasteels 1993). In contrast, defensive strategies in marine arthropods were generally considered to depend on camouflage, behaviour and/or their physical defensive weapons. *P. litorale* is the first marine arthropod in which defensive secretion and an allelochemical effect on a predator in its natural habitat were able to be demonstrated.

In searching for chemical defence in other marine arthropods, other pycnogonid species may be good candidates. All of them show relatively low motility and are morphologically unprotected, as their cuticles are not calcified. Unfortunately, information about pycnogonids as food for other animals is poor. From the few records available, Arnaud and Bamber (1987) concluded that pycnogonids are ingested more incidentally than actively, and that they do not constitute an appreciable part of any predator's diet. Krapp and Nieder (1993) found some evidence that five pycnogonid species (not *P. litorale*) from the northern Mediterranean are eaten by bleniid fishes. However, recent feeding experiments in our laboratory with *Anoplodactylus lentus* (Fig. 10.1a) from the North American Atlantic coast and *Ammothella biunguiculata* from the Sea of Japan suggest chemical defence against crustacean predators in these pycnogonids as well. Both species are unpalatable to *C. maenas*. The nature of the feeding deterrent compounds in these cases is not yet clear. A further indication for chemical defence in pycnogonids is the presence of numerous epidermal glands in all species (Helfer and Schlottke 1935). The function of these glands could as yet not be explained satisfactorily. It seems possible that they have defensive functions in other pycnogonid species as well. Thus, defensive secretion might well be a common feature to many (all?) pycnogonids.

10.4.2 Zooecdysteroids as Feeding Deterrents

Apart from the classical hormonal functions, there exist few reports on semiochemical functions of ES in arthropods. In crustaceans, a sex pheromone function of 20E was postulated by Kittredge and Takahashi (1972). However, this could not be verified by Atema and Gagosian (1973), Adelung et al. (1980) and Seifert (1982). Spencer and Case (1984) found sensory responses towards 20E at a level of 10^{-13} M in the chemoreceptors of spiny lobsters. However, as this is true for both sexes as well as for juveniles, it cannot corroborate the pheromonal hypothesis. Clear evidence for ES as components of a sex pheromone was given by Taylor et al. (1991) in two tick species.

In arthropods the use of ES as interspecific semiochemicals, the so-called allelochemicals, is as yet restricted to *P. litorale*. At first view, ES look like universal protection against all kinds of arthropods. They are active as hormones in concentrations of 10^{-8}–10^{-7} M (Kaeuser et al. 1990) and thus are potentially dangerous to the consumer. The use of phytoecdysteroids as feeding deterrents has already been suggested by Gailbraith and Horn (1966). Many plants contain ES (Lafont and Horn 1989), including ecdysone, 20E and ponasterone A, which are biologically active in all arthropods. ES levels in plants are generally several orders of magnitude higher than in arthropods and reach up to 3.2% of dry weight (Bandara et al. 1989). They can act as chemical defence against "unadapted" phytophagous insects by disturbing the insect's larval development when ingested (Lafont et al. 1991; Sláma et al. 1993). However, the ES are not necessarily a 100% protection. Some phytophagous insects have developed effective inactivation and excretion mechanisms in order to overcome the defence by ES (Hikino et al. 1975).

The use of ES as feeding deterrents in arthropods seems to be relatively rare. This may be due to the fact that the high ES concentrations required are potentially dangerous to the arthropod producer itself and therefore require specific adaptations to the physiology and anatomy of the producer. High levels comparable with those in pycnogonids (10^{-3} M) occur, e.g. in the eggs of ticks (5×10^{-6} M) (Connat et al. 1985) and locusts (10^{-4} M) (Hinton 1981), where a defensive function cannot be excluded. On the other hand, the use of ES as defensive compounds is likely in non-arthropod invertebrates, where they have no hormonal effects (as is the case in plants) and can therefore be contained at high levels. Gueriero and Pietra (1985) found high ES levels in the zoanthid *Gerardia savaglia*, where a defensive function against predatory crustaceans seems possible.

10.4.3 Evolution of Chemical Defence in Pycnogonids

The conditions for the use of ES as hormones and as interspecific messengers are similar: The ability to synthesize ES and the existence of specific ES receptors. The function of ES as allelochemicals in pycnogonids may have passed through the following evolutionary steps:

1. Synthesis of ES is a primitive feature of almost all invertebrates.
2. The use of ES as hormones was achieved by the common ancestors of all living arthropods. As yet, it has not been found in other organisms.
3. Defensive effects of ES presuppose their hormonal effects in arthropods, i.e. their danger to the hormonal equilibrium. The capability of the crustaceans to perceive ES in the food may be the result of a synevolution with a variety of ES-producing organisms. In this context, not only do pycnogonids come into question as ES producers, but also non-arthropod invertebrates and plants.
4. The defensive system in *P. litorale* may be a specific adaptation to the ES-perceiving capabilities of predatory crustaceans. The presence and maintenance of such high ES levels reflects the high predatory pressure by the crabs. Further investigations will show whether defensive secretion of ES occurs in other pycnogonids or in other arthropods as well, or whether this system in *P. litorale* is unique.

Acknowledgement. The author thanks Dr. Mona Ferguson for critically reading the manuscript.

References

Adelung D, Seifert P, Buchholz F (1980) Studies on the identification and the action of the sexual pheromone of the shore crab *Carcinus maenas*. Verh Dtsch Zool Ges 73:316
Arnaud F, Bamber RN (1987) The biology of Pycnogonida. Adv Mar Biol 24:1–96

Ashburner M (1980) Chromosomal action of ecdysone. Nature 285:435–436

Atema J, Gagosian RB (1973) Behavioral responses of male lobsters to ecdysones. Mar Behav Physiol 2:15–20

Bandara BMR, Jayasinghe L, Karunaratne V, Wannigama GP, Bokel M, Kraus W, Sotheeswaran S (1989) Ecdysterone from the stem of *Diploclisia glaucescens*. Phytochemistry 28:1073–1075

Behrens W (1981) Morphologische und hormonphysiologische Untersuchungen bei einem Pantopoden, *Pycnogonum litorale* Ström. Diss, Universität Ulm, Ulm

Behrens W, Bückmann D (1983) Ecdysteroids in the pycnogonid *Pycnogonum litorale* (Ström) (Arthropoda, Pantopoda). Gen Comp Endocrinol 51:8–14

Bergmann W, McLean MJ, Lester D (1943) Contributions to the study of marine products. XIII. Sterols from various invertebrates. J Org Chem 8:21–282

Blum MS (1981) Chemical defense of arthropods. Academic Press, New York, pp 1–36

Bückmann D, Tomaschko K-H (1992) 20-hydroxyecdysone stimulates molting in pycnogonid larvae (Arthropoda, Pantopoda). Gen Comp Endocrinol 88:261–266

Bückmann D, Starnecker G, Tomaschko K-H, Wilhelm E, Lafont R, Girault JP (1986) Isolation and identification of major ecdysteroids from the pycnogonid *Pycnogonum litorale* Ström (Arthropoda, Pantopoda). J Comp Physiol B 156:759–765

Butenandt A, Karlson P (1954) Über die Isolierung eines Metamorphosehormons der Insekten in kristalliner Form. Z Naturforsch 9b:389–391

Case J, Gwilliam GF (1961) Amino acid sensitivity of the dactyl chemoreceptors of *Carcinus maenas*. Biol Bull 121:449–458

Clayton RB (1964) The utilization of sterols by insects. J Lipid Res 5:3–19

Connat JL, Diehl PA, Gfeller H, Morici M (1985) Ecdysteroids in females and eggs of the Ixodid tick *Amblyomma hebraeum*. Int J Invertebrate Reprod Dev 8:103–116

Crothers JH (1968) The biology of the shore crab *Carcinus maenas* (L.) 2. The life of the adult crab. Field Stud 2:579–614

Delbeque J-P, Weidner K, Hoffmann KH (1990) Alternative sites for ecdysteroid production in insects. Inv Reprod Dev 18:29–42

Deroe C, Pasteels JM (1982) Distribution of adult defense glands in chrysomelids (Coleoptera: Chrysomelidae) and its significance in the evolution of defense mechanisms within the family. J Chem Ecol 8:67–82

Dettner K (1993) Defensive secretions and exocrine glands in free living staphylinid beetles – their bearing on phylogeny (Coleoptera: Staphylinidae). Biochem Syst Ecol 21:143–162

Dettner K, Schwinger G (1980) Defensive substances from pygidial glands of water beetles. Biochem Syst Ecol 8:89–95

Faulkner DJ (1994) Marine natural products. Nat Prod Rep 11:355–394

Franke S, Käuser G (1989) Occurrence and hormonal role of ecdysteroids in nonarthropods. In: Koolman J (ed) Ecdysone, from chemistry to mode of action. Thieme, Stuttgart, pp 296–307

Gailbraith MN, Horn DHS (1966) An insect-moulting hormone from a plant. Chem Commun 905–906

Garcia M, Girault J-P, Lafont R (1986) Ecdysteroid metabolism in the terrestrial snail *Cepaea nemoralis* (L.). Int J Invertebrate Reprod Dev 9:43–58

Gnatzy W, Schmidt M, Römbke J (1984) Are the funnel-canal organs the "campaniform sensilla" of the shore crab *Carcinus maenas* (Crustacea, Decapoda)? Zoomorphology 104:11–20

Goad LJ (1978) The sterols of marine invertebrates: composition, biosynthesis, and metabolites. In: Scheuer PJ (ed) Marine natural products, chemical and biological perspectives, II. Academic press, New York, pp 75–173

Guerriero A, Pietra F (1985) Isolation, in large amounts, of the rare plant ecdysteroid ajugasterone-C from the Mediterranean zoanthid *Gerardia savaglia*. Comp Biochem Physiol 80B:277–278

Habermehl G, Christ B, Krebs HC (1976) Die freien Steroide in *Metridium senile*. Naturwissenschaften 1:42

Hedgpeth JW (1947) On the evolutionary significance of the Pycnogonida. Smithson Misc Collect 106:1–54

Helfer H, Schlottke E (1935) Pantopoda. In: Bronns HG (ed) Klassen und Ordnungen des Tierreichs, vol 5, IV. Abt, 2. Buch. Leipziger Akademische Verlagsgesellschaft, Leipzig, pp 75–79

Hikino H, Ohizumi Y, Takemoto T (1975) Detoxication mechanism of *Bombyx mori* against exogenous phytoecdysone ecdysterone. J Insect Physiol 21:1953–1963

Hinton E (1981) Biology of insect eggs, vol 1. Pergamon Press, Oxford

Hoffmann JA, Lagueux M (1985) Endocrine aspects of embryonic development in insects. In: Kerkut GA, Gilbert LI (eds) Comprehensive insect physiology, biochemistry and pharmacology, vol 1. Pergamon, Elmsford, New York, pp 435–460

Jegla TC (1990) Evidence for ecdysteroids as molting hormones in Chelicerata, Crustacea, and Myriapoda. In: Gupta AP (ed) Morphogenetic hormones of arthropods, vol 1. Rutgers University Press, London, pp 229–273

Juberthie C, Bonaric JC (1992) Ecdysial glands and ecdysteroids in terrestrial Chelicerata. In: Gupta AP (ed) Morphogenetic hormones of arthropods, vol 1. Rutgers University Press, London, pp 271–305

Kaeuser G, Koolman J, Karlson P (1990) Mode of action of molting hormones in insects. In: Gupta AP (ed) Morphogenetic hormones of arthropods, vol 1. Rutgers University Press, London, pp 362–387

Kittredge JS, Takahashi FT (1972) The evolution of sex pheromone communication in the Arthropoda. J Theor Biol 35:467–471

Krapp F, Nieder J (1993) Pycnogonids as prey of blenniid fishes. Cah Biol Mar 34:383–386

Lafont R, Horn DHS (1989) Phytoecdysteroids: structures and occurrence. In: Koolman J (ed) Ecdysone, from chemistry to mode of action. Thieme, Stuttgart, pp 39–64

Lafont R, Wilson ID (1997) The ecdysone handbook. The Chromatographic Society, Suite 4, Clarendon Chambers, 32 Clarendon Street Nottingham NG1 5JD

Lafont R, Bouthier A, Wilson ID (1991) Phytoecdysteroids: structures, occurrence, biosynthesis and possible ecological significance. In: Hrdy I (ed) Insect chemical ecology. SPB Academic Publ, The Hague, pp 197–214

Lokensgard J, Smith RL, Eisner T, Meinwald J (1993) Pregnanes from defensive glands of a belastomid bug. Experientia 49:75–176

Lotz G, Bückmann D (1968) Die Häutungen und die Exuvie von *Pycnogonum litorale* (Ström) (Pantopoda). Zool Jahrb Anat 85:529–536

Maroy P, Kaufmann G, Dübendorfer A (1988) Embryonic ecdysteroids of *Drosophila melanogaster*. J Insect Physiol 34:633–637

Müller H-G (1993) World catalogue and bibliography of the recent Pycnogonida. Wissenschaftlicher Verlag, Laboratory for Tropical Ecosystems Research & Information Service, PO Box 2268, D-35532 Wetzlar. Floppy disk (IBM-DOS)

Müller KJ, Walossek D (1986) Arthropod larvae from the upper Cambrian of Sweden. Trans R Soc Edinb Earth Sci 77:157–179

Pasteels JM (1993) The value of defensive compounds as taxonomic characters in the classification of leaf beetles. Biochem Syst Ecol 21:135–142

Pawlik JR (1993) Marine invertebrate chemical defenses. Chem Rev 93:1911–1922

Savolainen V, Wuest J, Lafont R, Connat J-L (1991) Induction of supermolting in female ticks *Ornithodoros moubata* by ingestion of 20-hydroxyecdysone-22-acetate. Gen Comp Endocrinol 82:221

Seifert P (1982) Studies on the sex pheromone of the shore crab, *Carcinus maenas*, with special regard to the ecdysone excretion. Ophelia 21:147–158

Sláma K, Abukakirov NK, Gorovits MB, Baltaev UA, Saatov Z (1993) Hormonal activity of ecdysteroids from certain Asiatic plants. Insect Biochem Mol Biol 23:181–185

Spencer M, Case JF (1984) Exogenous ecdysteroids elicit low-threshold sensory responses in spiny lobsters. J Exp Zool 229:163–166

Spindler K-D, Keller R, O'Connor JD (1981) The role of ecdysteroids in crustacean moulting cycle. In: Hoffmann JA (ed) Progress in ecdysone research. Elsevier/North-Holland, Amsterdam, pp 247–280

Steel CGH, Vafopoulou X (1989) Ecdysteroid titer profiles during growth and development of arthropods. In: Koolman J (ed) Ecdysone, from chemistry to mode of action. Thieme, Stuttgart, pp 221–231

Takahashi FT, Kittredge JS (1973) Suppression of the feeding response in the crab *Pachygrapsus crassipes*: pheromone induction. Tex Rep Biol Med 31:403–408

Taylor P, Phillips JS, Sonenshine DE, Hanson FE (1991) Ecdysteroids as a component of the genital sex pheromone in two species of hard ticks, *Dermacentor variabilis* and *Dermacentor andersoni* (Acari: Ixodidae). Exp Appl Acarol 12:275–296

Tomaschko K-H (1987) Biologische Wirksamkeit und Titerverlauf der Ecdysteroide von *Pycnogonum litorale* (Ström) (Arthropoda, Pantopoda). Verh Dtsch Zool Ges 80:182

Tomaschko K-H (1994a) Ecdysteroids from *Pycnogonum litorale* (Arthropoda, Pantopoda) act as chemical defense against *Carcinus maenas* (Crustacea, Decapoda). J Chem Ecol 20:1445–1455

Tomaschko K-H (1994b) Defensive secretion of ecdysteroids in *Pycnogonum litorale* (Arthropoda, Pantopoda). Z Naturforsch (C) 49:367–371

Tomaschko K-H (1995) Autoradiographic and morphological investigations of the defensive ecdysteroid glands in adult *Pycnogonum litorale* (Arthropoda: Pantopoda). Eur J Entomol 92:105–112

Tomaschko K-H, Bückmann D (1990) Metabolism and excretion of ecdysteroids in *Pycnogonum litorale* (Arthropoda, Pantopoda). Invertebr Reprod Dev 18:130–131

Tomaschko K-H, Bückmann D (1993) Excessive abundance and dynamics of unusual ecdysteroids in *Pycnogonum litorale* (Ström) (Arthropoda, Pantopoda). Gen Comp Endocrinol 90:296–305

Tomaschko K-H, Guckler R, Bückmann D (1994) Ecdysteroids as feeding inhibitors in decapod crustaceans. Abstr 17th Conf of European Comparative Endocrinologists, Cordoba, 109 pp

Tomaschko K-H, Guckler R, Bückmann D (1995) A new bioassay for the investigation of a membrane associated ecdysteroid receptor in decapod crustaceans. Neth J Zool 45:93–97

Tomaschko K-H, Wilhelm E, Bückmann D (1997) Growth and reproduction of *Pycnogonum litorale* Ström (Pantopoda) in the laboratory. Mar Biol (in press)

Weigert O (1995) Die Aufnahme von Vorstufen, die Synthese und die Abgabe von Ecdysteroiden bei dem Pantopoden *Pycnogonum litorale* (Ström). Diplomarbeit Universität Ulm, Ulm

Wilhelm E, Bückmann D, Tomaschko K-H (1997) The life cycle and the population dynamics of *Pycnogonum litorale* Ström (Pantopoda) in the natural habitat. Mar Biol (in press)

Part D

Phytophages and Their Enemies: Interactions Between Aphids and Their Predators and Parasitoids

11 Growth and Development in Parasitoid Wasps: Adaptation to Variable Host Resources

M. Mackauer, R. Sequeira, and M. Otto

11.1 Introduction

Fitness returns to a female parasitoid depend on how she distributes offspring over available hosts and host patches (reviewed in Godfray 1994), and on how immature parasitoids adapt to variation in host resources. The kind and the amount of resources provided by the host determine its quality for parasitoid survival and development (Mackauer and Sequeira 1993). In idiobionts, species that develop in non-feeding host stages such as eggs and pupae, the host represents a "closed" system. The amount of available resources depends on host size, which can be assessed by the female at oviposition, regardless of any physiological changes that may occur later during the interaction (Strand 1986; King 1990). In contrast, koinobionts exploit hosts that continue to feed, grow and metamorphose after parasitization; the host represents an "open" system. Because quality is determined by the host's nutrition and growth potential after parasitization, which vary with circumstances during the interaction (Sequeira and Mackauer 1992a, 1994), it cannot be evaluated from age and size differences alone.

Life-history responses reflect the optimal balance between adaptation and constraint (Sibly and Calow 1986). Especially in koinobiont parasitoids, host choice often reflects the female's ability to find and oviposit in or on a host, rather than to provide optimum resources for immatures. The mother's choice is generally irreversible; only in a few species with a predatory lifestyle are hosts selected by the externally feeding larvae as opposed to the free-living adults (Gauld and Bolton 1988). Immature parasitoids should be expected, therefore, to evolve ontogenetic responses to resource constraints. Larval growth rate, development time and adult body size are the principal components of resource-utilization strategies in hymenopteran parasitoids.

We address two related aspects of host-parasitoid interactions. First, we examine patterns of adaptive host use in idiobiont and koinobiont species; we ask if host quality is association-specific. Second, we compare the growth rates of male and female parasitoids developing in the same kinds of hosts; because females gain weight faster than males under the same conditions, the degree of sexual dimorphism varies with host quality. It is suggested that idiobiont and koinobiont resource-utilization strategies do not, in principle, differ at the hypothetical upper end of the host-size spectrum. We review the recent litera-

Ecological Studies, Vol. 130
Dettner et al. (eds.) Vertical Food Web Interactions

ture, with emphasis on aphid-parasitoid-hyperparasitoid associations. As examples of koinobiont parasitoids we use species of aphidiid wasps, which are solitary endoparasitoids of aphids (Mackauer and Chow 1986). As an example of idiobiont parasitoids we use *Dendrocerus carpenteri*, a solitary ectoparasitoid of aphidiid wasps inside mummified aphids (Sullivan 1988).

11.2 Models of Parasitoid Growth

Mackauer and Sequeira (1993) reviewed patterns of parasitoid growth and development under different host constraints. They proposed three different models of parasitoid adaptation to host resources. The immatures of idiobiont species typically grow at a constant rate in a closed resource environment (model 1). Koinobionts, in contrast, can adapt to dynamic changes in an open resource environment by varying the ratio of the growth rate to the development rate (models 2 and 3).

Let host quality, Q, be defined as the total amount of host resources available during the interaction, with the amount of resources available at oviposition denoted by R_o and the amount acquired during parasitoid development denoted by R_d. Note that R_o is determined by the state of the host at parasitization and, in that sense, is fixed, whereas R_d depends on the dynamics of resource acquisition until host death at t_d:

$$Q = f(R_o + R_d). \tag{1}$$

In normally feeding aphids, the increase in mass with age follows a sigmoid curve which reaches a plateau marked by the beginning of reproduction (Mackauer 1986; Sequeira and Mackauer 1992a). Consequently, both R_o and R_d are functions of host age at oviposition, t_o, under most conditions (Fig. 11.1). Parameter values can be obtained from a mathematical equation fitted to the sigmoid growth curve of unparasitized aphids; R_o is estimated by the definite integral of the growth function from $t = 0$ to t_o, and R_d by the definite integral from t_o to t_d (R. Sequeira and M. Mackauer, unpubl.). Predicted values of R_d may be over- or underestimated, however, if parasitism affects host feeding and resource acquisition (Cloutier and Mackauer 1979, 1980; Bai and Mackauer 1992). For idiobionts developing in a closed resource system, $R_d \approx 0$ and Eq. (1) simplifies to $Q = f(R_o)$.

In many host-parasitoid associations, parasitoid body size in terms of dry mass, DM, is a nonnegative, but not necessarily linear, function of host quality. Sequeira and Mackauer (1992a) showed that growth in aphidiid wasps follows a so-called J-curve. Mass increased exponentially between oviposition and the time the last-instar larva killed the host; from its maximum at pupation, mass declined again between pupation and adult eclosion. Causton (1977) provided a derivation of the mean relative growth rate, MRGR; the equation makes no assumption about the dynamics of growth and can be applied to arbitrary intervals:

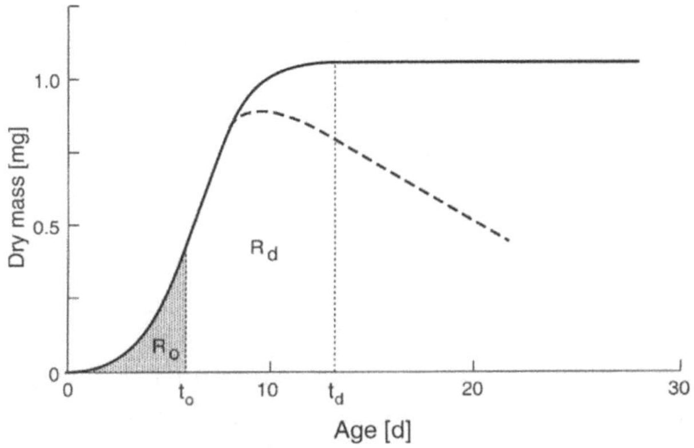

Fig. 11.1. Hypothetical growth curve of pea aphid, *Acyrthosiphon pisum*, in relation to age. Dry mass reaches a plateau at the onset of reproduction. An aphid parasitized at age t_o contains current resources (R_o, *shaded area*) and can acquire additional resources (R_d) prior to death at time t_d. Parasitized aphids may grow slower, in terms of dry mass, than unparasitized aphids, which is indicated by the *dashed line*. (Redrawn from Sequeira and Mackauer 1992a)

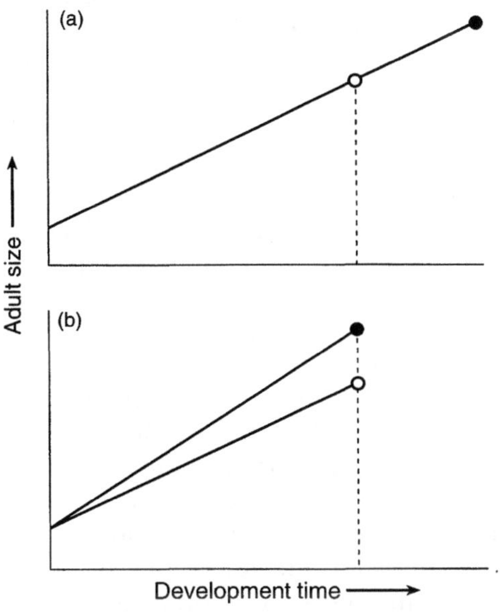

Fig. 11.2. Relationship between development time and adult body size. **a** Both species (sexes) grow at the same rate, but one grows longer for a gain in size. **b** One species (sex) grows faster than the other and attains a larger size without longer development

$$\text{MRGR} = \frac{\log_e \text{DM}_2 - \log_e \text{DM}_1}{t_2 - t_1}, \qquad (2)$$

where DM_1 and DM_2 are, respectively, dry mass at the beginning and end of the interval from t_1 to t_2. For species growing at a constant rate, any gain in mass can be achieved only by growing longer (Fig. 11.2a). Alternatively, a gain in size without a correlated increase in development time requires a higher growth rate (Fig. 11.2b).

11.3 Patterns of Resource Utilization

11.3.1 Idiobiont Parasitoids

Idiobionts are adapted to exploit fixed or rapidly diminishing host resources. Host quality is indexed largely by host size rather than host age (Mackauer and Sequeira 1993). In general, phylogenetic differences between hosts may often be of little consequence in egg parasitoids (Strand 1986; Bai et al. 1994; Kazmer and Luck 1995) but are significant in pupal and ectoparasitoids (Mackauer and Sequeira 1993). Adaptive responses of idiobionts to variable host resources include the production of more daughters and, in gregarious species, larger clutches in relatively young and/or large hosts (Waage and Ng 1984), little or no flexibility in the growth rate resulting in a positive correlation between the amount of host resources and development time (Salt 1940), and different utilization of host resources by male and female parasitoids (Bai et al. 1994; Urano and Hijii 1995).

Urano and Hijii (1995) compared patterns of resource utilization and sex allocation between *Atanycolus initiator* and *Spathius brevicaudis*, two solitary ectoparasitoids of subcortical beetles infesting Japanese pine trees. The two parasitoids have similar host ranges but differ in size and ovipositor length, with *A. initiator* being about seven times larger (in wet body mass) than *S. brevicaudis*. Parasitoid body size was positively correlated with host size, but the two species responded differently to the resources available in hosts of the same size, *Shirahoshizo* larvae. These larvae are relatively small hosts for *A. initiator* but relatively large hosts for *S. brevicaudis*. Whereas the relationship between host and parasitoid size was linear in *A. initiator*, it was logarithmic (increasing at a decreasing rate) in *S. brevicaudis*. The slope of the regression differed significantly between males and females on large, but not small, hosts; this shows that the sexes grew at different rates. As the size of the host larvae increased, a decreasing proportion of host mass was utilized by the smaller *S. brevicaudis*, especially males. By contrast, both sexes of *A. initiator* consumed most host resources, excepting the largest hosts.

The relationship between host size and host quality may be confounded by other factors such as physiological state. Evidence comes from studies of *Dendrocerus carpenteri*, a generalist hyperparasitoid of the prepupal and pupal stages of aphidiid wasps inside mummified aphids. The solitary larvae feed externally on the host (Bennett and Sullivan 1978). Mackauer and Lardner (1995) reported that females hyperparasitized small and large mummified pea aphids equally, independent of the sex of the immature *Aphidius ervi* inside the mummies. Mummy size had a significant influence on offspring sex allocation by *D. carpenteri*, however. Mated females deposited more fertilized eggs (daughters) if the mummies were large and more unfertilized eggs (sons) if the mummies were small, with the exact pattern depending on encounter sequence (Chow and Mackauer 1996). To assess the relative importance of the size of the primary parasitoid (a correlate of mummy size; Henkelman 1979) as opposed to its age, M. Otto (unpubl.) compared patterns of growth and development of *D. carpenteri* on eight host types: 9-day-old (prepupae), 11-day-old (early pupae) and 13-day-old (late pupae close to adult emergence) *A. ervi* inside pea aphids, *Acyrthosiphon pisum*, which were parasitized in the first (small mummies) and third nymphal (large mummies) instar; two groups of hyperparasitoids were reared on *A. ervi* pupae (10 days old) developing in English grain aphid, *Sitobion avenae*. Adult size of *D. carpenteri*, as indexed by the length of the hind tibia, was positively correlated with mummy size, whereas the growth rate was independent of mummy size and aphid species. Females were larger than males, but both sexes gained in size approximately at the same rate per unit of time. Consistent with model 1 of Mackauer and Sequeira (1993), *D. carpenteri* attained a larger body size only by growing longer (Fig. 11.3). The pea aphid is a much larger aphid than *S. avenae*, which was reflected in the sizes of mummies and immature *A. ervi*. Hyperparasitoids grew at the same rate on early pupal stages of *A. ervi* despite differences in host

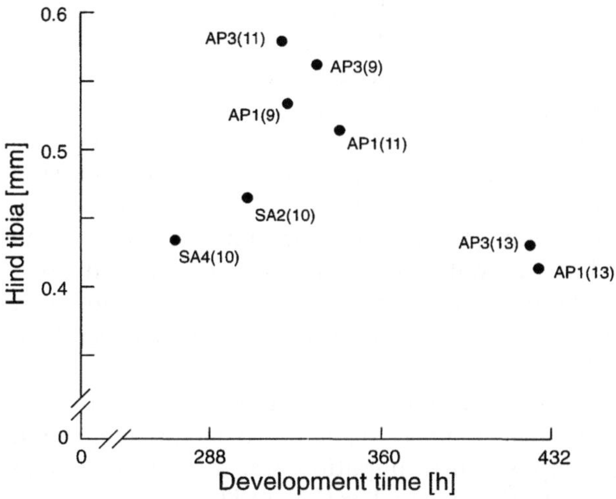

Fig. 11.3. Influence of host type on development time and body size in female *Dendrocerus carpenteri*, an idiobiont ectoparasitoid developing on immature *Aphidius ervi* inside mummified aphids. Body size is indexed by the length of the hind tibia. Hyperparasitoids grew at approximately the same rate on prepupal and pupal hosts but needed longer, without gaining in size, to eclose on late-pupal stages. *AP*, *Acyrthosiphon pisum*; *SA*, *Sitobion avenae*. The *first* and, in *brackets*, *second number* following the abbreviated aphid name are, respectively, aphid instar at parasitization by *A. ervi* and age of *A. ervi* when hyperparasitized. (Data from M. Otto, unpubl.)

size and aphid species, which shows that these host types were qualitatively the same. In contrast, *D. carpenteri* grew much slower on late than early pupae of *A. ervi* inside pea aphids although these types were the same size. Growth was reduced in hyperparasitoids developing on 13-day-old *A. ervi*, in which parts of the exoskeleton were already well developed. Larvae, apparently, were unable to consume all the host resources which remained in the mummy, a constraint analogous to small host size in *S. avenae*.

11.3.2 Koinobiont Parasitoids

The koinobiont strategy of resource utilization depends on the continued survival and growth of the host after parasitization, especially if the host at the time of parasitization is much smaller than the adult parasitoid. A flexible growth rate is one of the key adaptive features of koinobiont life-history responses that facilitates the exploitation of the host's growth potential (Mackauer and Sequeira 1993; Harvey et al. 1994).

Some of the best-studied examples of adaptive responses to variable host resources are provided by aphidiid wasps (Liu 1985; Mackauer 1986; Sequeira and Mackauer 1992a,b, 1993, 1994). For example, *Monoctonus paulensis* parasitizes a range of aphid species but selectively attacks the smallest available hosts (Calvert and van den Bosch 1972). Recent studies by R. Sequeira and M. Mackauer (unpubl.) showed that adult body size and development time varied with host type. The mean relative growth rate, as estimated by the ratio of adult DM to development time, suggests that *M. paulensis* can utilize *S. avenae*, a relatively small host with low post-parasitization growth potential, more efficiently than larger hosts, such as *A. pisum*, with greater growth potential. Development time from oviposition to host death varied with aphid instar; it was longer in parasitoids developing in fourth- (L4) than second-instar (L2) pea aphids. Total development time increased in proportion to body mass, however, such that the mean growth rate was the same in both host types. As Table 11.1 shows, host quality was not only a function of size and post-parasitization growth potential but was also influenced by taxonomic identity, a broad measure of physiological and biochemical differences between host types.

In *Ephedrus californicus* developing in the four nymphal instars of the pea aphid, adult size increased with host instar from L2 to L4 (Sequeira and Mackauer 1993). Time from oviposition to pupation and total development time were both greater in L2 and L3 than in smaller (L1) and larger (L4) aphids. In relatively large hosts (L4), resources were sufficient for the parasitoid simultaneously to maximize body size and minimize development time; however, in small hosts (L1), available resources were utilized primarily to maximize body size. In mechanistic terms, the onset of logarithmic growth during early larval development of *E. californicus*, presumably, was delayed in L2 and L3 aphids until the host had attained a larger size.

Table 11.1. Relationship between growth potential of different aphids and mean relative growth rate of the parasitoid *Monoctonus paulensis*. (Data from R. Sequeira and M. Mackauer, unpubl.)

Host aphid				Parasitoid MRGR ($\log_e \mu g\,day^{-1}$)	
Species	Instar	Dry mass (μg)	R_d (μg)	Male	Female
S. avenae	L2	31.2 ± 6.4	1168	0.2840 ± 0.0015	0.2944 ± 0.0016
A. pisum	L2	56.6 ± 6.2	3920	0.3099 ± 0.0007	0.3169 ± 0.0006
A. pisum	L4	182.6 ± 23.2	7007	0.3118 ± 0.0014	0.3186 ± 0.0009
M. creelii (green)	L2	82.3 ± 13.2	8254	0.3107 ± 0.0028	0.3156 ± 0.0031
M. creelii (pink)	L2	87.9 ± 12.2	8714	0.3219 ± 0.0037	0.3293 ± 0.0012

Parasitoids were reared on five host types (*Sitobion avenae*, *Acyrthosiphon pisum*, and a red and green colour morph of *Macrosiphum creelii*) at 20 ± 0.5 °C, 50–60% RH and continuous light. Aphid instar and dry mass were determined at the time of parasitization, t_o. Values for which variation is shown are means (±SEM).
Abbreviations: R_d, post-parasitization growth potential (see Fig. 11.1); MRGR, mean relative growth rate, is estimated by the ratio of the $\log_e$-transformed DM at eclosion and total development time, with dry mass at oviposition assumed to have a unit value of 1.

Unlike *E. californicus*, the growth rate of *A. ervi* developing in pea aphids varied nonlinearly in the four nymphal instars; the highest and lowest growth rates were achieved in L2 and L3 instars (Sequeira and Mackauer 1992a). Development time increased with adult size in L1 and L3 hosts but decreased with no change in adult size in L4 hosts (Sequeira and Mackauer 1992b). Thus, in L2 and L3 hosts, the optimization of adult body size took precedence over the minimization of development time, whereas in L4 hosts the additional resources were utilized to minimize development time. Furthermore, the time from oviposition to pupation remained constant across host instars, which may indicate a fundamental difference in developmental strategies between *A. ervi* and *E. californicus* (Mackauer and Sequeira 1993).

Liu (1985) examined life-history responses of *Aphidius sonchi* developing in two nymphal instars of the sowthistle aphid, *Hyperomyzus lactucae*. Adult wasps from L3 hosts were larger, lived longer and contained more mature eggs at eclosion than their counterparts from L1 hosts. The time from ovipostion to pupation did not differ between these groups, however, which suggests that immature *A. sonchi* adjusted the rates of growth and of development in response to the host's size and growth potential. The ability of *Aphidius* wasps to exploit host resources acquired after parasitization is perhaps best epitomized by *Aphidius smithi*. Very small but viable and fertile offspring developed from eggs deposited in pea aphid embryos while inside their mother (Mackauer and Kambhampati 1988), evidence that critical biomass is not a condition for successful moulting to the adult stage in this species.

Host-parasitoid developmental interactions involving koinobionts other than aphidiid wasps were reviewed by Vinson (1990), Mackauer and Sequeira (1993) and Harvey et al. (1994). Physiological synchrony of growth between, for example, *Microplitis* species and their lepidopteran hosts conforms largely to model 2 of Mackauer and Sequeira (1993). In hosts that at oviposition are similar in size to, or larger than, the parasitoid, much of the available resources may be inaccessible to the immature parasitoid; this could account for arrested development and/or low growth rates of early-instar larvae. Usually, adult size of the parasitoid increased with host age, but development was often prolonged in younger hosts (Smith et al. 1994; Croft and Copland 1995). Using *Venturia canescens*, a solitary parasitoid of pyralid moths, Harvey et al. (1994) demonstrated flexible growth rates of the type involving arrested development in young or small hosts, followed by constant growth to adulthood. The period of growth, from oviposition to pupation of the wasp, decreased progressively in older hosts. Examination of their results reveals that parasitoids grew at the same rate in L2, L3 and L4 hosts, but larvae developing in L5 hosts gained mass much faster. Fifth-instar hosts pupated soon after parasitization, however, and may be physiologically different from earlier instars, which would caution against any direct comparisons.

11.4 Sexual Dimorphism in Parasitoid Growth

Females are larger than males in most species of parasitoid wasps (Hurlbutt 1987). Hymenoptera are haplo-diploid, which enables mated females to determine offspring sex by controlling fertilization (Cook 1993). If females gain relatively more than males in fitness from increased size, a correlate of fecundity, wasps should allocate more fertilized eggs to large hosts and more unfertilized eggs to small hosts (Charnov 1982). Implicit in this prediction is the assumption, generally unstated, that both sexes grow at the same rate but a female develops longer, and so needs more resources, to attain pupation weight (Fig. 11.2a). For example, Charnov et al. (1981) reported that the smaller males of *Lariophagus distinguendus*, a pteromalid parasitoid of grain weevils, *Sitophilus granarius*, emerged earlier than the larger females from same-sized hosts. The evidence for other idiobionts is less clear, however. In *Muscidifurax raptor*, a solitary parasitoid of housefly pupae, development time was influenced by host size in females but not males; females eclosed later from large rather than small pupae (Seidl and King 1993).

Dimorphism can also result from sex-specific trade-offs between development time and size at maturity, or from one sex exploiting host resources more efficiently than the other. Mackauer (1996) compared adult dry mass between males and females of *A. ervi*, *A. smithi*, *E. californicus* and *M. paulensis* developing in same-sized pea aphids under similar conditions. In all four species, females grew to a larger size than males, independent of host size at parasitization. The relationship between male and female mass was allometric rather than isometric, however. Parasitoids developing in small aphids differed most in mass, with dimorphism declining at a decelerating rate with an increase in host size at oviposition. Development time did not differ between the sexes in *A. ervi* and *A. smithi* (Mackauer and Henkelman 1975; Sequeira

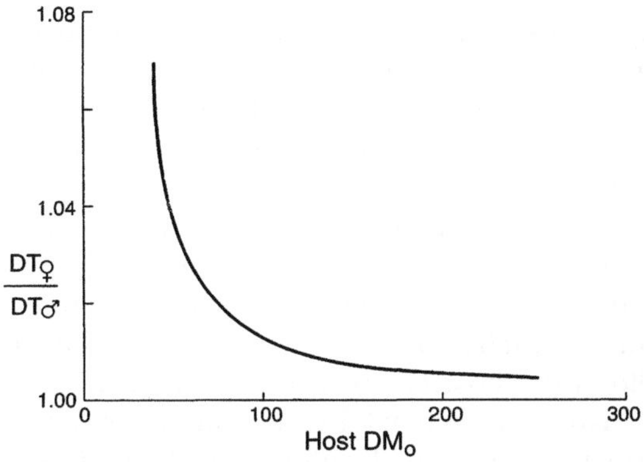

Fig. 11.4. Sexual dimorphism of development time (*DT*) in *Ephedrus californicus* developing in pea aphid, *Acyrthosiphon pisum*. Females developing in small hosts required relatively more time than males from oviposition to eclosion, but both sexes eclosed at the same time in large hosts. (Data from Sequeira and Mackauer 1993)

and Mackauer 1992b), which suggests that females gained mass faster per unit of time. In *E. californicus* this intrinsic difference was accentuated by sex-specific trade-offs between adult size and development time (Sequeira and Mackauer 1993). Without nutritional constraints, both sexes eclosed at the same time but females gained proportionally more in mass from developing in large hosts. In very small aphids females developed slower than males (Fig. 11.4), investing additional development time for potential gains in size and size-dependent fitness returns. A flexible response to host constraints allows *E. californicus* to exploit small aphids that have limited growth potential but are easier to attack and parasitize. Because the fitness costs of a longer time-to-adult on overall reproductive achievement may be severe (Lewontin 1965), a minimum size apparently is critical for females but less so for males (van den Assem et al. 1989).

11.5 Conclusions

Koinobiosis, presumably, has evolved from idiobiosis as an alternative strategy that allowed parasitoid wasps to exploit exposed hosts at earlier stages of development (Gauld 1988). Small hosts are generally easier to capture and subdue than large ones, but they may not be optimal for immature development due to the limited amount of resources they contain. Consequently, if hosts vary in quality as a function of size, a potential conflict exists between the interests of the female to minimize the costs of finding and subduing hosts and the interests of the immature parasitoid to gain the maximum amount of resources for growth and development (Mackauer 1973).

We have shown that idiobionts and koinobionts respond differently to resource constraints. Developing in a closed or fast diminishing resource environment, idiobionts should always grow at the physiological maximum to reduce post-parasitization mortality risk. At a given temperature, size and age differences between hosts translate into different slopes of the growth trajectory. By comparison, the principal ontogenetic response of koinobionts to variation in host quality is the ability of immature stages to vary the growth rate in relation to the development rate. Development appears to be size-dependent, which allows the parasitoid to delay exponential growth in low-quality hosts until the nutritional or physiological state of the host becomes more favourable (Mackauer and Sequeira 1993). Commonly, delayed parasitoid development simply allows the host to grow to a larger size (and acquire more resources) before parasitism affects vital metabolic processes.

For example, host size and growth potential of aphids are dynamically related, both being functions of host age at oviposition (Fig. 11.5). Current host resources represented by size at parasitization, R_o, increase exponentially with host age, t_o, whereas future host resources decline at an increasing rate with host age. Aphidiid wasps can be often larger than their hosts at the time of parasitization. The importance of future host resources, R_d, in relation to present resources, R_o, thus can be expected to decrease with increasing R_o. Furthermore, as the interval between the time of oviposition and host death declines, $t_o \rightarrow t_d$, the idiobiont and koinobiont resource-utilization strategies are likely to become indistinguishable. At the hypothetical upper end of the host-size spectrum, a koinobiont grows at the physiological maximum without benefit from trade-offs between size and development time.

Many studies have shown that fitness in insect parasitoids is closely linked to adult body size and size-related traits (King 1987, 1989; van den Assem et al. 1989; Bai et al. 1994; Visser 1994; Kazmer and Luck 1995). In aphidiid wasps such as *M. paulensis* optimization of body size involves growing as large as possible (Table 11.1). In contrast, in *E. californicus* adult size reflects the optimal balance between the ecological consequences of growing larger (for increased oviposition success) and completing development earlier.

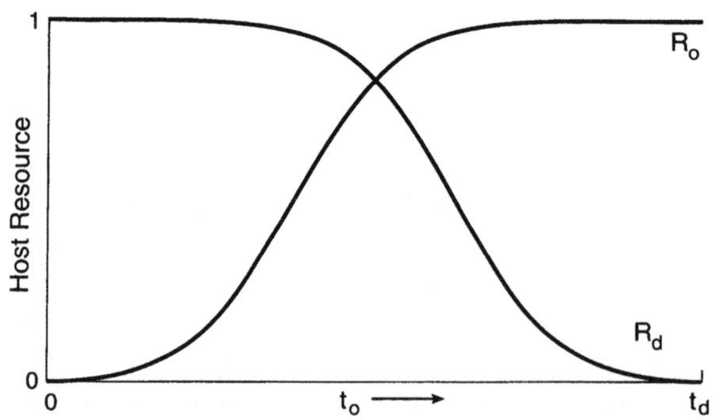

Fig. 11.5. Dynamics of current (R_o) and future (R_d) resources in hosts parasitized by a koinobiont in relation to host age at oviposition (t_o). As the internal between the time t_o and the time of host death, t_d, declines, the koinobiont resource-utilization strategy converges to the idiobiont strategy

Body size, apparently, is under different selection in males and females. Aphidiid females gained more mass per unit of time than males in hosts of equal quality (Mackauer and Henkelman 1975; Sequeira and Mackauer 1992b). This difference could result from sex-specific differences in resource utilization, with diploid females being metabolically more efficient than haploid males. An alternative, but less likely, explanation is that female larvae stimulate aphid feeding and resource acquisition more than male larvae (Cloutier and Mackauer 1979, 1980). Sexual dimorphism in size was allometric rather than isometric, however, i.e. the difference in body size between males and females declined with increasing size and, implicitly, host quality (Mackauer 1996). Such intrinsic dimorphism may be accentuated by sex-specific ontogenetic responses to variable host resources (Mackauer and Sequeira 1993). In species that depend on size for oviposition success, such as *E. californicus*, attaining a mimimum body size is more important for females than males, which would explain why females, but not males, develop longer in suboptimal hosts for a potential gain in mass (Fig. 11.4).

The mechanisms underlying dimorphism in adult size are probably complex, with differences in resource utilization between males and females representing only one factor. Such intrinsic sexual differences may be confounded by the female's strategy for offspring and offspring sex allocation, which is influenced by the population mating structure (Bulmer 1983; LaBarbera 1989). These questions have not received much attention, and available data are few and often contradictory. For example, evolutionary host-choice models are unable to explain a male-biased dimorphism in size, as observed in many Ichneumonidae (Hurlbutt 1987). For males to be on average larger than females, males would need more or better resources for larval development to grow either longer (Fig. 11.2a) or gain size faster (Fig. 11.2b). No evidence is available, however, that, unlike other parasitoids, ichneumonid mothers allocate unfertilized eggs disproportionately to high-quality large hosts and fertilized eggs to low-quality small hosts. Moreover, any gains in size achieved at the cost of longer development must be balanced against reduced mating success of late, as opposed to early, eclosing males (Singer 1982). We suggest that hypotheses about the evolution of offspring sex allocation to hosts varying in quality need to take into account intrinsic differences in resource utilization between male and female parasitoid wasps.

Acknowledgement. We thank the Natural Sciences and Engineering Research Council of Canada for financial support.

References

Bai B, Mackauer M (1992) Influence of hyperparasitism on development rate and adult size in a solitary parasitoid wasp, *Aphidius ervi*. Funct Ecol 6:302–307

Bai B, Luck RF, Forster L, Stephens B, Janssen JAM (1994) The effect of host size on quality attributes of the egg parasitoid, *Trichogramma pretiosusm*. Entomol Exp Appl 64:37–48
Bennett AW, Sullivan DJ (1978) Defensive behavior against tertiary parasitism by the larva of *Dendrocerus carpenteri*, an aphid hyperparasitoid. J NY Entomol Soc 86:153–160
Bulmer MG (1983) Models for the evolution of protandry in insects. Theor Popul Biol 23:314–322
Calvert DJ, van den Bosch R (1972) Behavior and biology of *Monoctonus paulensis* (Hymenoptera: Braconidae), a parasite of dactynotine aphids. Ann Entomol Soc Am 65:773–779
Causton DR (1977) A biologist's mathematics. E Arnold, London
Charnov EL (1982) The theory of sex allocation. Princeton University Press, Princeton
Charnov EL, Los-den Hartogh RL, Jones WT, van den Assem J (1981) Sex ratio evolution in a variable environment. Nature (Lond) 289:27–33
Chow A, Mackauer M (1996) Sequential allocation of offspring sexes in the hyperparasitoid wasp, *Dendrocerus carpenteri*. Anim Behav 51:859–870
Cloutier C, Mackauer M (1979) The effect of parasitism by *Aphidius smithi* (Hymenoptera: Aphidiidae) on the food budget of the pea aphid, *Acyrthosiphon pisum*. Can J Zool 57:1605–1611
Cloutier C, Mackauer M (1980) The effect of superparasitism by *Aphidius smithi* (Hymenoptera: Aphidiidae) on the food budget of the pea aphid, *Acyrthosiphon pisum*. Can J Zool 58:241–244
Cook JM (1993) Sex determination in the Hymenoptera: a review of models and evidence. Heredity 71:421–435
Croft P, Copland MJW (1995) The effect of host instar on the size and sex ratio of the endoparasitoid *Dacnusa sibirica*. Entomol Exp Appl 74:121–124
Gauld ID (1988) Evolutionary patterns of host utilization by ichneumonoid parasitoids (Hymenoptera: Ichneumonidae and Braconidae). Biol J Linn Soc 35:351–377
Gauld I, Bolton B (eds) (1988) The Hymenoptera. Oxford University Press, Oxford
Godfray HCJ (1994) Parasitoids. Behavioral and evolutionary ecology. Princeton University Press, Princeton
Harvey JA, Harvey IF, Thompson DJ (1994) Flexible larval growth allows use of a range of host sizes by a parasitoid wasp. Ecology 75:1420–1428
Henkelman DH (1979) A study of weight variation in *Aphidius smithi* (Hymenoptera: Aphidiidae), a parasite of the pea aphid, *Acyrthosiphon pisum* (Homoptera: Aphididae). MSc Thesis, Simon Fraser University, Burnaby, British Columbia
Hurlbutt B (1987) Sexual size dimorphism in parasitoid wasps. Biol J Linn Soc 30:63–89
Kazmer DJ, Luck RF (1995) Field tests of the size-fitness hypothesis in the egg parasitoid *Trichogramma pretiosum*. Ecology 76:412–425
King BH (1987) Offspring sex ratios in parasitoid wasps. Q Rev Biol 62:367–396
King BH (1989) Host-size dependent sex ratios among parasitoid wasps: does host growth matter? Oecologia 78:420–426
King BH (1990) Sex ratio manipulation by the parasitoid wasp *Spalangia cameroni* in response to host age: a test of the host-size model. Evol Ecol 4:149–156
LaBarbera M (1989) Analyzing body size as a factor in ecology and evolution. Annu Rev Ecol Syst 20:97–117
Lewontin RC (1965) Selection of colonizing ability. In: Baker HG, Stebbins GL (eds) The genetics of colonizing species. Academic Press, New York, pp 79–91
Liu SS (1985) Development, adult size and fecundity of *Aphidius sonchi* reared in two instars of its aphid host, *Hyperomyzus lactucae*. Entomol Exp Appl 37:41–48
Mackauer M (1973) Host selection and host suitability in *Aphidius smithi* (Hymenoptera: Aphidiidae). In: Lowe AD (ed) Perspectives in aphid biology. Bull Entomol Soc NZ 2:20–29
Mackauer M (1986) Growth and developmental interactions in some aphids and their hymenopterous parasites. J Insect Physiol 32:275–280
Mackauer M (1996) Sexual size dimorphism in solitary parasitoid wasps: influence of host quality. Oikos 76:265–272

Mackauer M, Chow FJ (1986) Parasites and parasite impact. In: McLean GD, Garrett RG, Ruesink WG (eds) Plant virus epidemics: monitoring, modelling and predicting outbreaks. Academic Press, Sidney, pp 95–118
Mackauer M, Henkelman DH (1975) Effect of light-dark cycles on adult emergence in the aphid parasite *Aphidius smithi.* Can J Zool 53:1201–1206
Mackauer M, Kambhampati S (1988) Parasitism of aphid embryos by *Aphidius smithi*: some effects of extremely small host size. Entomol Exp Appl 49:167–173
Mackauer M, Lardner RM (1995) Sex-ratio bias in an aphid parasitoid-hyperparasitoid association: a test of two hypotheses. Ecol Entomol 20:118–124
Mackauer M, Sequeira R (1993) Patterns of development in insect parasites. In: Beckage NE, Thompson SN, Federici BA (eds) Parasites and pathogens of insects, vol 1. Parasites. Academic Press, San Diego, pp 1–23
Salt G (1940) Experimental studies in insect parasitism VII. The effects of different hosts on the parasite *Trichogramma evanescens* Westwood. Proc R Entomol Soc (Lond) A 15:81–95
Seidl SE, King B (1993) Sex-ratio manipulation by the parasitoid wasp *Muscidifurax raptor* in response to host size. Evolution 47:1876–1882
Sequeira R, Mackauer M (1992a) The nutritional ecology of an insect host-parasitoid association: the pea aphid-*Aphidius ervi* system. Ecology 73:183–189
Sequeira R, Mackauer (1992b) Covariance of adult size and development time in the parasitoid wasp *Aphidius ervi* in relation to the size of its host, *Acyrthosiphon pisum.* Evol Ecol 6:34–44
Sequeira R, Mackauer M (1993) The nutritional ecology of a parasitoid wasp, *Ephedrus californicus* Baker (Hymenoptera: Aphidiidae). Can Entomol 125:423–430
Sequeira R, Mackauer M (1994) Variation in selected life-history parameters of the parasitoid wasp, *Aphidius ervi.* Entomol Exp Appl 71:15–22
Sibly RM, Calow P (1986) Physiological ecology of animals. Blackwell, Oxford
Singer MC (1982) Sexual selection for small size in male butterflies. Am Nat 119:440–443
Smith HA, Capinera JL, Pena JE, Linbo-Terhaar B (1994) Parasitism of pickleworm and melonworm (Lepidoptera: Pyralidae) by *Cardiochiles diaphaniae* (Hymenoptera: Braconidae). Environ Entomol 23:1283–1293
Strand MR (1986) The physiological interactions of parasitoids with their hosts and their influence on reproductive strategies. In: Waage J, Greathead D (eds) Insect parasitoids. Academic Press, London, pp 97–136
Sullivan DJ (1988) Hyperparasites. In: Minks AK, Harrewijn P (eds) Aphids: their biology, natural enemies and control. World crop pests, vol 2B. Elsevier, Amsterdam, pp 189–203
Urano T, Hijii N (1995) Resource utilization and sex allocation in response to host size in two ectoparasitoid wasps on subcortical beetles. Entomol Exp Appl 74:23–35
van den Assem J, van Iersel JJA, Los-den Hartogh RL (1989) Is being large more important for female than for male parasitic wasps? Behaviour 108:160–195
Vinson SB (1990) Physiological interactions between the host genus *Heliothis* and its guild of parasites. Arch Insect Biochem Physiol 13:63–81
Visser ME (1994) The importance of being large: the relationship between size and fitness in females of the parasitoid *Aphaereta minuta* (Hymenoptera: Braconidae). J Anim Ecol 63:963–978
Waage JK, Ng SM (1984) The reproductive strategy of a parasitic wasp. I. Optimal progeny and sex allocation in *Trichogramma evanescens.* J Anim Ecol 53:401–415

12 Patch Quality and Fitness in Predatory Ladybirds

A.F.G. DIXON

12.1 Introduction

Ladybirds are regarded with affection by people who have no interest in, or even an aversion to insects in general. This is reflected in the prevalence of holy attributives in the colloquial names, in many languages, of these much-befriended insects. That the first great success in biological control came in 1889 with the introduction of the Australian ladybird (*Rodolia cardinalis*) into California to combat a coccid scale insect that was threatening the existence of the Californian citrus industry further popularised ladybirds.

In spite of the great and long-standing interest in ladybirds, there has been little success in accounting for why these insects have generally been more successfully employed in the biological control of coccids than aphids. It is clear that adult ladybirds select the patches of prey in which to lay their eggs, and their larvae, because of poor powers of dispersal, are committed to staying within a patch. Nevertheless, most of the work on foraging behaviour has been carried out on larvae. These studies, not surprisingly, have revealed that the behaviour of larvae tends to keep them within a patch of prey (Ferran and Dixon 1993). A reason for the preponderance of studies on larval behaviour is that larvae are more voracious and easier to study than adults. Recently, however, there has been an increase in the awareness of the importance of adult decision-making in determining fitness. This has resulted in a number of experimental and theoretical studies on the foraging behaviour of adults (Hemptinne et al. 1992, 1993; Kindlmann and Dixon 1993), which have greatly increased our understanding of ladybird-aphid systems. In addition, studies on the vulnerability of ladybird eggs to cannibalism and predation by other species of ladybirds (Agarwala and Dixon 1993b,c) are beginning to reveal the interspecific relationships within the guilds of ladybirds attacking aphids.

The aim of this chapter is to review the advantages, in terms of fitness, of adult ladybirds assessing patch quality, offer an explanation for the ineffectiveness of aphidophagous ladybirds in biological control and account for the evolution of cannibalism in ladybirds and the effect it may have had on interspecific relationships in ladybird guilds.

Ecological Studies, Vol. 130
Dettner et al. (eds.) Vertical Food Web Interactions

12.2 Ladybird Aphid Systems

As in other predator prey systems (Nicholson 1933), aphids are patchily distributed in space and concentrated into "groups" or "active centres", as Elton (1949) calls them. In addition, the aphids in these active centres characteristically show dramatic changes in abundance in time, even in the absence of natural enemies, with the different species of aphids peaking in abundance at different times. That is, the prey of aphidophagous ladybirds occurs in active centres or patches, which vary in quality in space and time. Laboratory studies have shown that the survival of the first instar larvae of ladybirds is very dependent on the abundance of young aphids and that adult ladybirds are likely to remain, mature and lay eggs in patches only where their capture of prey per unit time is above a certain critical threshold. At this lower critical threshold there are sufficient young aphids for the survival of the first-instar larvae (Dixon 1959).

Field studies have confirmed the existence of a lower critical threshold in aphid abundance for egg-laying in ladybirds (Wratten 1973; Honek 1978; Mills 1979). The time from egg hatch to pupation in ladybirds spans more than one generation of aphids and is often similar in duration to the period for which patches of aphids contain sufficient prey to sustain ladybird larvae. In addition to feeding on aphids, ladybird larvae and adults will readily eat conspecific eggs and larvae, in both the laboratory and the field (Mills 1982b; Osawa 1989). The above defines ladybird aphid systems and highlights those features that are likely to be critical in determining ladybird fitness.

12.2.1 Optimal Foraging by Ladybirds

A simulation model of a ladybird aphid system, which takes the minimum aphid population density requirements of the first-instar larvae of ladybirds and the risk of cannibalism into account, indicates that the best strategy is for ladybirds to lay a few eggs at the beginning of the development of aphid colonies/patches (Fig. 12.1; Kindlmann and Dixon 1993). If they lay their eggs later, the larvae will not mature before the aphids become scarce (Fig. 12.1a). In addition, if many eggs are laid, the larvae reduce the rate of population increase of the aphids and cause an earlier decline in their abundance. If this happens the larvae resort to cannibalism to survive. This results in the survival of a few small adults, which have a low potential fecundity (Fig. 12.1c; Dixon and Guo 1993). That is, if ladybirds are to maximise their fitness, they should lay a few eggs early in the development of an aphid colony (Fig. 12.1b). It is assumed that the lower critical threshold for oviposition, relative developmental times of prey and predator, and incidence of cannibalism have been shaped by selection acting at the individual level.

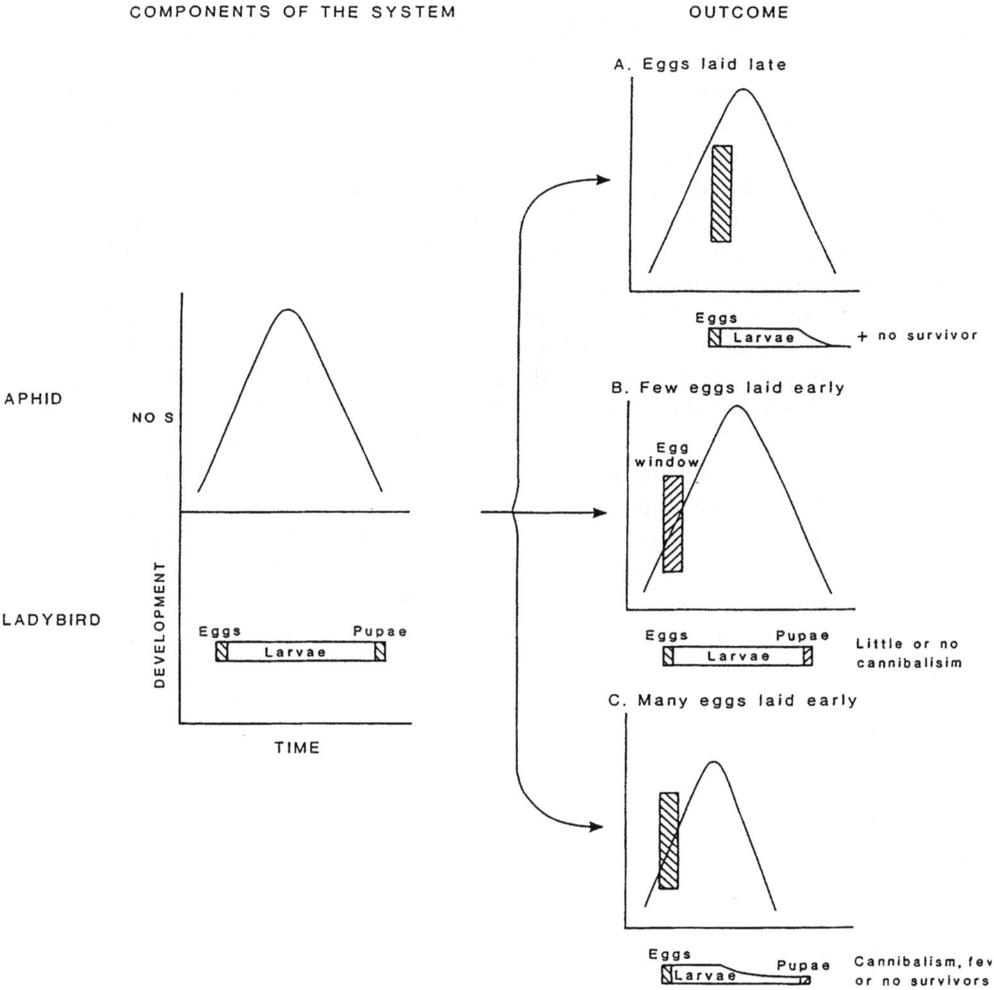

Fig. 12.1A–C. Graphical presentation of the components of the ladybird-aphid system: temporal changes in the abundance of aphids and the relative developmental time of the ladybird, and the outcome if the eggs are laid late (**A**), a few eggs are laid early (**B**) or many eggs are laid early (**C**)

12.2.2 Assessment of Patch Quality

12.2.2.1 Field Observations

What evidence is there that ladybirds forage optimally? The number of eggs that the two-spot ladybird, *Adalia bipunctata* (L.), lays per unit area of lime tree foliage relative to the aphid population density on lime trees in the field has been determined by Mills (1979) and Wratten (1973). The relationship for

42 trees in 1 year revealed by the more detailed study of Mills is given in Fig. 12.2. This relationship is made up of three components: the aggregative response of adults to aphid density, the functional response determining the number of aphids consumed per adult and the reproductive numerical response that determines the conversion of aphids consumed to eggs produced. The intercept on the abscissa of the relationship in Fig. 12.2 represents the density at which sufficient aphids are encountered, captured and eaten to cover the maintenance costs of ladybirds; the lower critical threshold for oviposition. The slope of the relationship represents the increase in coccinellid oviposition: since the slope is significantly less than 1, the combined oviposition increases at a decreasing rate, indicating that this aphid is most strongly exploited for egg-laying at the lower densities. From the numbers of adult ladybirds observed on the trees, Mills (1979) calculated the fraction of the total numbers of adult coccinellids and aphids that occurred on each tree and the aggregative index (Hassell and May 1973). This indicates that adults are aggregating in patches of high aphid abundance, but the response is very weak and the aggregative index of 0.52 suggests that it contributes little, if anything, to stability (Hassell and May 1974). Adult seven-spot ladybirds, *Coccinella septempunctata* L., show a similar very weak aggregative response to patches of cereal aphids of different densities in the field (Rautapää 1976). This has

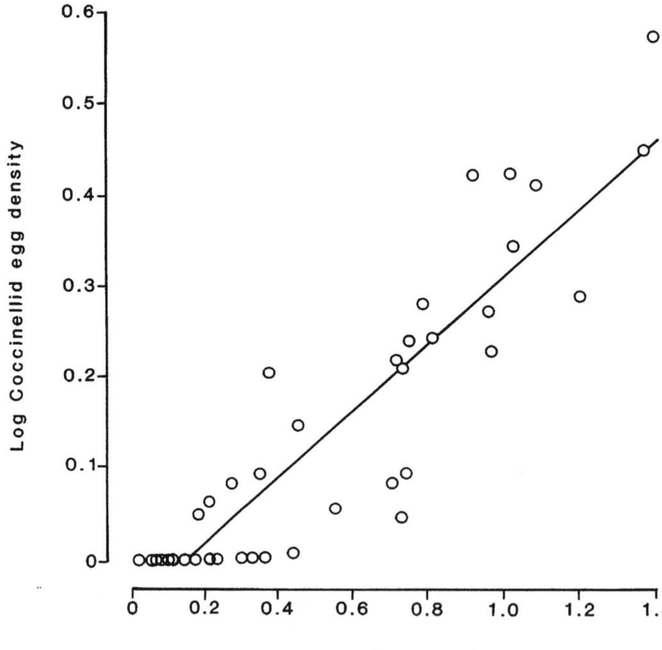

Fig. 12.2. Relationship between the logarithms of the numbers of two-spot ladybird eggs and lime aphid numbers, both per unit area of foliage, on 42 lime trees in 1977. (After Mills 1979)

been interpreted as a consequence of the poor ability of ladybirds in locating the few patches that support high aphid abundance (Mills 1979) and the effect of satiation (Mills 1982a).

The two-spot ladybird lays eggs on apple trees, nettles and wheat, in this sequence, as the aphid populations develop on these plants (Hemptinne 1989). In these different habitats and on lime trees (Figs. 12.2, 12.3) most of the eggs are laid over a short period of time, the *egg window*, before each of these aphids peaks in abundance (Hemptinne et al. 1992). As individual ladybirds live and oviposit for very much longer (Table 12.1) than the duration for which patches of aphids can sustain ladybirds, it is likely that individual ladybirds exploit several patches of aphids, in space and time. That is, the *egg window* is not a consequence of ladybirds only being able to oviposit over a short period of time. They would also appear to be potentially capable of laying considerably more eggs in a patch, but in order to do this the beetles would have to remain in a patch for longer, as the number of eggs they can produce per day is limited (Dixon 1959; Fig. 12.4A) by their capacity to convert aphids into eggs.

Dividing the number of eggs laid by the number of adults of *A. bipunctata* per unit area gives an indication of the number of eggs laid per adult during the *egg window* (Mills 1982b; Fig. 12.5). Assuming an equal sex ratio and that females laid eggs before leaving a tree, and accepting that the maximum clutch

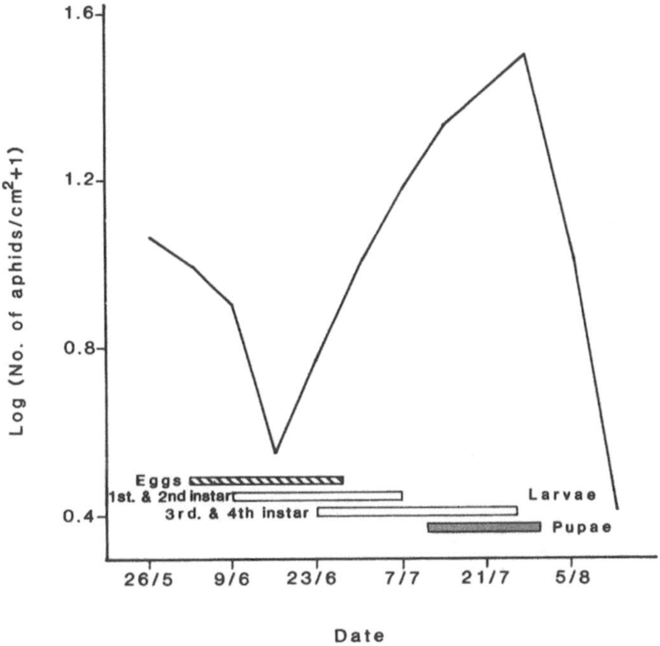

Fig. 12.3. Occurrence in time of the eggs, small and large larvae, and pupae of the two-spot laybird relative to changes in the abundance of the lime aphid on a lime tree in 1977. (After Mills 1979)

Table 12.1. Adult longevity and total fecundity of three species of aphidophagous ladybirds

Species	Longevity (days)	Fecundity (no. of eggs)	Reference
Adalia bipunctata	76–135	250–1466	Blackman (1967) Ellingsen (1969) Hamalainen et al. (1975)
Propylea japonica	46–116	404–1481	Hukusima and Komada (1972) Kawauchi (1985)
Harmonia axyridis	50–170	778–3819	Hukusima and Kamei (1970)

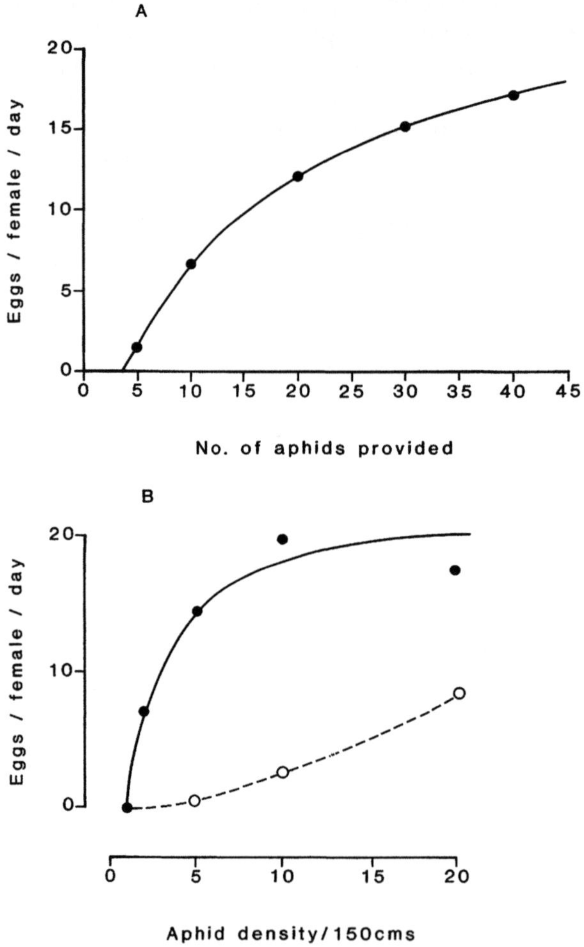

Fig. 12.4A,B. Numbers of eggs laid per day by ten-spot ladybirds fed different numbers of aphids per day (A), and the number of eggs laid daily by two-spot ladybirds fed different numbers of aphids daily and kept on their own (●—●) or with a conspecific larva (○---○) (**B**). (After Dixon 1959; Hemptinne et al. 1992)

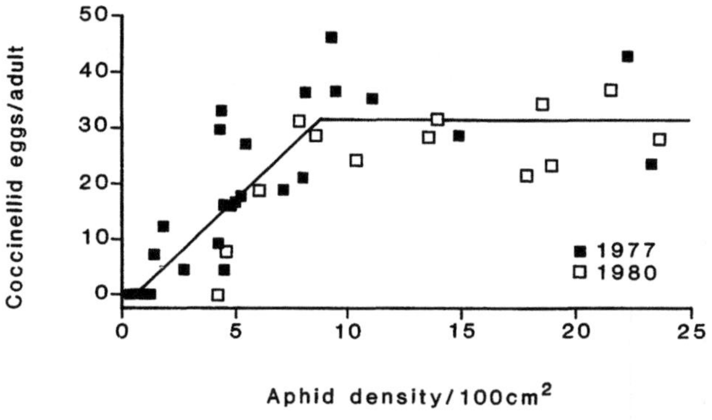

Fig. 12.5. Relationship between the number of eggs laid per adult two-spot ladybird and lime aphid population density in the field in 2 years. (After Mills 1982b)

size for this species is 20–30 eggs (Stewart et al. 1991a,b), it would appear that females lay a maximum of two to three clutches over a wide range of aphid population densities. If by staying longer in a patch females could lay more eggs, it is relevant to ask: what causes the adults to cease oviposition and leave a patch, especially as the aphids are still increasing in abundance (Fig. 12.3)?

12.2.2.2 Laboratory Observations

In the laboratory, the fecundity of females of *A. bipunctata* and *Adalia decempunctata* (L.), kept on their own, rises at a decreasing rate towards an upper asymptote as prey density increases (Dixon 1959; Hemptinne et al. 1992; Fig. 12.4A). In the presence of conspecific larvae, however, they are extremely reluctant to oviposit (Fig. 12.4B) and are 1.6 times more active (Fig. 12.6). The higher level of activity in the presence of conspecific larvae would, if not constrained, result in their leaving the patch, even though aphids are abundant. Interestingly, they do not respond in this way to the presence of similar-sized larvae of a different but closely related (same genus) species of ladybird or to the odour of their own larvae (Hemptinne et al. 1992). The same marked reluctance to lay eggs is also observed when gravid females of both *A. bipunctata* and *C. septempunctata* are offered substrates for oviposition that have recently been traversed by conspecific, but *not* heterospecific larvae. Chloroform extracts of conspecific larval tracks are equally effective in inhibiting oviposition (Doumbia et al. 1997). That is, females appear to respond to chemical cues in the tracks left by conspecific larvae, and these cues act as an oviposition-inhibiting pheromone. The net effect of the responses of adults to the presence of conspecific larvae is that they are highly likely to cease oviposition and leave the patch.

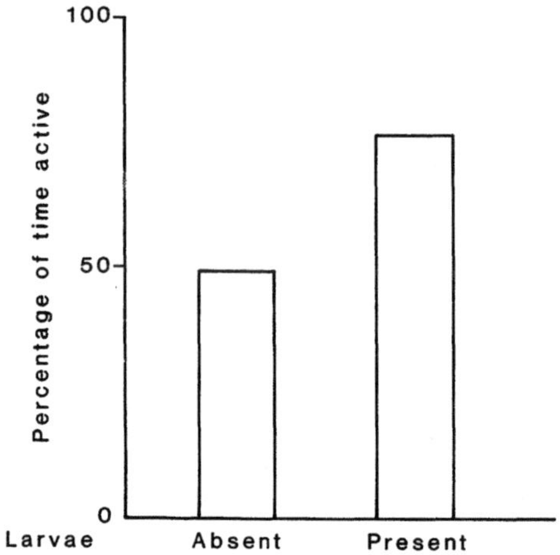

Fig. 12.6. Percentage of the time that two-spot ladybird females are active when kept in Petri dishes with and without conspecific larvae

12.2.3 Cannibalism

Eggs and larvae are at risk from cannibalism (Mills 1982b; Osawa 1989, 1992). Therefore, the adaptive significance of the response of ovipositing adults to larval tracks is that it may enable them to avoid areas already occupied by larvae and so reduce the risk of their eggs being eaten. What evidence is there for this? In the field, the mortality of eggs due to cannibalism increases dramatically as the density of ladybird egg/larvae increases (Mills 1982b; Fig. 12.7).

In the laboratory, the risk of egg and early larval cannibalism is inversely related to aphid population density (Agarwala and Dixon, 1993b) and the reluctance to lay eggs is strongest in areas where the concentration of larval tracks is greatest (Doumbia et al. 1997). That is, as eggs hatch, the increasing numbers of larvae result in an increased risk of egg and early larval cannibalism. This poses a very serious and increasing risk to the potential fitness of adults that continue to oviposit in the patch. The field data (Figs. 12.5, 12.7), however, tend to indicate that aphid population density has little or no effect on the incidence of cannibalism (Mills 1982b). Therefore, in terms of fitness, adults that assess the quality of patches and avoid those that are already being exploited by conspecific larvae are likely to be at a selective advantage over those that do not.

12.2.3.1 Prudent Predators

As indicated in Section 12.2.1, ladybirds, by virtue of their abundance in patches, can influence the future supply of prey. Populations of ladybirds can maximise their equilibrium density by managing the prey population in each patch so as to maximise prey growth rate (Fig. 12.1B; Slobodkin 1968, 1977).

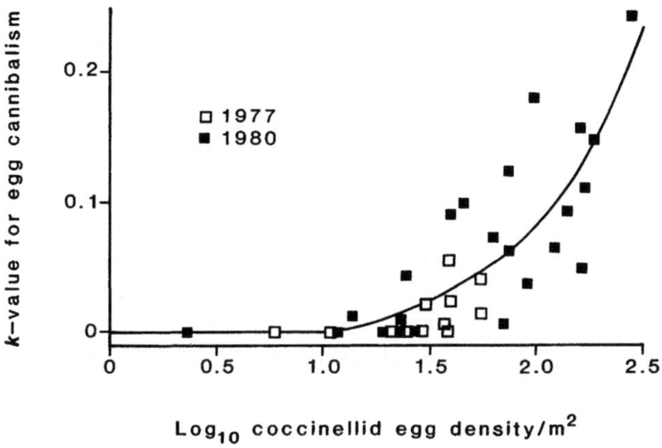

Fig. 12.7. Relationship between egg mortality (k) and two-spot ladybird egg density on lime trees in 1977 and 1980 [$k = 0.002\ (E-10.93)^{0.862}$; E = egg density; $r^2 = 0.67$; $P < 0.001$]. (After Mills 1982b)

However, such prudent predators are vulnerable to cheaters, as it sets the table for predators that do not curb their foraging efforts. That is, prudent predation is not an evolutionary stable strategy (Maynard Smith 1964, 1982; Hardin 1968).

In curtailing their oviposition and leaving a patch already exploited by larvae, ladybirds would appear to be safeguarding the future supply of prey for their larvae. That is, they appear to be behaving as *prudent predators*. However, this prudent behaviour is clearly a consequence of selection acting at the individual level. Adults that "cheat" and continue to oviposit in a patch that contains conspecific larvae are very likely to have a lower fitness, due to egg cannibalism, than adults that seek other unexploited patches.

12.2.3.2 Evolution of Cannibalism

Selection is likely to favour adults that oviposit only where prey is sufficiently abundant to sustain their young larvae. Any eggs they encounter they should destroy, providing they are not their own, as they are likely to be those of intra- or interspecific competitors. This prediction has support in that hungry females of *A. bipunctata* show more reluctance to eat their own eggs than those of other females, whereas males show no reluctance to eat the eggs they have sired (Agarwala and Dixon 1993a). As males spend most of their time searching for females and quickly leave an area after mating (Hemptinne et al. 1996), they are unlikely to encounter eggs they have sired. Young larvae also show a reluctance to eat sibling eggs. That is, females and young larvae appear to be able to avoid eating their own eggs or sibs, but eat those of other individuals of the same species. Eggs are potentially a high-quality food, which can be used by adults for egg production and by larvae to increase their chance of survival.

The trend in patch quality in time is dependent on a number of factors, e.g. changes in host-plant quality affecting aphid abundance, and the numbers of conspecific and heterospecific natural enemies. That is, for the late larval stages of a natural enemy with a long development time, the quality of the patch is less certain than for the early larval stages. Under these circumstances, selection is likely to favour cannibalism. Should the aphids become too scarce for the larvae to survive, then those that resort to cannibalism are more likely to survive. In addition, cannibalism reduces the rate of decline in aphid abundance and further increases the chances of survival of the cannibal in the longer term. However, the latter is a consequence and not the selective advantage of cannibalism, which is the acquisition of food leading to an increase in a larva's fitness. Cannibalism, however, involves risk of injury or death. This possibly accounts for why most cannibals attack the more vulnerable moulting, smaller, moribund or prepupal larvae rather than fully active larvae. Although advantageous to larvae under certain conditions, this form of cannibalism is clearly disadvantageous for a mother, as it reduces her fitness.

As in other predators, cannibalism is advantageous for the cannibal (Nicholson 1933; Fox 1975; Duelli 1981; Agarwala 1991), and not only when prey is scarce. Cannibalism has also been very important in shaping the attack strategy of ladybirds. That is, the apparent prudent oviposition behaviour of ladybirds can be ascribed to selection acting via cannibalism at the individual level. By not ovipositing in patches containing larvae, adults increase their fitness.

12.3 Aphidophaga Guilds

Many species and groups of insects are predators of aphids. It is not unusual for a population of aphids to be simultaneously attacked by several species of aphidophaga. Even ladybirds are not specific to a particular species of aphid. They show varying degrees of overlap in their aphid preferences (Honek 1985) and specificity for particular habitats (Majerus 1991). In its preferred habitat, a ladybird's most serious competitor for aphids and threat to its fitness is likely to be its own species. This possibly accounts for why ovipositing ladybirds do not respond to the presence of larvae of other species (Sect. 12.2.2.2).

As indicated earlier (p. 213), larvae and adults should eat the eggs and young larvae of other aphidophaga, as they pose a threat to their fitness. This possibly accounts for why eggs of some species appear to be well defended chemically against predation (Agarwala and Dixon 1993b,c). Ladybirds use bitter-tasting alkaloids as a defence against predation by ants and birds (Rothschild 1961; Pasteels et al. 1973; Mueller et al. 1984; Marples et al. 1989). Eggs of both *A. lipunctata* and *C. septempunctata* are less readily eaten by the larvae and adults of other species of ladybirds than by their own species (Agarwala and Dixon 1993b), that is, ladybirds are defended not only against vertebrate predators but also insect predators.

Eggs can be afforded some protection from cannibalism by painting them with an extract of another species' eggs (Agarwala and Dixon 1993b). Using clusters of eggs made up of different proportions of two species of ladybirds, it has been shown that the level of inhibition of predation is a function of the proportion of the eggs that belong to another species. It would appear that the defence is based on gustatory factors rather than colour and that the level of deterrence associated with clusters of eggs is greater than that associated with single eggs. In this context it is interesting to note that alkaloids are present in all stages in ladybirds and their effectiveness in deterring ants is concentration-dependent (Pasteels et al. 1973). The distasteful substance, therefore, is likely to contain the same alkaloids used by adults for defence against predators. Individual eggs, even those that are toxic (Agarwala and Dixon 1993c), are vulnerable to predation. Eggs in clusters, however, are less likely to be attacked and damaged. As the effectiveness of the deterrent substances is dose-dependent, an important factor in the evolution of egg-clustering in ladybirds could be that it is the most cost-effective way of reducing the incidence of predation.

12.4 Biological Control

Predators are considered to be less effective biocontrol agents than parasites (Taylor 1935). This is supported by many observations: out of 93 cases of "substantial" or "complete" biological control reported by van den Bosch and Messenger (1973), only 10 cases of substantial and 2 cases of complete control involved predators only. The reductions in host/prey density below the enemy free value (q value) by parasitoids are reported to be an order of magnitude greater than by predators (Kindlmann and Dixon 1997). The famous exception is the control of the cottony-cushion scale, *Icerya purchasi*, by the coccidophagous ladybird beetle predator, *R. cardinalis*. This outstanding success resulted in the widespread and haphazard use of natural enemies for "biocontrol". Several other species of coccidophagous ladybirds have also proved effective biocontrol agents (Taylor 1935; Clausen 1940; Bartlett 1978), but few or none of the aphidophagous species (Clausen 1940). Is it possible to account for the very different effectiveness of these two groups of ladybirds?

Mills (1982a) states that coccidophagous species are generally small and feed almost continuously, whereas the larger aphidophagous species are characterised by long periods of inactivity due to satiation. He suggests that in feeding on immobile prey, like coccids, there are advantages in reducing the proportion of each prey item eaten to that which is readily extracted, easily assimilated and rapidly digested. A reduction in digestion time would allow more continuous feeding, while a reduction in the proportion of each prey item eaten would increase the potential impact on the prey population. In contrast, aphidophagous species tend to eat all of each prey item, and this nonselective feeding results in frequent satiation. Thus, the success of coccidophagous species in biological control is attributed to their more optimal use of prey.

What evidence is there that the incidence of satiation is different in these two groups of ladybirds? Taylor (1935), in accounting for the success of *Cryptognatha nodiceps* in controlling the scale insect, *Aspidotus destructor*, states that both larvae and adults of this ladybird feed continuously during the day. However, colleagues who have worked with both aphidophagous and coccidophagous species have not noticed a marked difference in the incidence of satiation (J.-L. Hemptinne and A. Magro, pers. comm.), and Iperti et al. (1977) recorded that coccidophagous species are less voracious than aphidophagous species.

If coccidophagous species make more optimal use of their prey, they should, all other things being equal, have faster rates of development than aphidophagous species. However, the reverse appears to hold. It is difficult to compare species because developmental times are affected by temperature and food quality/quantity (Dixon 1959). This can partly be overcome by using the fastest developmental time recorded for each species. The data appear to indicate that coccidophagous species fairly consistently take considerably longer to complete their development (Fig. 12.8) or at least do not develop faster than aphidophagous species. This is contrary to what one would expect if coccidophagous species feed continuously and are less often satiated than aphidophagous ladybirds. If coccidophagous species do have a significantly *longer* developmental time, then it raises an interesting question, which is considered in Section 12.5. Interestingly in this respect, larvae of the coccidophagous species, *Exochomus quadripustulatus*, fed aphids retain the typical development time of coccidophagous species (Table 12.2). However, the adults fed scale produce fewer eggs than those fed aphids (Radwan and Lövei 1983).

If coccidophagous ladybirds do not show a more optimal use of prey, how can we account for their undoubted success in biological control? In the case of *R. cardinalis*, an egg is laid under an adult female scale or its ovisac and the larva *completes* its development by consuming the scale and its eggs (Clausen 1940). It is likely that the presence of a larva under a scale deters other ladybirds from ovipositing (Hemptinne et al. 1993). That is, the larvae of this predator do not have to seek for and capture active prey and the generation times of *Rodolia* and its prey are comparable. *C. nodiceps* has likewise been very successful in dramatically reducing the abundance of the coconut scale, *A. destructor*, in Fiji (Taylor 1935). In this system the adult ladybird eats the adult coccid after extracting it from its scale, leaves the ovisac undamaged and lays an egg in the empty scale. The larva eats the contents of the ovisac, i.e. eggs and

Table 12.2. The fastest times for development from first instar to adult emergence from the pupa that have been recorded for aphidophagous and coccidophagous species of ladybirds

Aphidophaga			Coccidophaga		
Species	Development time (Days)	Reference	Species	Development time (Days)	Reference
Adalia bipunctata	11	Okrouhla et al. (1983)	*Azya trinitatis*	28	Taylor (1935)
Calvia quatuordecimguttata	12	Lamana and Miller (1995)	*Chilocorus bipustulatus*	22	Podoler and Henen (1983)
Coccinella novemnotata	9	McMullen (1967)	*Chilocorus kuwanae*	24	Podoler and Henen (1983)
Coccinella septempunctata	8	Okrouhla et al. (1983)	*Chilocorus nigritus*	22	Tirumala (1954)
Coccinella transversoguttata	10	Obrycki and Tauber (1981)	*Coelophora quadrivitta*	10	Chazeau (1981)
Cheilomenes sulphurea	13	Okrouhla et al. (1983)	*Cryptognatha nodiceps*	17	Taylor (1935)
Coleomegilla maculata	11	Obrycki and Tauber (1978)	*Cryptognatha simillima*	22	Taylor (1935)
Eriops connexa	9	Miller and Panstian (1992)	*Cryptolaemus montrouzieri*	17	Ramesh Babu and Azam (1987)
Harmonia axyridis	11	Hukusima and Kamei (1970)	*Diomus hennesseyi*	19	Kanika-Kiamfu et al. (1992)
Hippodamia convergens	9	Miller (1992)	*Exochomus flaviventris*	26	Kiyindou (1989)
Hippodamia parenthesis	10	Orr and Obrycki (1990)	*Exochomus quadripustulatus*[a]	28	Radwan and Lövei (1983)
Hippodamia sinuata	10	Michels and Behle (1991)	*Hyperaspis jucunda*	21	Nsiama She et al. (1984)
Hippodamia tredecimpunctata tibialis	14	Okrouhla et al. (1983)	*Hyperaspis lateralis*	25	McKenzie (1932)
			Hyperaspis marmottani	23	Umeh (1982)
Lioadalia flavomaculata	9	Okrouhla et al. (1983)	*Hyperaspis raynevali*	20	Kiyindou and Fabres (1987)
Menochilus sexmaculatus	12	Okrouhla et al. (1983)	*Hyperaspis senegalensis*	26	Fabres and Kiyindou (1985)
Propylea quatuordecimpunctata	9	Quilici (1981)	*Pharosymnus numidicus*	18	Kehat (1968)
Scymnodes frontalis	18	Gibson et al. (1992)	*Sycmnus aeneipennis*	27	Taylor (1935)
Scymnodes lividigaster	17	Anderson (1981)	*Sycmnus reunioni*	17	Izhevsky and Orlinsky (1988)
			Sycmnus severini	13	Taylor (1935)
			Sycmnus sp.	20	Taylor (1935)

[a] Coccidophagous species reared on aphids.

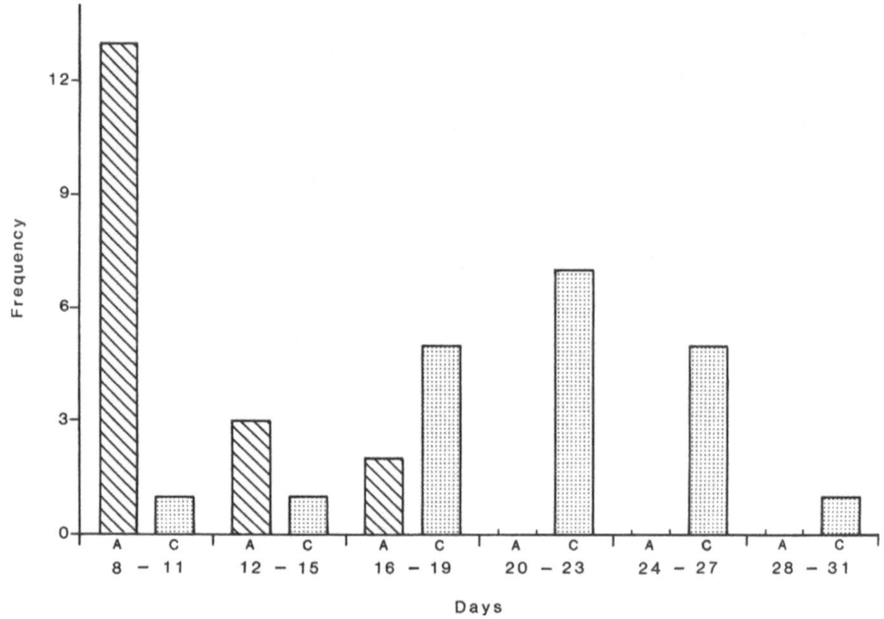

Fig. 12.8. Frequency distributions of the fastest developmental times recorded for aphidophagous (□) and coccidophagous (▧) species of ladybird (cf. Table 12.2)

young scale, and then disperses and completes its development by eating mainly young stages of scale. That is, initially the young larvae do not have to search for prey. In addition, the total developmental time of the ladybird (19–25 days) is considerably shorter than that of the scale (32–35 days) (Taylor 1935).

Ladybird/aphid and ladybird/coccid systems differ strikingly in two features. In the latter system the generation times of the predator and prey are comparable, whereas in the former the prey has a much shorter developmental time than the predator. Secondly, the larvae of coccidophagous species usually do not have to search for prey, especially during the early larval stages, which is in marked contrast to the situation in aphidophagous species, where the larvae from birth search for active prey. That is, in being able to mature or reach an advanced stage of development by eating one prey item, coccidophagous species are able to track changes in the abundance of their prey more closely and, as a consequence, are more effective in reducing the abundance of their prey than aphidophagous species (Kindlmann and Dixon 1997). In addition, in aphidophagous species there is both a lower critical threshold of prey abundance and an upper critical threshold of larval abundance, which further constrains the effectiveness of aphidophagous species, by curtailing their numerical reproductive response.

12.5 Evolution of Life-History Traits

If the marked apparent difference in the developmental times of aphidophagous and coccidophagous ladybirds is confirmed, then it is likely to be a consequence of the different selection pressures operating in the two systems. It is known that shorter developmental times can be selected for in ladybirds (Rodriguez-Saonia and Miller 1995) and the ratio of egg to adult weights is correlated with developmental time (Stewart et al. 1991b). That is, species laying relatively large eggs have a short development time and vice versa.

As the larvae of aphidophagous species scramble for resources, that is, they pursue and subdue active prey, they need to be relatively large. This is supported by the observation that there appears to be a lower limit to egg size in aphidophagous species (Stewart et al. 1991b). A consequence of this is that the first instar larvae are large relative to the adult, and their developmental time is shorter than it would be if they were smaller. In coccidophagous species the eggs are deposited under individual scales or their ovisacs, which, compared to aphids, are relatively long-lived, and the ladybird larvae do not have to search for resources. In this case, the prediction is that the relative size of the eggs could be small. If this is so, then the prediction is that coccidophagous species should have relatively long developmental times (cf. Fig. 12.8). That is, characteristics of their prey have possibly been important in shaping the reproductive strategies of aphidophagous and coccidophagous ladybirds.

12.6 Conclusions

Both aphidophagous and coccidophagous ladybirds refrain from ovipositing in patches of prey already being exploited by conspecific larvae. In responding in this way they are likely to increase their fitness, as fewer of their offspring will be eaten by older larvae. However, aphidophagous and coccidophagous predator-prey systems differ in two respects. In the former, the prey is smaller and has a much shorter developmental time than the predator, which from the time of egg hatch has to search for active prey. As a consequence, their effectiveness in exploiting prey populations is severely constrained by bottlenecks in prey abundance. In the coccidophagous predator-prey system, the predator and prey are of similar size and have similar developmental periods, and there are sufficient resources in one prey item for complete or nearly complete development of the predator. As a consequence, coccidophagous predators can more closely track the abundance of their prey. That is, coccidophagous species are more likely to be effective biological control agents than aphidophagous species, which is what has been observed. It is likely that the development times and mobility of the prey, aphids and coccids, have been important in shaping the reproductive strategies of these ladybirds.

Acknowledgements. I am indebted to Jean-Louis Hemptinne and Pavel Kindlmann, whose experimental and mathematical expertise and enthusiasm made our ladybird studies possible, and to Jean-Louis for reading and commenting on the manuscript.

References

Agarwala BK (1991) Why do ladybirds (Coleoptera: Coccinellidae) cannibalize? J Biosci 16:103–109

Agarwala BK, Dixon AFG (1993a) Kin recognition: egg and larval cannibalism in *Adalia bipunctata* (Coleoptera: Coccinellidae). Eur J Entomol 90:45–50

Agarwala BK, Dixon AFG (1993b) Laboratory study of cannibalism and interspecific predation in ladybirds. Ecol Entomol 17:303–309

Agarwala BK, Dixon AFG (1993c) Why do ladybirds lay eggs in clusters? Funct Ecol 7:541–548

Anderson JME (1981) Biology and distribution of *Scymnodes lividigaster* (Muls.) and *Leptothea galbula* (Muls.), Australian ladybirds (Coleoptera: Coccinellidae). Proc Linn Soc N S W 105:1–15

Bartlett BR (1978) Coccidae. In: Clausen CP (ed) Introduced parasites and predators of arthropod pests and weeds: a world review. Agricultural Research Service Handbook No. 480, USDA, pp 57–74

Blackman RL (1967) The effects of different foods on *Adalia bipunctata* L. and *Coccinella 7-punctata* L. Ann Appl Biol 59:207–219

Chazeau J (1981) Données sur la Biologie de *Coelophora quadrivittata* [Col.: Coccinellidae], prédateur de *Coccus viridis* [Hom.: Coccidae] en Nouvelle-Calédonie. Entomophaga 26:301–312

Clausen CP (1940) Entomophagous insects. McGraw-Hill, New York

Dixon AFG (1959) An experimental study of the searching behaviour of the predatory coccinellid beetle *Adalia decempunctata* (L.) J Anim Ecol 28:259–281

Dixon AFG, Guo Y (1993) Egg and cluster size in ladybird beetles (Coleoptera: Coccinellidae): the direct and indirect effects of aphid abundance. Eur J Entomol 90:457–463

Doumbia M, Hemptinne J-L, Dixon AFG (1997) Assessment of patch quality by ladybirds: role of larval tracks. Oecologia (submitted)

Duelli P (1981) Is larval cannibalism in lacewings adaptive? (Neuroptera: Chrysopidae). Res Popul Ecol 23:193–209

Ellingsen I-J (1969) Fecundity, aphid consumption and survival of the aphid predator *Adalia bipunctata* L. (Col., Coccinellidae). Nor Entomol Tidsskr 16:91–95

Elton CS (1949) Population interspersion: an essay on animal community patterns. J Ecol 37:1–23

Fabres G, Kiyindou A (1985) Comparison du potentiel biotique de deux coccinelles (*Exochomus flaviventris* et *Hyperaspis senegalensis hottentotta*, Col. Coccinellidae) prédatrices de *Phenococcus manihoti* (Hom., Pseudococcidae) au Congo. Acta Oecol Oecol Appl 6:339–348

Ferran A, Dixon AFG (1993) Foraging behaviour of ladybird larvae (Coleoptera: Coccinellidae). Eur J Entomol 90:383–402

Fox LR (1975) Cannibalism in natural populations. Annu Rev Ecol Syst 6:87–106

Gibson RL, Elliot NC, Schaefer P (1992) Life history and development of *Scymnus frontalis* (Coleoptera: Coccinellidae) on four aphid species. J Kans Entomol Soc 65:410–415

Hamalainen M, Markkula M, Raij T (1975) Fecundity and larval voracity of four lady beetle species (Col., Coccinellidae). Ann Entomol Fenn 41:124–127

Hardin G (1968) The tragedy of the commons. Science 162:1243–1247
Hassell MP, May RM (1973) Stability in insect host-parasite models. J Anim Ecol 42:693–726
Hassell MP, May RM (1974) Aggregation of predators and insect parasites and its effect on stability. J Anim Ecol 43:567–594
Hemptinne J-L (1989) Ecophysiologie d'*Adalia bipunctata* (L.) (Coleoptera: Coccinellidae). Thèse de doctorat, Université Libre de Bruxelles, Bruxelles
Hemptinne J-L, Dixon AFG, Coffin J (1992) Attack strategy of ladybird bettles (Coccinellidae): factor shaping their numerical response. Oecologia 90:238–245
Hemptinne J-L, Dixon AFG, Doucet J-L, Petersen J-E (1993) Optimal foraging by hoverflies (Diptera: Syrphidae) and ladybirds (Coleoptera: Coccinellidae): mechanisms. Eur J Entomol 90:451–455
Hemptinne J-L, Dixon AFG, Lognay G (1996) Searching behaviour and mate recognition by males of the two-spot ladybird beetle, *Adalia bipunctata* (L.) Ecol Entomol 21:165–170
Honek A (1978) Trophic regulation of postdiapause ovariole maturation in *Coccinella septempunctata* [Col.: Coccinellidae]. Entomophaga 23:213–216
Honek A (1985) Habitat preferences of aphidophagous coccinellids [Coleoptera]. Entomophaga 30:253–264
Hukusima S, Kamei M (1970) Effects of various species of aphids as food on development, fecundity and longevity of *Harmonia axyridis* Pallas (Coleoptera: Coccinellidae). Res Bull Fac Agric Gifu Univ 29:53–66
Hukusima S, Komada N (1972) Longevity and fecundity of overwintered adults of *Propylea japonica* Thunberg (Coleoptera: Coccinellidae). Res Bull Fac Agric Gifu Univ 33:83–87
Iperti G, Katsoyannos P, Laudeho Y (1977) Étude comparative de l'anatomie des coccinelles aphidiphages et coccidiphages et appartenance d'*Exochomus quadripustulatus* L. à l'un de ces groupes entomophages (Col. Coccinellidae). Ann Soc Entomol Fr NS 13:427–437
Izhevsky SS, Orlinsky AD (1988) Life history of the imported *Scymnus* (*Nephus*) *reunioni* [Col.: Coccinellidae] predator of mealybugs. Entomophaga 33:101–114
Kanika-Kiamfu J, Kiyindou A, Brun J, Iperti G (1992) Comparison des potentialités biologiques de trois coccinelles prédatrices de la cocchenille farineuse du manioc *Phenaccocus manihoti* (Hom. Pseudococcidae). Entomophaga 37:277–282
Kawauchi S (1985) Comparative studies on the fecundity of three aphidophagous coccinellids (Coleoptera: Coccinellidae). Jpn J Appl Entomol Zool 29:203–209
Kehat M (1968) The feeding behaviour of *Pharoscymnus numidicus* (Coccinellidae), predator of the date palm scale, *Parlatoria Blanchardi*. Entomol Exp Appl 11:30–42
Kindlmann P, Dixon AFG (1993) Optimal foraging in ladybird beetles (Coleoptera: Coccinellidae) and its consequences for their use in biological control. Eur J Entomol 90:443–450
Kindlmann P, Dixon AFG (1997) GTRS – Determinants of prey abundance in insect predator-prey interactions. Evol Ecol (submitted)
Kiyindou A (1989) Seuil thermique de développement de trois coccinelles prédatrices de la cochenille du manioc au Congo. Entomophaga 34:409–417
Kiyindou A, Fabres G (1987) Étude de la capacité d'accroissement chez *Hyperaspis raynevali* [Col.: Coccinellidae] prédateur introduit au Congo pour la régulation des populations de *Phenacoccus manihoti* [Hom.: Pseudococcidae]. Entomophaga 32:181–189
Lamana ML, Miller JC (1995) Temperature-dependent development in a polymorphic lady beetle *Calvia quatuordecimguttata* (Coleoptera: Coccinellidae). Ann Entomol Soc Am 88:785–790
Majerus MEN (1991) Habitat and host plant preferences of British ladybirds. Entomol Mon Mag 127:167–175
Marples NM, Brakfield PM, Cowie RJ (1989) Differences between the 7-spot and 2-spot ladybird beetles (Coccinellidae) in their toxic effects on a bird predator. Ecol Entomol 14:79–84
Maynard Smith, J. (1964) Group selection and kin selection. Nature 201:1145–1147
Maynard Smith J (1982) The evolution of social behaviour – a classification of models. In: King's College Sociobiology Group, Current problems in sociobiology. Cambridge University Press, Cambridge, pp 29–44

McKenzie HL (1932) Biology and feeding habits of *Hyperaspis lateralis*. Univ Calif Publ Entomol 6:10–17

McMullen RD (1967) The effects of photoperiod, temperature, and food supply on rate of development and diapause in *Coccinella novemnotata*. Can Entomol 99:578–586

Michels GJ, Behle RW (1991) Effects of two prey species on the development of *Hippodamia sinnuta* (Coleoptera: Coccinellidae) larvae at constant temperatures. J Econ Entomol 84:1480–1484

Miller JC (1992) Temperature-dependent development of the convergent ladybird (Coleoptera: Coccinellidae). Environ Entomol 21:197–201

Miller JC, Paustian JW (1992) Temperature-dependent development of *Eriops connexa* (Coleoptera: Coccinellidae). Environ Entomol 21:1139–1142

Mills NJ (1979) *Adalia bipunctata (L.)* as a generalist predator of aphids. PhD Thesis, University of East Anglia, Norwich

Mills NJ (1982a) Satiation and the functional response: a test of a new model. Ecol Entomol 7:305–315

Mills, N.J. (1982b). Voracity, cannibalism and coccinellid predation. *Ann. Appl. Biol.* 144–148.

Mueller RH, Thompson ME, Dipardo RM (1984) "Stereo-" and regioselective total synthesis of the hydropyrido [2,1,6-de] quinolizine ladybug defensive alkaloids. J Org Chem 49:2217–2213

Nicholson AJ (1933) The balance of animal populations. J Anim Ecol 2:132–178

Nsiama She HD, Odebiyi JA, Herren HR (1994) The biology of *Hyperaspis jucunda* [Col.: Coccinellidae] and exotic predator of the cassava mealybug *Phenacoccus manihoti* [Hom.: Pseudococcidae] in southern Nigeria. Entomophaga 29:87–93

Obrycki JJ, Tauber MJ (1978) Thermal requirements for development of *Coleomegilla maculata* (Coleoptera: Coccinellidae) and its parasite *Perilitus coccinellae* (Hymenoptera: Braconidae). Can Entomol 110:407–412

Obrycki JJ, Tauber MJ (1981) Phenology of three coccinellid species, thermal requirements for development. Ann Entomol Soc Am 74:31–36

Okrouhlá M, Chakrabarti S, Hodek I (1983) Developmental rate and feeding capacity in *Cheilomenes sulphurea* (Coleoptera: Coccinellidae). Vestn Cesk Spol Zool 47:105–117

Orr CJ, Obrycki JJ (1990) Thermal and dietary requirements for development of *Hippodamia parenthesis* (Coleoptera:Coccinellidae). Environ Entomol 19:1523–1527

Osawa N (1989) Sibling and non-sibling cannibalism by larvae of a ladybird beetle *Harmonia axyridis* Pallas (Coleoptera: Coccinellidae) in the field. Res Popul Ecol 31:153–160

Osawa N (1992) Sibling cannibalism in the ladybird beetle *Harmonia axyridis*: Fitness consequences for mother and offspring. Res Popul Ecol 34:45–55

Pasteels JM, Deroe C, Tursch B, Braekman JC, Daloze D, Hootele C (1973) Distribution et activités des alcaloides détensifs des Coccinellidae. J Insect Physiol 19:1771–1784

Podoler H, Henen J (1983) A comparative study of the effects of constant temperatures on development time and survival of two coccinellid beetles of the genus *Chilocorus*. Phytoparasitica 11:167–176

Quilici S (1981) Etude biologique de *Propylea quatuordecimpunctata* L. Efficacité prédatrice compareé de trois types de coccinelles aphidophages en lutte biologique contre les pucerons sous serre. Thèse Doct 3 ème Cycle, Univ Paris IV, Paris

Radwan Z, Lövei GL (1983) Aphids as prey for the coccinellid *Exochomus quadripustulatus*. Entomol Exp Appl 34:283–286

Ramesh Babu T, Azam KM (1987) Biology of *Cryptolaemus montrouzieri* Mulsant [Coccinellidae: Coleoptera] in relation with temperature. Entomophaga 32:381–386

Rautapää J (1976) Population dynamics of cereal aphids and method of predicting population trends. Ann Agric Fenn 15:272–293

Rodriguez-Saonia C, Miller JC (1995) Life history traits in *Hippodamia convergens* (Coleoptera: Coccinellidae) after selection for fast development. Biol Control 5:389–396

Rothschild M (1961) Defensive odours and Müllerian mimicry among insects. Trans R Entomol Soc Lond 113:101–122

Slobodkin LB (1968) How to be a predator. Am Zool 8:45–51

Slobodkin LB (1974) Prudent predation does not require group selection. Am Nat 108:665–673
Stewart LA, Dixon AFG, Ruzicka Z, Iperti G (1991a) Clutch and egg size in ladybird beetles. Entomophaga 36:329–333
Stewart LA, Hemptinne J-L, Dixon AFG (1991b) Reproductive tactics of ladybird beeltles: relationship between egg size, ovariole number and developmental time. Funct Ecol 3:380–385
Taylor THC (1935) The campaign against *Aspidiotus destructor*, Sign., in Fiji. Bull Entomol Res 26:1–102
Tirumala R (1954) Attempts at the utilisation of *Chilocorus nigritus* Fab. (Coleoptera, Coccinellidae) in the Madras State. Indian J Entomol 16:205–209
Umeh EDNN (1982) Biological studies on *Hyperaspis marmottani* Fairm. (Col.: Coccinellidae), a predator of the cassava mealybug *Phenacoccus manihoti* Mat-Ferr. (Hom., Pseudococcidae). Z Angew Entomol 94:530–532
van den Bosch R, Messenger PS (1973) Biological control. Intertext Books, New York
Wratten SD (1973) The effectiveness of the coccinellid beetle, *Adalia bipunctata* (L.), as a predator of the lime aphid, *Eucallipterus tiliae* L. J Anim Ecol 42:785–802

13 Interactions Between Ants and Aphid Parasitoids: Patterns and Consequences for Resource Utilization

W. Völkl

13.1 Introduction

Many insect species of various taxa produce carbohydrate-rich excretions called honeydew, which is regularly collected by ants (for reviews, see Way 1963; Buckley 1987; Hölldobler and Wilson 1990). Both partners may derive benefits from this mutualistic association: the ants obtain an important source of nutrients; in return, they may act as effective guards for the tended species by warding off predators and parasitoids (e.g. Way 1954, 1963; Banks 1962; Pierce and Easteal 1986; Völkl 1992; Jiggins et al. 1993). However, a number of recent studies have shown that this protection is incomplete, and that ants cannot always provide an enemy-free space for their mutualistic partners. A number of predators and parasitoids have evolved morphological and/or behavioural adaptations to gain access to ant-attended resources (e.g. Eisner et al. 1978; Maschwitz et al. 1984, 1988; Mason et al. 1991; Völkl 1992, 1995).

In temperate regions, aphids are the most important insect group having developed a mutualistic relationship with ants (Zwölfer 1958; Buckley 1987; Hölldobler and Wilson 1990). Aphids are attacked by a diverse complex of predators (Dixon 1985) and by parasitoids of the hymenopterous family Aphidiidae (Mackauer and Stary 1967; Starý 1970). Aphidiid wasps develop exclusively as solitary endoparasitoids of aphids (Starý 1970, 1988). The host range of about 45 out of some 120 European species covers ant-attended aphid species (Börner 1952; Starý 1966; Mackauer and Starý 1967). Females of these species interact with ants when foraging and ovipositing, and consequently ant attendance has a considerable influence on their resource exploitation and thus on population dynamics. For example, *Lysiphlebus cardui* – a specialized parasitoid of the black bean aphid, *Aphis fabae* – benefits from ant attendance by achieving higher parasitization rates (Völkl 1994), while *Trioxys angelicae* – another wasp species attacking *A. fabae* – is not able to exploit ant-attended resources (Völkl and Mackauer 1993).

The present study compares the ant-parasitoid relationships for various aphidiid species parasitizing ant-attended resources and analyses some factors that influence these interactions (Fig. 13.1). First, the patterns of interactions between foraging female wasps and honeydew-collecting ant workers are described. Second, the attempt is made to analyse which strategies have been evolved to escape ant aggression. Third is the search for environmental factors

Ecological Studies, Vol. 130
Dettner et al. (eds.) Vertical Food Web Interactions

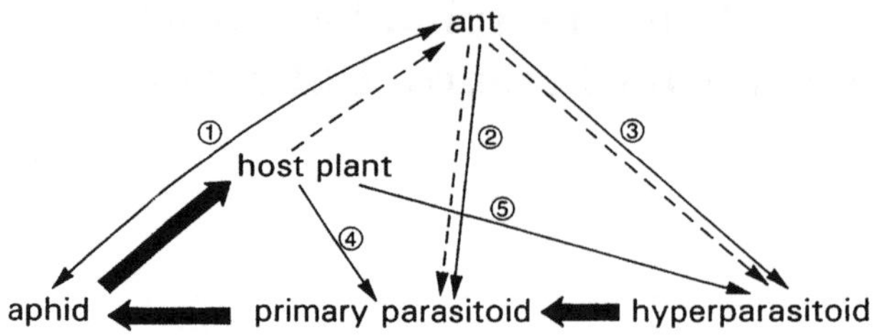

Fig. 13.1. Interactions within the aphid-parasitoid food web. *Bold black arrows* Trophic relationships; *thin arrows*: *1* ant-aphid mutualism; *2, 3* interactions between ants and primary parasitoids or hyperparasitoids; *4, 5* host-plant structure may influence primary parasitoid and hyperparasitoid foraging; *thin broken arrows* host-plant structure may influence ant behaviour and ant-parasitoid interactions

influencing the outcome of ant-parasitoid interactions, and demonstration of the consequences for aphidiid resource utilization.

13.2 Patterns of Interactions

The analysis of interactions between ant workers and foraging aphidiid females for 34 European aphidiid species and the available information for three East Asian and one North American species (Mason 1922; Takada 1983; Takada and Hashimoto 1985) revealed three general types of ant-parasitoid interactions (Table 13.1). (1) Parasitoids are treated aggressively by all tested ant species: foraging parasitoid females are treated extremely aggressively and generally killed if captured by an ant worker, independent of the ant species involved. Ants usually perceive these wasps quickly even when they approach from behind. (2) Parasitoids are treated aggressively only by particular ant species. Foraging females of a number of aphidiid species are attacked and killed only by particular ant species, while other ant species either ignore the parasitoid females (e.g. *Aclitus sappaphis*, *Paralipsis eikoae*) or respond less aggressively to an encounter with a foraging parasitoid female (e.g. *Pauesia* spp.). In the latter case, parasitoids are usually not perceived when approaching from behind. They are repelled by the ants from the vicinity of the aphid colonies but not vigorously attacked. (3) Parasitoids are generally treated non-aggressively. Foraging aphidiid females are generally disregarded by ant workers, independent of the ant species. Females of these species are able to forage and oviposit unmolested within ant-attended aphid colonies.

The responses of ant workers towards foraging males and towards adults emerging from the mummy yielded essentially the same results (Weisser and Völkl 1996; W. Völkl, unpubl.). By contrast, it has never been observed in my experience that ants, independent of the ant species, preyed within a natural system selectively on premature parasitoid stages, i.e. on parasitized aphids or

Table 13.1. Interaction types among various aphidiid and ant species

Parasitoid species	Ant species[a]				
	Ln	Lfl	Lfu	M	Fp
Ephedrus persicae Froggatt	a	–	a	a	(–)
Ephedrus plagiator (Nees)	a	–	a	a	a
Praon abjectum (Haliday)	a	(–)	a	a	a
Praon volucre (Haliday)	a	–	(–)	a	(–)
Aphidius tanacetarius Mackauer	a	–	–	b	b
Aphidius colemani Viereck	a	–	–	a	a
Aphidius matricariae (Haliday)	a	–	(–)	a	(–)
Euaphidius setiger Mackauer	a	–	a	a	a
Euaphidius cingulatus (Ruthe)	a	–	b	–	–
Lysiphlebus fabarum (Marshall)	c	–	c	c	c
Lysiphlebus cardui (Marshall)	c	–	–	c	c
Lysiphlebus hirticornis Mackauer	c	–	–	c	c
Lysiphlebus melandriicola Stary	c	–	–	c	(–)
Lysiphlebus confusus Tremblay & Eady	c	–	c	(–)	(–)
Lysiphlebus testaceipes (Cresson)	c	–	–	c	(–)
Lysiphlebus fritzmuelleri Mackauer	c	–	–	(–)	(–)
Lysiphlebus macrocornis Mackauer	c	c	–	c	–
Adialytus ambiguus (Haliday)	c	–	–	(–)	–
Pauesia silvestris Stary	a	–	b	–	b
Pauesia pini (Haliday)	a	–	b	–	b
Pauesia pinicollis Stary	a	–	b	–	b
Pauesia picta (Haliday)	a	–	b	–	b
Pauesia laricis (Haliday)	a	–	b	–	b
Xenostigmus bifasciatus (Haliday)[b]	(–)	–	–	–	b
Diaeretiella rapae Mc'Intosh	a	–	–	a	(–)
Paralipsis enervis (Nees)	c	a	–	c	–
Paralipsis eikoae Yasumatsu[c]	c	–	–	a[d]	–
Aclitus sappahis Takada & Shiga[c]	a	–	–	c[d]	–
Aclitus obscuripennis (Förster)	c	(–)	–	c	–
Trioxys angelicae (Haliday)	a	–	(–)	a	a
Trioxys betulae Marshall	a	–	(–)	–	a
Trioxys falcatus Mackauer	a	–	a	a	a
Trioxys acalephae (Marshall)	a	–	(–)	a	(–)
Trioxys pallidus (Haliday)	a	–	(–)	(–)	a
Protaphidius wismanni (Ratzeburg)	–	–	b	–	–
Protaphidius nawaii Yasumatsu[e]	b	–	–	–	–
Monoctonus cerasi (Marshall)	a	–	–	–	(–)
Lipolexis gracilis (Förster)	b	(–)	–	b	(–)

[a] Ln = *Lasius niger*, Lfl = *Lasius flavus*, Lfu = *Lasius fuliginosus*, M = *Myrmica* spp., Fp = *Formica polyctena*. [b] Mason (1922). [c] Takada and Hashimoto (1985). [d] The ant species was *Pheidole fervida* Fr. Smith (Takada and Hashimoto 1985). [e] Takada (1983). All other data: present study. (For methods, see Völkl and Mackauer 1993).

a = aggressive ant response, parasitoid shows no adaptation to ant aggression. b = aggressive ant response, parasitoid displays general anti-predator behaviour. c = non-aggressive ant-response, parasitoid shows specific adaptations (chemical and behavioural mimicry). (–) = interactions have not been studied. – = interactions are very unlikely in the field.

parasitoid mummies. These responses my differ in exotic ant-parasitoid relationships. Van den Bosch et al. (1979) observed a selective predation of *Trioxys pallidus*-parasitized walnut aphids by the introduced Argentine ant *Iridomyrmex humilis* in California, and Stechmann et al. (1996) reported predation of *Aphidius colemani* mummies by *Solenopsis geminata* in Tonga.

13.3 Evolutionary Strategies and Resource Utilization in Ant-Aphidiid Interactions

13.3.1 Parasitoid Species Without Adaptations to Ant Attendance

About 40% of the aphidiid species shown in Table 13.1 are attacked by all ant species that were tested. Their only chance to escape aggression is to flee immediately, e.g. by dropping off the plant. By this means, they are widely restricted to unattended aphid colonies in their resource utilization (Völkl 1992; Mackauer and Völkl 1993). For example, the average number of *T. angelicae* mummies per *A. fabae* colony was significantly lower in ant-attended colonies than in unattended ones (Fig. 13.2). This general pattern was independent of the host plant, although *T. angelicae* had a relatively higher oviposition success in ant-attended colonies on goosefoot (*Chenopodium* spp.) than on creeping thistle (*Cirsium arvense*) and spindle bush (*Evonymus europaeus*). The differences in parasitization success between host plants may

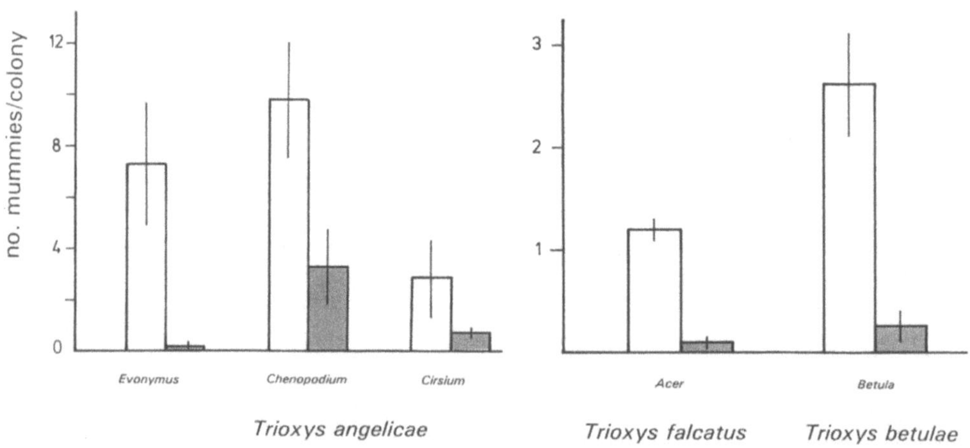

Fig. 13.2. Patterns of parasitism by *Trioxys angelicae* on three different host plants during 1986 to 1988, by *Trioxys falcatus* on *Acer* spp. during 1993/1994 and by *Trioxys betulae* on *Betula pendula* during 1993 in the presence (*shaded bars*) and absence (*open bars*) of ants. Data are pooled for all sampling dates and ant species. Values give mean ± SE. All samples (sample size: n > 50 for each combination) were collected in the vicinity of Bayreuth, northern Bavaria, Germany

be mainly attributed to two factors. First, *A. fabae* forms on average larger colonies on thistle (335 ± 215 aphids/colony; n = 332) and goosefoot (287 ± 221 aphids/colony; n = 267) than on spindle bush leaves (36 ± 12 aphids/colony; n = 198). These large aphid colonies cannot be guarded by ants as effectively as small ones. Second, the higher structural diversity of the feeding sites of *A. fabae* (leaves + flowers) on goosefoot plants in comparison to the feeding sites on thistle (stem) may lead to an increased probability of a successful oviposition before an ant attack. A similar drastic reduction of parasitoid numbers in ant-attended colonies appeared also for *Trioxys falcatus*, a specialised parasitoid of *Periphyllus* spp. on *Acer* spp., and for *Trioxys betulae*, which exclusively parasitizes *Symydobius oblongus*, a heavily tended callaphidid species dwelling on *Betula pendula* (Fig. 13.2).

13.3.2 Species That Avoid Ant Aggression

About 60% of the studied aphidiid species have evolved mechanisms to avoid or escape aggression by at least some ant species. These anti-predator defenses may be either general behavioural patterns, like walking away or dropping off, which do not result in a non-recognition by ants, or specific adaptations, like chemical or behavioural mimicry or camouflage resulting in non-recognition.

13.3.2.1 Species with General Behavioural Adaptations

A number of aphidiid species prevent ant aggression by an avoidance behaviour. These species are characterized by their quick, agile and flexible movements including the ability to run backwards as quickly as forwards. Furthermore, they have comparably large cup-shaped eyes which provide a wide visual field and allow them to see backwards. Typical representatives of this type are members of the genus *Pauesia*, which attack conifer aphids of the genus *Cinara*. Many *Cinara* species are attended by red wood ants (*Formica* spp.), which are commonly assumed to provide some protection against the aphid's natural enemies (Scheurer 1964, 1971; Fossel 1972). For example, *Formica polyctena* workers repelled quickly moving females of *Pauesia pinicollis*, a common parasitoid of *Cinara pinea* on *Pinus sylvestris*, from the vicinity of the aphid colony by a short pursuit (Fig. 13.3). This pursuit, however, was given up after the *P. pinicollis* female had retreated quickly backwards to a needle, or had dropped off. *P. pinicollis* females often perceived honeydew-collecting or foraging *F. polyctena* workers at a distance of more than 1 cm and thus were able to avoid contact. Having perceived an ant, or after a pursuit, *P. pinicollis* females remained on average for 2 min on a needle ("waiting behaviour") and kept a "security distance" of 1.5 cm to the nearest worker's head. Such non-moving, and also very slow-moving, parasitoids were usually disregarded by *F. polyctena* workers. If ant workers abandoned the tended aphid for a few seconds, or if they turned their abdomen towards the waiting *P. pinicollis*

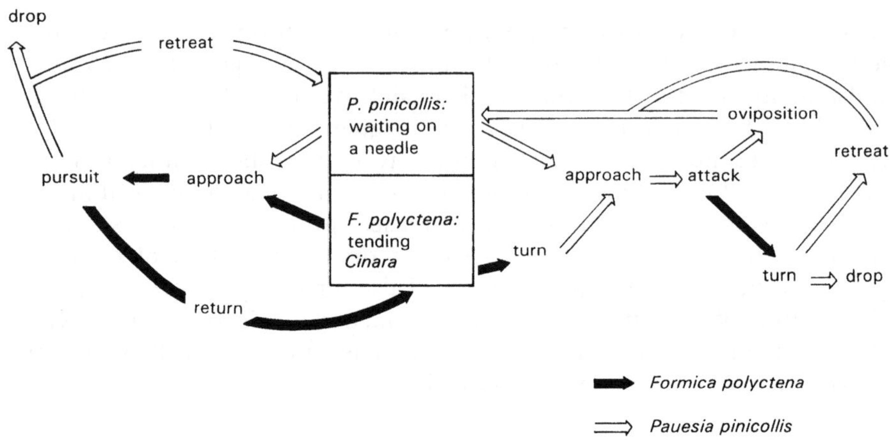

Fig. 13.3. A scheme of a typical sequence of interactions between *Pauesia pinicollis* and *Formica polyctena*

female, the parasitoids quickly approached the *C. pinea* colony and tried to oviposit before an ant returned, or before the ant turned again (Fig. 13.3). A convergent parasitoid behaviour and similar patterns of interactions were found in a number of other *Pauesia* species and within other parasitoid genera, like *Aphidius* and *Protaphidius* (Table 13.1).

There is considerable evidence that some *Pauesia* species may learn during interaction with *F. polyctena* workers. Foraging *P. pinicollis* females changed their behaviour after some ant encounters. (1) They slowed down their search speed; (2) they considerably reduced the "security distance" when waiting on needles; and (3) they approached ants slowly also from the side and even from the front. Furthermore, females retreated to a significantly lower proportion in response to a quick ant approach and remained instead motionless, thereby preventing attacks. A similar effect of experience-related change in behaviour during ant-parasitoid interactions was observed for *Pauesia picta*, a parasitoid of *C. pinea*, and for *Pauesia pini*, a common parasitoid of various *Cinara* species on pine, spruce and larch. The benefit of this "learning" was increased oviposition success in the presence of ants: in *P. pinicollis*, the oviposition rate of experienced females was significantly higher (2.43 ± 0.52 eggs/h; $n = 36$) than that of naive females (0.85 ± 0.6 eggs/h; $n = 36$).

The success of a general anti-predator behaviour depends on the aggressiveness of the ant species involved. The quick and aggressive workers of *Lasius niger* caught and killed parasitoids (Table 13.1), which tried to obviate their aggression by an agile behaviour or at least prevented successful ovipositions. By contrast, parasitoid species applying avoidance behaviour succeeded in coexisting with less aggressive and/or less agile ant species, like *Lasius fuliginosus*, *Formica polyctena or Myrmica* spp. (Table 13.1). Parasitoid species which were able to coexist with *Lasius niger* have always evolved additional adaptations to escape attacks by this aggressive ant (see below).

The patterns of parasitism in the field for species that escape ant aggression by agile behaviour widely resembled ant aggressiveness. *Euaphidius cingulatus*, a specialized parasitoid of *Pterocomma* spp. on *Salix* spp. achieved a high parasitization success when attacking *L. fuliginosus*-attended *Pterocomma salicis* on *Salix* sp. (Fig. 13.4). By contrast, parasitism was significantly reduced in unattended *P. salicis* colonies and virtually absent in *L. niger*-attended *P. salicis* colonies. Similarly, the parasitization rates of *Aphidius tanacetarius* were comparably high in unattended and *Myrmica rugulosa*-attended colonies of *Metopeurum fuscoviride* on *Tanacetum vulgare*, while parasitism by this parasitoid species was negligible in *L. niger*-attended colonies (Mackauer and Völkl 1993). The average number of *Pauesia* mummies was also highest on *Pinus sylvestris* twigs with *F. polyctena*-attended colonies of *Cinara pini* or *Cinara pinea*. The average number of *Pauesia* mummies on unattended twigs was considerably lower, and *L. niger*-attended twigs yielded no mummies at all. However, the mummies of the different *Pauesia* species are difficult to distinguish, and exact countings for the different species were not possible due to high hyperparasitism. Therefore, there is currently no direct evidence that all *Pauesia* species benefit from red wood ants.

There is also some evidence that *E. cingulatus* and *Pauesia* spp. benefit from low temperatures and low insolation. Females of these species are very active

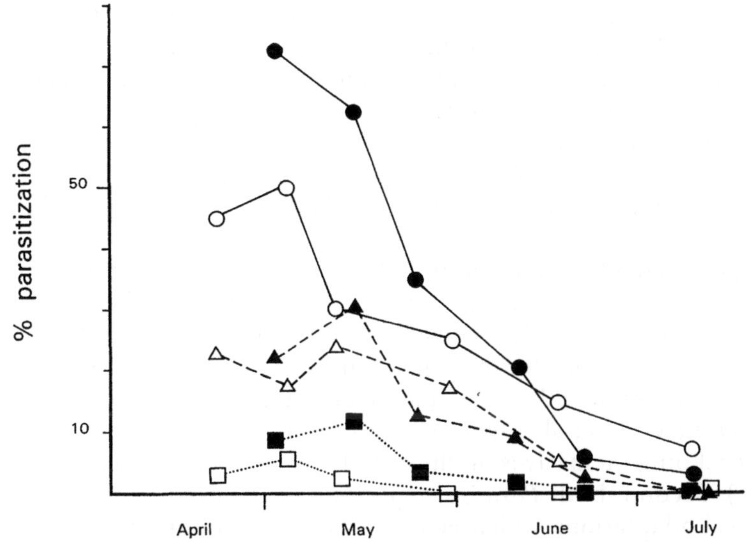

Fig. 13.4. Seasonal patterns of parasitism by *Euaphidius cingulatus* in colonies of *Pterocomma salicis* on *Salix* sp. at a sampling site near Bayreuth, northern Bavaria, Germany, in 1993 and 1994. *Circles*, *Lasius fuliginosus*-attended colonies; *squares*, *Lasius niger*-attended colonies; *triangles* unattended colonies; *open signs* 1993; *filled signs* 1994

and agile even at temperatures below 10 °C and without sunshine, while ants are quite inactive under these conditions, often sitting motionless within the aphid colony. Thus, parasitoid females may benefit from the ants by a reduction of aphid defence behaviour but are not disturbed in their foraging activities by ant workers. Indeed, the parasitization rates within *L. fuliginosus*-attended *P. salicis* colonies during peak infestation in April/May were negatively correlated with daily mean temperatures ($r_S = -0.886$, $n = 12$, $p < 0.001$; Fig. 13.4).

13.3.2.2 Parasitoid Species with Specific Adaptations to Ant Attendance

Some aphidiid species have evolved specific adaptations to obtain access to ant-attended resources. These species were tolerated by all tested ant species and generally disregarded by honeydew-collecting ant workers (Table 13.1). When foraging within an ant-attended aphid colony, parasitoid females usually display slow "calm" movements and show an almost cryptic behaviour. Their normal response to an encounter with an ant worker is either cowering or disregarding (Völkl and Mackauer 1993). For *L. cardui*, there is considerable evidence that this species mimics the epicuticular hydrocarbon profile of its host, *A. fabae cirsiiacanthoidis*, to fool honeydew-collecting ant workers (Liepert and Dettner 1993; Völkl and Mackauer 1993), thereby applying a "wolf-in-sheep's-clothing" strategy. This could explain why different, and unrelated, ant species respond similarly to both the parasitoid and the host aphid (Völkl and Mackauer 1993), why *L. cardui* females are not attacked by ants during eclosion from the mummy (W. Völkl, unpubl.), and why caged females of *L. cardui* attempt to oviposit in conspecific wasps when deprived of hosts (Starý and Völkl 1988). The strategy of chemical mimicry is mainly applied by representatives of the genus *Lysiphlebus*, *Adialytus* (Mackauer and Völkl 1993; Liepert and Dettner 1995) and *Aclitus* (Takada and Hashimoto 1985; W. Völkl, unpubl.).

The most advanced adaptation to ant-attended resources has been found in *Paralipsis enervis* and *Paralipsis eikoae*, two specialized parasitoids of root aphids. Both species are exclusively associated with the ant *L. niger* and have evolved a number of adaptations for this coexistence. Firstly, to escape aggression by *L. niger*, they mimic either the epicuticular hydrocarbon pattern of their host (*P. enervis*; Völkl et al. 1996) or they actively acquire the odour of their host ant by rubbing themselves on the ant's body (*P. eikoae*; Takada and Hashimoto 1985). Second, both *Paralipsis* species have evolved behavioural and morphological adaptations which enable a tactile communication with and regurgitation by *L. niger* (Manéval 1940; Takada and Hashimoto 1985; Völkl et al. 1996). They both engage in mutual antennating (antennal tapping) with an encountered *L. niger* worker and elicit regurgitation of liquid food via trophallaxis by the ant worker. As a morphological adaptation to this trophallaxis, both species have elongated glossae (Takada and Hashimoto

1985). For *P. enervis*, however, neither mutual antennating nor trophallaxis was observed during encounters with workers of *Tetramorium caespitum* and *Myrmica laevinodis*, although neither ant species attacks the parasitoid. By contrast, *L. flavus* workers began to engage in mutual antennation, but subsequently attacked and killed *P. enervis* (Völkl et al. 1996). This different ant behaviour resulted in a different parasitoid longevity. *P. enervis* survived for only 10 min in the presence of *L. flavus* due to ant aggression. Survival increased to approx. 1 day in the presence of host aphids and *T. caespitum* or *M. laevinodis*, or without ants due to lacking trophallaxis. Survival increased significantly to more than 5 days in the presence of *L. niger*, which regularly provided food to the parasitoids (Völkl et al. 1996). These differences in ant-parasitoid interactions and the resulting positive consequences for parasitoid longevity, and thus for parasitoid fitness, may provide an explanation for the close and specific association between *P. enervis* and *L. niger*.

Aphidiid species that gain access to ant-attended resources by non-recognition often reach extremely high parasitization rates within their host colonies. For example, the parasitization rates of *L. cardui* increase rapidly during the season and can reach up to 100% parasitism in ant-attended colonies during midsummer. By contrast, unattended colonies are usually colonized later in the season and parasitized less heavily (Fig. 13.5). Similar

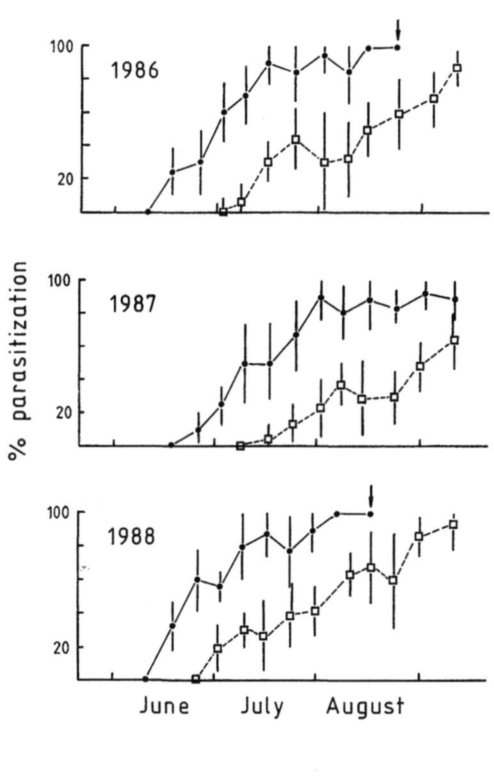

Fig. 13.5. Seasonal patterns of parasitism by *Lysiphlebus cardui* in ant-attended (*filled circles*) and unattended (*open squares*) colonies of *Aphis fabae cirsiiacanthoidis* on *Cirsium arvense*. *Arrows* indicate the eradication of all *A. fabae* colonies by *L. cardui*. (Völkl 1992)

patterns of parasitism and differences between ant-attended and unattended colonies were found for the closely related species, *Lysiphlebus hirticornis* (Mackauer and Völkl 1993) and *Lysiphlebus fabarum* (Starý 1970).

13.3.3 Benefits from Ant Attendance for Aphidiid Wasps

The progeny of aphidiid wasps is attacked by a variety of hyperparasitoid species which attack either the primary parasitoid's larvae within the living aphid or the aphid mummy (Sullivan 1988). In the absence of ant attendance, hyperparasitoids ususally cause a high mortality among aphidiids (Sullivan 1988; Mackauer and Völkl 1993). By contrast, hyperparasitism may be significantly reduced in ant-attended colonies, since females of most hyperparasitoid species have difficulties in laying eggs within ant-attended aphid colonies (Hübner and Völkl 1996). For example, *L. cardui* suffered from an average rate

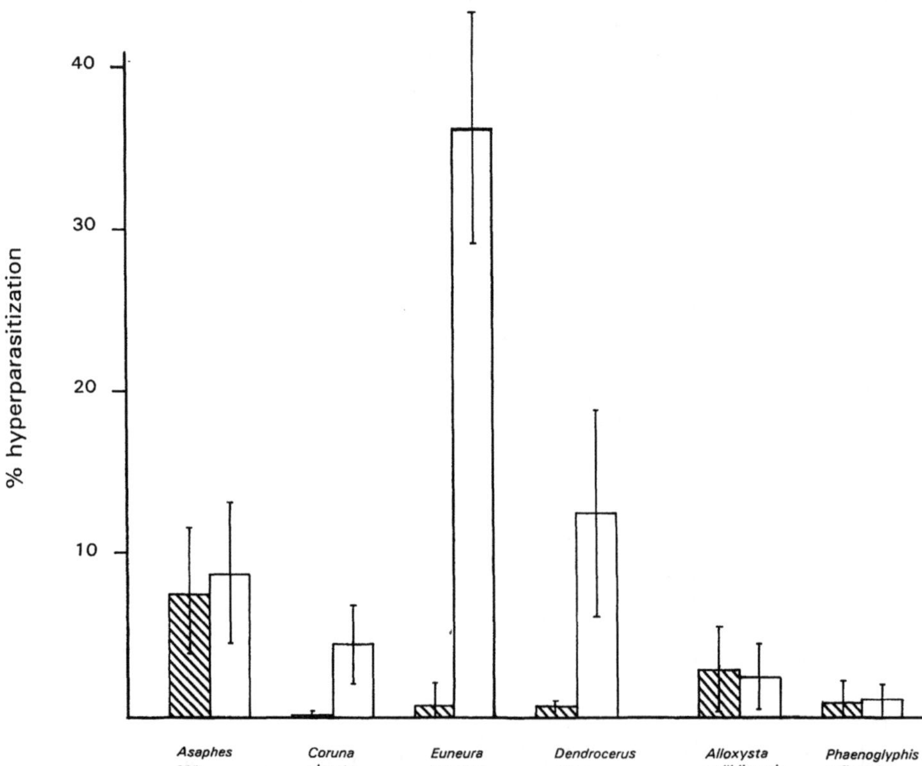

Fig. 13.6. Frequency of hyperparasitism in *Cinara pini* mummies on *Pinus sylvestris* (primary parasitoid: *Pauesia* spp.). *Hatched bars* Mummies having developed in ant-attended colonies on bark; *open bars* mummies having developed on needles or in unattended colonies on bark. All samples were collected in coniferous forests of the Fichtelgebirge near Bad Berneck, northern Bavaria, Germany

of hyperparasitism of 70% in unattended colonies of *A. fabae*, while this rate dropped to an average of 17% in ant-attended colonies (Völkl 1992). Similar benefits were found for *L. hirticornis* attacking *M. fuscoviride* on *T. vulgare* (Mackauer and Völkl 1993) and for *Pauesia* spp. attacking *C. pini* on *P. sylvestris* (Fig. 13.6). Parasitized individuals of *C. pini* either remain in the colony and mummify on the bark, or they move to the outer part of a neighbouring needle for mummification. However, *F. polyctena* workers usually only tend aphids - and mummies - on the bark, while they do not care for aphids and mummies on pine needles. *Pauesia* mummies collected within the *C. pini* colony were only slightly hyperparasitized, mainly by *Asaphes* spp. (7.6%) and *Alloxysta pallidicornis* (2.9%). *Euneura augarus*, a specialized hyperparasitoid of conifer aphids, caused only 0.8% mortality in mummies inside colonies but 36.4% mortality outside the aphid colony (Fig. 13.6). Furthermore, wasps developing in ant-attended colonies may benefit from reduced predation, since many predators are attacked and repelled from the aphid colony by tending ants (e.g. Banks 1962; Bradley 1973; Jiggins et al. 1993).

13.3.4 Distribution in Habitats

Ant-parasitoid interactions may also effect the parasitoids' spatial distribution in their natural habitats. *Tanacetum vulgare*, for example, originally grew mainly in the floodplains along riverbanks, but is now common in secondary habitats like ruderals, roadsides and railway embankments. In the natural (primary) habitat, most colonies of the monophagous aphid *M. fuscoviride* are either unattended or *Myrmica laevindodis*-attended, while *L. niger*-attended colonies comprise the major proportion in secondary habitats. *M. fuscoviride* is attacked by two monophagous aphidiid species, *A. tanacetarius* and *L. hirticornis*. In the Rhine valley between Bonn and Sinzig (western Germany), *A. tanacetarius*, which is attacked by *L. niger* (Table 13.1), was common in the primary habitats on the floodplains along the Rhine shore, but rare in nearby secondary habitats. The opposite situation was found for *L. hirticornis*. This species has difficulties in dealing with unattended *M. fuscoviride* due to their vigorous defense. *L. hirticornis* was extremely common in *L. niger*-attended colonies in nearby secondary habitats where it reached high densities (Mackauer and Völkl 1993) but rare along the Rhine shore, where only few *M. fuscoviride* colonies were attended by *L. niger*.

13.4 Conclusions

Parasitoids that exploit ant-attended resources need strategies to overcome ant aggression. In this respect, we may broadly distinguish between escape strategies and protective strategies (Hübner and Völkl 1996). Escape strategies include avoidance behaviour and are often combined with morphological adaptations like jumping ability or a well-armoured body. Species applying escape strategies are often treated aggressively by ants and either repelled or killed. Encyrtid and pteromalid psyllid parasitoids or hyperparasitoids, for example, escape ant aggression by their jumping ability, while some eucharitid parasitoids of ants (Wojcik 1989) are partially protected by their well-armoured body. Others, like the aphid hyperparasitoid *Alloxysta brevis*, use chemical defense (Völkl et al. 1994). Aphidiid wasps have no such morphological adaptations to escape or survive ant aggression. Species that have no adaptations to obviate ant aggression, like *Trioxys* and *Ephedrus* spp., have to avoid ant-attended colonies and try to forage for unattended colonies. Other species, like *Pauesia* spp., obviate ant aggression by general behavioural antipredator adaptations like quick movements. The success of this strategy depends on various factors. First, different ant species exhibit different degrees of aggressiveness (Table 13.1). Escape strategies of *Pauesia* spp. failed in *L. niger*-attended colonies, while they provided access to *F. polyctena* colonies. Also, the same ant species may display different responses when tending different aphid species, as found for *F. polyctena*, which was more aggressive in colonies of *C. pini* than of *C. pinea* (Kroupa 1995). Furthermore, ant-aphid mutualism and thus ant aggressiveness may be influenced by host-plant quality (Breton and Addicott 1992), by the presence of alternative honeydew sources (Addicott 1978; Del-Claro and Oliveira 1993) or by seasonal patterns in the ants' carbohydrate demand. Second, plant architecture and the host plant's structural diversity may influence the interactions by providing refuges for the parasitoid (Fig. 13.3). Third, only species which lay their eggs very quickly may be able to circumvent ant guards and parasitize tended hosts. This has been shown for *Pauesia* species, but also for braconid parasitoids of lycaenid butterflies (Fiedler 1995). In contrast, species with very long handling times, like the aphid hyperparasitoids *Dendrocerus carpenteri* and *Syrphophagus aphidivorus* and the psyllid parasitoid *Syrphophagus taeniatus*, are virtually unable to parasitize an ant-attended host despite their effective escape behaviour (Novak 1994; Hübner and Völkl 1996).

Protective strategies which result in non-aggression include chemical or behavioural mimicry. Species applying one of these strategies are either prevented from or protected against ant aggression. Protective strategies are often independent of the involved ant species and allow the parasitoid to forage freely within ant-attended resources (Vander Meer et al. 1989; Völkl and Mackauer 1993). In aphid parasitoids, chemical mimicry has been evolved in at least three genera (*Aclitus*, *Lysiphlebus*, *Paralipsis*). The hosts of all these species display only few or no defense behaviours, and all parasitoid species are characterized by long stings.

There is no clear pattern among aphid parasitoids for different evolutionary strategies correlated with the parasitoids' degree of specialization. All strategies occurred within both specialists and generalists. Among the species presented in Table 13.1, 11 generalist and 13 specialists are heavily attacked by *L. niger*, while 6 specialists and 6 generalists are protected by chemical mimicry. Furthermore, there is a slight trend for parasitoids of obligate myrmecophilous aphids to develop better adaptations to ant attendance than parasitoids of facultatively myrmecophilous aphids or parasitoids with a broad host range. Five parasitoids of obligate myrmecophilous aphids are heavily attacked by *L. niger*, while eight species are disregarded. In contrast, 17 species parasitizing facultatively myrmecophilous aphids are attacked by *L.* niger, while only 6 species are disregarded.

There is a clear pattern of resource utilization when comparing species without adaptations, with general adaptations and with specific adaptations (Table 13.2). In laboratory experiments, species without adaptations were unable to forage successfully in ant-attended colonies. By contrast, species which were able to obviate ant aggression benefited from increased oviposition success. Their oviposition rate, with one exception, was higher in the presence of non-aggressive ants than in their absence. The main reason for this greater success was a reduced aphid defense behaviour against parasitoid attacks in the presence of the ants. However, species with specific adaptations resulting in non-aggression achieved a substantially higher oviposition rate than species with general anti-predator behaviour. The main reason is that the latter have to spend a considerable amount of time for "ant handling" while the former are able to lay eggs continuously (Völkl 1994).

All species that gained access to ant-attended resources derived similar benefits. First, their oviposition success was considerably higher (Table 13.2).

Table 13.2. Comparison of oviposition rates (mean number of ovipositions/h) for various aphid parasitoid species in the presence and absence of ants. Within species, means sharing the same letter do not differ significantly at $p < 0.05$ (Mann Whitney U-test). For methods, see Mackauer and Völkl (1993) and Völkl and Mackauer (1993)

Parasitoid species	Host	Host plant	Oviposition rate	
			Ant present	Ant absent
Trioxys angelicae[a,d]	*Aphis fabae*	*Cirsium arvense*	0a	58.2 ± 22.2b
Trioxys betuale[b,e]	*Symydobius oblongus*	*Betula pendula*	0a	5.3 ± 2.3b
Aphidius tanacetarius[c,f]	*Metopeurum fuscoviride*	*Tanacetum vulgare*	52.0 ± 14.6a	17.4 ± 6.6b
Pauesia picta[b,e]	*Cinara pinea*	*Pinus sylvestris*	1.9 ± 0.5a	1.3 ± 0.3b
Pauesia pinicollis[b,e]	*Cinara pinea*	*Pinus sylvestris*	2.4 ± 0.5a	1.1 ± 0.3b
Pauesia pini[b,e]	*Cinara piceicola*	*Picea abies*	3.7 ± 0.3a	1.8 ± 0.2b
Lysiphlebus cardui[a,d]	*Aphis fabae*	*Cirsium arvense*	24.0 ± 6.6a	17.4 ± 3.6b
Lysiphlebus testaceipes[a,d]	*Aphis fabae*	*Cirsium arvense*	35.4 ± 19.2a	42.6 ± 13.8a
Lysiphlebus hirticornis[a,e]	*Metopeurum fuscoviride*	*Tanacetum vulgare*	34.1 ± 12.1a	6.3 ± 2.5b

Ant species: [a] *L. niger*. [b] *F. polyctena*. [c] *M. rugulosa*.
[d] Völkl and Mackauer (1993).
[e] W. Völkl, unpubl.
[f] Mackauer and Völkl (1993).

Second, hyperparasitization was generally significantly reduced in ant-attended aphids compared to unattended ones. Third, ant-attended colonies persist longer and grow larger than unattended ones (Addicott 1979; Bristow 1984; Völkl 1990). They should therefore represent a much more stable and predictable resource for foraging parasitoid females. These factors have an important influence on the aphids' and parasitoids' population dynamics (Völkl 1992; Mackauer and Völkl 1993) and may also have been an important factor influencing the evolution of egg-laying behaviour (Weisser et al. 1994).

Acknowledgements. This contribution is dedicated to Helmut Zwölfer, from whose knowledge, help and critical discussions I have benefited very much during the past years. The comments of K. Dettner, K. Fiedler, K.H. Hoffmann, G. Hübner, M. Mackauer, M. Romstöck-Völkl, B. Stadler, P. Starý, K.H. Tomaschko and W. Weisser on earlier drafts helped substantially to improve the manuscript.

References

Addicott JH (1978) Competition for mutualists: aphids and ants. Can J Zool 56:2093–2096

Addicott JH (1979) A multispecies aphid-ant association: density dependence and species-specific effects. Can J Zool 57:558–569

Banks CJ (1962) Effect of the ant, *Lasius niger*, on insects preying on small populations of *Aphis fabae* Scop. on bean plants. Ann Appl Biol 50:669–679

Börner CB (1952) Aphidae Europae Centralis. Mitt Thür Bot Ges 4:1–184

Bradley GA (1973) Effect of *Formica obsuripes* (Hymenoptera: Formicidae) on the predator-prey relationship between *Hyperaspis congressis* (Coleoptera: Coccinellidae) and *Toumeyella numismaticum* (Homoptera: Coccidae). Can Entomol 105:1113–1118

Breton LM, Addicott JF (1992) Does host-plant quality mediate aphid-ant mutualism? Oikos 63:253–259

Bristow C (1984) Differential benefits from ant-attendance to two species of Homoptera on New York ironweed. J Anim Ecol 53:715–726

Buckley R (1987) Interactions involving plants, Homoptera, and ants. Annu Rev Ecol Syst 18:111–135

Del-Claro K, Oliveira PS (1993) Ant-homopteran interaction: do alternative sugar sources distract tending ants? Oikos 68:202–206

Dixon AFG (1985) Aphid ecology. Blackie, Glasgow

Eisner T, Hicks K, Eisner M, Robson DS (1978) "Wolf-in-sheep's-clothing" strategy of a predaceous insect larva. Science 199:790–794

Fiedler K (1995) Lebenszyklen tropischer Bläulinge – von Interaktionen mit Ameisen geprägt. Rundgespr Kom Ökol 10:199–214

Fossel A (1972) Die Populationsdichte einiger Honigtauerzeuger und ihre Abhängigkeit von der Betreuung durch Ameisen. Waldhygiene 9:185–191

Hölldobler B, Wilson EO (1990) The ants. Springer, Berlin Heidelberg New York

Hübner G, Völkl W (1996) Behavioral strategies of aphid hyperparasitoids to escape aggression by honeydew-collecting ants. J Insect Behav 9:143–157

Jiggins C, Majerus MEN, Gough U (1993) Ant defence of colonies of *Aphis fabae* Scopoli (Hemiptera: Aphididae), against predation by ladybirds. Br J Entomol Nat Hist 6:129–137

Kroupa A (1995) Der Einfluß von Pflanzenstruktur, Wirtsart und Ameisenbelauf auf das Fouragierverhalten von *Pauesia silvestris* und *Pauesia pinicollis* (Hymenoptera: Aphidiidae) an Waldkiefern. Diploma Thesis, University of Bayreuth, Bayreuth

Liepert C, Dettner K (1993) Recognition of aphid parasitoids by honeydew-collecting ants: the role of cuticular lipids in a chemical mimicry system. J Chem Ecol 19:2143–2153
Liepert C, Dettner K (1995) Chemical mimicry of aphid parasitoids of the genus *Lysiphlebus* (Hymenoptera, Aphidiidae). Abstr XII Meet ISCE, Oct 2–6, 1995, Chile, p 58
Mackauer M, Stary P (1967) World Aphidiidae. Le Francois, Paris
Mackauer M, Völkl W (1993) Regulation of aphid populations by aphidiid wasps: does aphidiid foraging behaviour or hyperparasitism limit impact? Oecologia 94:339–350
Manéval H (1940) Observations sur un Aphidiidae (Hym.) myrmécophile. Description du genre et de l'éspèce. Bull Soc Linn Lyon 9:9–14
Maschwitz U, Schroth M, Manel M, Tho YP (1984) Lycaenids parasitizing symbiotic ant-plant partnerships. Oecologia 64:78–80
Maschwitz U, Nässig WA, Dumpert K, Fiedler K (1988) Larval carnivory and myrmecoxeny, and imaginal myrmecophily in miletine lycaenids (Lepidoptera: Lycaenidae) on the Malay peninsula. Tyô to Ga 39:167–181
Mason AC (1922) Life history studies of some Florida aphids. Fla Entomol 5:53–65
Mason RT, Fales HM, Eisner M, Eisner T (1991) Wax of a whitefly and its utilization by a chrysopid larva. Naturwissenschaften 78:28–30
Novak H (1994) The influence of an attendance or larval parasitism in hawthorn psyllids (Homoptera: Psyllidae). Oecologia 99:72–78
Pierce NE, Easteal S (1986) The selective advantage of attendant ants for the larvae of a lycaenid butterfly, *Glaucopsyche lygdamus*. J Anim Ecol 55:451–462
Scheurer S (1964) Zur Biologie einiger Fichten bewohnender Lachnidenarten (Homoptera, Aphidina). Z Angew Entomol 53:153–178
Scheurer S (1971) Der Einfluß der Ameisen und der natürlichen Feinde auf einige an *Pinus sylvestris* lebende Cinarinen in der Dübener Heide. Pol Pismo Entomol 41:197–229
Starý P (1966) Aphid parasites of Czechoslovakia. Academia, Praha
Starý P (1970) Biology of aphid parasites, with respect to integrated control. Ser Entomol 6. Dr W Junk, Den Haag
Starý P (1988) Aphidiidae. In: Minks AK, Harrewijn P (eds) Aphids. Their biology, natural enemies and control, vol 2B. Elsevier, Amsterdam, pp 171–184
Starý P, Völkl W (1988) Aggregations of aphid parasitoid adults (Hymenoptera, Aphidiidae). Z Angew Entomol 105:270–279
Stechmann DH, Völkl W, Starý P (1996) Ants as a critical factor in the biological control of the banana aphid *Pentalonia nigronervosa* in Oceania. J Appl Entomol 120:119–123
Sullivan DJ (1988) Hyperparasites. In: Minks AK, Harrewijn P (eds) Aphids. Their biology, natural enemies and control, vol 2B. Elsevier, Amsterdam, pp 189–203
Takada H (1983) Redescription and biological notes on *Protaphidius nawaii*. Kontyu 51:112–121
Takada H, Hashimoto Y (1985) Association of the root aphid parasitoids *Aclitus sappaphis* and *Paralipsis eikoae* (Hymenoptera, Aphidiidae) with the aphid-attending ants *Pheidole fervida* and *Lasius niger* (Hymenoptera, Formicidae). Kontyu 53:150–160
van den Bosch R, Hom R, Matteson P, Frazer BD, Messenger PS, Davis CS (1979) Biological control of the walnut aphid in California: impact of the parasite, *Trioxys pallidus*. Hilgardia 47:1–13
Vander Meer RK, Jouvenaz DP, Wojcik DP (1989) Chemical mimicry in a parasitoid (Hymenoptera: Eucharitidae) of fire ants (Hymenoptera: Formicidae). J Chem Ecol 15:2247–2261
Völkl W (1990) Fortpflanzungsstrategien von Blattlausparasitoiden (Hymenoptera, Aphidiidae): Konsequenzen ihrer Interaktionen mit Wirten und Ameisen. Diss University of Bayreuth, Bayreuth
Völkl W (1992) Aphids or their parasitoids: who actually benefits from ant-attendance? J Anim Ecol 61:273–281
Völkl W (1994) The effect of ant-attendance on the foraging behaviour of the aphid parasitoid *Lysiphlebus cardui*. Oikos 70:149–155
Völkl W (1995) Behavioural and morphological adaptations of the coccinellid *Platynaspis luteorubra* for exploiting ant-attended resources. J Insect Behav 8:653–670

Völkl W, Mackauer M (1993) Interactions between ants and parasitoid wasps foraging for *Aphis fabae* spp. *cirsiiacanthoidis* on thistles. J Insect Behav 6:301–312

Völkl W, Hübner G, Dettner K (1994) Interactions between *Alloxysta brevis* (Hymenoptera, Cynipoidea, Alloxystidae) and honeydew collecting ants; how an aphid hyperparasitoid overcomes ant aggression by chemical defense. J Chem Ecol 20:2621–2635

Völkl W, Liepert C, Birnbach R, Hübner G, Dettner K (1996) Chemical and tactile communication between the root aphid parasitoid *Paralipsis enervis* and trophobiotic ants: consequences for parasitoid survival. Experientia 52:731–738

Way MJ (1954) Studies on the association of the ant, *Oecophylla longinoda* (Latr.) (Formicidae) with the scale insect *Saissetia zanzibarensis* Williams (Coccidae). Bull Entomol Res 45:113–154

Way MJ (1963) Mutualism between ants and honeydew-producing homoptera. Annu Rev Entomol 8:307–344

Weisser WW, Völkl W (1996) Dispersal in the aphid parasitoid, *Lysiphlebus cardui*. J Appl Entomol 120 (in press)

Weisser WW, Houston AI, Völkl W (1994) Foraging strategies in solitary parasitoids: the trade-off between female and offspring mortality. Evol Ecol 8:587–597

Wojcik DP (1989) Behavioral interactions between ants and their parasites. Fla Entomol 72:43–51

Zwölfer H (1958) Zur Systematik, Biologie und Ökologie unterirdisch lebender Aphiden (Hom., Aphidoidea) (Anoeciinae, Tetraneurini, Pemphigini und Fordinae). Z Angew Entomol 43:1–52

14 The Relative Importance of Host Plants, Natural Enemies and Ants in the Evolution of Life-History Characters in Aphids

B. STADLER

14.1 Introduction

Many aphid species show tremendous fluctuations in numbers both within and between habitats or successive years (Redfearn and Pimm 1988). Their population dynamics seem mainly to be a consequence of exploiting resources that are highly variable in space and time. Environmental variability includes changes in host-plant quality, weather conditions, relationship to ants, and predator or parasitoid pressure. However, to what extent these factors contribute to the selection of life-history traits such as size at birth, length of survival and number of offspring, or to behavioural patterns, is uncertain (Morris 1992). Although behaviour and life history can be studied independently, much behaviour is directly linked to life-history characters (e.g. allocation processes, survival) and therefore should be included in any analysis of life-history strategies in variable environments.

An implicit assumption in life-history studies is that multiplication in number is advantageous. This is especially so for parthenogenetically reproducing aphids since reproduction is a dominant feature in their mode of life. However, fitness in the Darwinian sense also implies success over ecological and geological time (Wilson 1990). Therefore, if a clone becomes extinct it has zero fitness, no matter how many offspring it produced in earlier generations. Consequently, in the aphid case, success is not easy to identify since many aphid species show tremendous rates of increase when conditions are favourable (Dixon 1985) but there are also numerous reports of large population fluctuations and local extinctions when environmental conditions become adverse (e.g. Addicott 1978a; Morris 1992; Dennis and Sotherton 1994).

This paradigm makes it legitimate to ask whether it is always advantageous for an aphid clone to produce offspring at the physiological maximum possible at a particular time (sensu Kindlmann and Dixon 1989). A clone might also increase its overall fitness if it invests less energy in offspring, that is, shows a reduced numerical response even though environmental quality is high. More energy could then be allocated to the soma, enhancing its future survival over an unpredictable, suboptimal period and resuming offspring production if environmental conditions become better again. Host-plant quality and/or distribution/fragmentation (Cappuccino 1988; Dixon et al. 1993) are important features of environmental quality for aphids; but host-plant characters alone

Ecological Studies, Vol. 130
Dettner et al. (eds.) Vertical Food Web Interactions

might not sufficiently explain realised life-history traits of aphids in all circumstances, especially if aphid-ant relationships are involved or if natural enemies inflict a significant selection pressure. Whether such relationships are sufficiently strong or systematic to cause life-history variation in aphids is an important question; there appears to be no definite answer, but the conceptual tools necessary to answer the question can be identified.

This chapter first describes particular aspects of the plant-aphid and aphid-parasitoid/predator relationships with special reference to the aphid species *Uroleucon jaceae* (L.), which has been studied in our laboratory for several years. This aphid feeds almost exclusively on knapweed (*Centaurea jacea* L.) and is generally highly mobile but never tended by ants. Next specific aspects in the biology of aphids, important for aphid-ant relationships are highlighted, e.g. their degree of mobility, their strength of association with ants and life-history characters, from a review of the literature. The dynamic processes in aphid-ant relationships are described, focusing on the aphid species *Cinara pilicornis* (Hartig), which lives on Norway spruce [*Picea abies* (L.) Karst.] and is less mobile and facultatively tended by ants. Finally these findings are integrated in a description of the evolution of several life-history characters in terms of aphids' responses to their host plants, predators or ants. It is further concluded that investigations on life-history characters in clonal organisms, of which aphids are a good example, need to appreciate more deeply the dynamic relationships with the selective environment during the course of the complete life cycle, if the evolution of these traits in a heterogeneous environment is to be explained.

14.2 The Association Between Aphids and Their Host Plants

Aphids feed on the phloem sap of plants and differ from chewing herbivores in usually not damaging plant tissue, although necrotic reactions can occur. The chemical interaction with their hosts is believed to be more delicate compared to leaf feeders, because phloem sap does not contain the same concentrations of noxious chemicals as structural plant tissues (Raven 1983). However, the quantities of sap which pass through the digestive system of some species of aphids are considerable, e.g. adults of *Tuberolachnus salignus* (Gmelin) excrete 1.9 μl/h, which corresponds to 33% of their fresh mass (Mittler 1958). That is, although phloem sap contains only small amounts of noxious chemicals per unit volume, nevertheless, aphids are likely to have to cope in total with considerable quantities of these substances.

The study of complex relationships between plants and their associated herbivores has increased our understanding of the mechanisms and underlying processes (see, e.g. Harborne 1993; White 1993). Several review articles covering different aspects of plant-aphid interactions have been published (e.g. on host specificity, van Emden 1972; plant stress, Major 1990; host plant quality, Kidd 1985, Douglas 1993; plant chemistry, Dreyer and Campbell 1987,

Montllor 1991; physiological mechanisms, Whitham et al. 1991, Leather 1994; plant distribution, Cappuccino 1988, Dixon and Kindlmann 1990). Because of the wealth of information already available on these topics this chapter is restricted to the interactions between aphids and their host plants, corroborating the importance of plant chemistry/host quality and its variability in the evolution of life-history characters in aphids.

To investigate the feedback effects on the host plants, 40 seedlings of *C. jacea* were reared to the tillering stage (10–15 leaves developed) in pots filled with either high- or low-quality soil (Stadler 1992). After the initial growth period, 10 plants of each treatment were infested with a single adult aphid, which was allowed to reproduce freely. To other 10 plants of each treatment were used as controls and kept under identical abiotic conditions in a growth chamber (20 ± 2 °C; 65% RH; 18:6h photoperiod at 14000lx). Two months after infestation with aphids the plants were sampled and several measurements were made on the accumulated biomass and chemistry of the plants. During the experiment any alatae that developed were removed to prevent colonisation of the control plants.

During the 2-month feeding period, *U. jaceae* had a severe effect on the development of its host plant. Table 14.1 gives the biomass of leaves and roots of *C. jacea* and Table 14.2 the concentrations of amino acids as well as C/N ratios for plants grown in both high and low quality soil, or infested with aphids vs. control plants without aphids. Two results seem to be important. Infested hosts achieve lower growth rates than control plants and the nutritional supply of the plant also determines biomass accumulation and chemical profile. Aphid infestation and low-quality soil both resulted in a relative increase in root biomass. However, the effect of aphid feeding seems less severe if the host plants are grown in low-quality soil, as no statistically significant differences were found between infested and non-infested plants in the

Table 14.1. Biomass parameters for whole plants of *C. jacea* grown in high- and low-quality soil infested with *U. jaceae* for 2 months or grown without aphids (means ± SE)

Treatment	High quality			Low quality		
	Without aphids		With aphids	Without aphids		With aphids
n	10		10	10		9
Dry mass (g plant^{-1})						
Leaves	1.10 ± 0.06	*	0.62 ± 0.06	0.35 ± 0.03	*	0.18 ± 0.02
Roots	1.23 ± 0.09	*	0.93 ± 0.09	0.74 ± 0.05	*	0.58 ± 0.04
Total	2.33 ± 0.12	*	1.55 ± 0.14	1.09 ± 0.08	*	0.76 ± 0.06
Shoot/root	0.92 ± 0.06	*	0.68 ± 0.05	0.47 ± 0.03	*	0.32 ± 0.03
No. of leaves	13.70 ± 0.7	*	10.30 ± 0.75	9.60 ± 0.4	ns	8.22 ± 0.5
Length of leaves (cm)	12.90 ± 0.7	ns	12.97 ± 0.65	9.71 ± 0.5	ns	8.23 ± 0.5
Specific leaf mass (g m^{-2})	56.17 ± 1.6	*	45.10 ± 1.72	49.57 ± 1.7	ns	46.27 ± 2.0
Leaf area (cm^2)	196.36 ± 9.7	*	140.18 ± 15.56	69.06 ± 5.0	*	39.50 ± 3.5

Tests for significance refer to the Mann-Whitney U statistics for groups of plants with and without aphids. ns = not significant; * = $p < 0.05$; n = sample size.

Table 14.2. Contents of major free amino acids of leaves and C/N ratios for whole plants of *C. jacea* grown in high- and low-quality soil infested with *U. jaceae* for 2 months or grown without aphids (means ± SE)

Treatment	High quality Without aphids		With aphids	Low quality Without aphids		With aphids
n	10		10	10		9
Amino acids [μmol/gFM]						
asp	0.29 ± 0.05	ns	0.27 ± 0.06	0.07 ± 0.01	ns	0.08 ± 0.01
glu	0.71 ± 0.09	ns	0.52 ± 0.08	0.44 ± 0.06	*	0.26 ± 0.02
asn	0.57 ± 0.15	*	0.20 ± 0.06	0.13 ± 0.02	ns	0.12 ± 0.03
ser	0.13 ± 0.02	ns	0.10 ± 0.02	0.11 ± 0.01	ns	0.07 ± 0.01
gln	0.14 ± 0.03	*	0.06 ± 0.01	0.05 ± 0.01	ns	0.04 ± 0.00
gly	0.03 ± 0.00	*	0.02 ± 0.00	0.02 ± 0.00	*	0.01 ± 0.00
thr	0.08 ± 0.01	ns	0.06 ± 0.01	0.06 ± 0.01	*	0.03 ± 0.01
ala	0.01 ± 0.00	ns	0.00 ± 0.00	0.01 ± 0.00	*	0.00 ± 0.00
arg	0.37 ± 0.04	*	0.23 ± 0.04	0.23 ± 0.01	*	0.14 ± 0.02
C/N ratio						
Leaves	20.72 ± 0.94	ns	21.53 ± 0.60	25.24 ± 0.84	ns	27.18 ± 1.24
Roots	23.01 ± 1.62	ns	26.24 ± 2.42	52.01 ± 3.35	*	66.49 ± 4.71

Tests for significance refer to the Mann-Whitney U statistics for groups of plants with and without aphids. ns = not significant; * = $p < 0.05$; n = sample size.

number of leaves, leaf length or leaf mass. Uninfested plants have higher amino acid concentrations than infested plants, and well-nourished plants store more free amino acids in their leaves than plants grown in poor-quality soil. No particular amino acid seems to be used preferentially by the aphids.

Only one comparison between infested and uninfested plants of C/N ratio is statistically significant (Table 14.2). However, the trend is similar to that described for amino acids and biomass with infested and low-quality plants (especially their roots) showing higher C/N ratios, due to low N contents.

Severe effects of aphids on plants have been reported for several tree-feeding aphid species. Loss in biomass of Douglas-fir was shown to be as high as 50–75% (Johnson 1965) and the chemistry of conifers is particularly variable (Kidd 1985; Kidd et al. 1990). Dixon (1971a) reports a 62% reduction in wood formation and a 40% reduction in leaf size in *Acer pseudosplantanus* (L.) when infested with *Drepanosiphum plantanoides* (Schr.). The net primary production of leaves of saplings of *Tilia vulgaris* Hayne infested with aphids the previous year is 1.6 times greater than the leaves of previously uninfested saplings (Dixon 1971b).

Aphids counter deterioration in host quality by an array of physiological or behavioural mechanisms such as reduced investment in gonads relative to soma (Wellings et al. 1980), prolonged developmental times (e.g. Dixon 1985) or reduced mobility (Stadler et al. 1994). Thus, soil quality and aphids, in dynamically affecting plant quality, also affect aphid physiology.

Plant distribution and fragmentation might also influence aphid life-history strategies (Dixon et al. 1987, 1993). In a multispecies comparison it has been demonstrated that aphids feeding on climax plants usually invest more in reproduction than aphids feeding on herbaceous plants, presumably because of the higher predictability of climax plants in space and time.

Summarising, plants constitute a highly selective environment which shapes the reproductive characters, migratory behaviour and physiology of aphids. The aim of the following sections is to evaluate the importance of the features associated with plants relative to competition, predation and parasitism or ant-tending in shaping the evolution of the life-history characters of aphids.

14.3 The Influence of Predators and Parasitoids on the Behavioural Traits of Aphids

Many aphid species defend themselves by kicking, droplet secretion or walking away when attacked by an insect predator or parasitoid (Klingauf 1967). The interaction of aphids with a natural enemy is specific and dependent on the relative sizes of the predator and prey, angle of attack (Dixon 1958), relative age (Weisser 1994), degree of hunger (Carter and Dixon 1982) or plant architecture (Carter et al. 1984; Völkl and Stadler 1996) and plant distribution (Kareiva 1987; Cappuccino 1988), to name just a few.

In spite of this knowledge, it is difficult to assess the degree to which predators influence the population dynamics or the evolution of reproductive patterns in aphids. Several authors attribute pronounced effects to predators such as coccinellids (Kareiva and Odell 1987; Ives et al. 1993), carabids (Hance 1987), syrphids (Sanders 1979; Furuta 1988), chrysopids (McEwen et al. 1993), spiders (Scheurer 1964) or birds (Smith 1966), while others question whether predators have a significant affect on aphid abundance, at least when aphid populations are increasing exponentially (Liebig 1987; Kindlmann and Dixon 1993).

The following section demonstrates that the aphid *U. jaceae* shows a different response to the generalist aphid predator *Coccinella septempunctata* L. and the parasitoid *Aphidius funebris* Mackauer, which specialises on aphids in the genus *Uroleucon*. Details of the experimental setup are described in Stadler (1989).

14.3.1 Defence Behaviour Shown by *U. jaceae* to Attacks by *C. septempunctata*

The search behaviour of a coccinellid larva is directed by plant structure, which results in larvae quickly locating aphid colonies. This relationship between searching behaviour and plant structure is independent of whether prey is recognised only after physical contact (Dixon 1959; Banks 1957) or in response to visible or olfactory cues (Stubbs 1980). Figure 14.1A shows the responses of different aphid instars and adult aphids to the presence of *C. septempunctata*. For each aphid developmental stage the defence reactions sums to 100%. The youngest instar larvae (L1) show no defence reactions. With increase in age, however, the responses become more differentiated. The proportion that show no reaction decreases and that showing droplet secretion, kicking (body shake), dropping or walking away increases. However, the only responses that saved the aphids from being eaten are walking away or dropping off the host plant. These reactions were mainly shown by L3, L4 and adult aphids, and

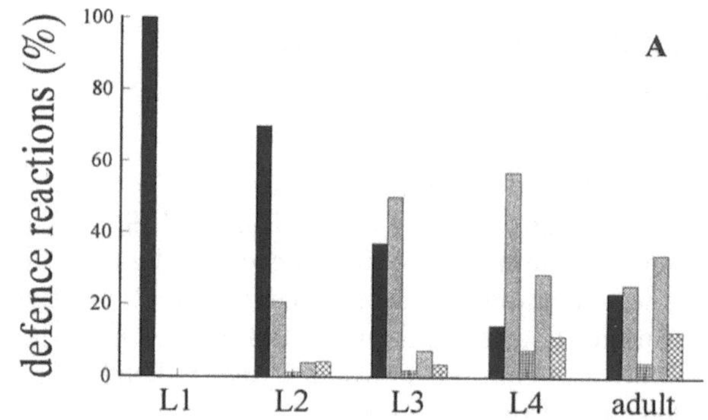

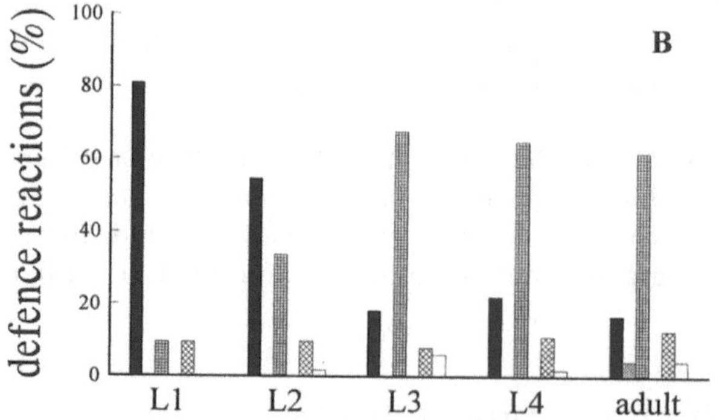

Fig. 14.1A,B. Defence reactions of *U. jaceae* towards attacks of A fourth instar larvae of *C. septempunctata* and B *A. funebris*. *L1*, *L2*, *L3*, *L4* First, second, third and fourth instar larvae. ■ no reaction; ▩ droplet secretion; ▩ kicking; ▩ dropping; ▩ walking; □ combinations of defence behaviours

rarely by L1 and L2 aphids ($Chi^2 = 36.20$; $df = 4$; $p < 0.001$; $n = 205$). Kicking and droplet secretion did not seem to affect predator efficiency.

14.3.2 Defence Behaviour Shown by *U. jaceae* to Attacks by *A. funebris*

In contrast to the defence shown by *U. jaceae* to coccinellids, the response to attacks of *A. funebris* is more diverse (Fig. 14.1B). As in their response to predators, young aphids show little reaction to parasites. In contrast, older aphids defend themselves by kicking, walking away, droplet secretion or a combination of kicking and droplet secretion. Again, L1 and L2 differ significantly from older aphids in their response to *A. funebris* ($Chi^2 = 65.793$; $df = 4$; $p < 0.001$; $n = 384$). The proportion that kicked increased from L1 to L3 and was the dominant response shown by L3, L4 and adult aphids. Interestingly, the dropping reaction was never used to escape ovipositing parasitoids.

14.4 Patterns of Aphid-Ant Associations

Information in the literature on more than 150 aphid species indicates that they show variation in the degree of mobility, ant-tending, intrinsic rate of increase (r_m) and colony structure (gregarious vs. more scattered). Unfortunately, not all of this information is available for all species, or available for comparable abiotic conditions. Especially r_m values differ greatly between studies and this restricts their usefulness in multispecies comparisons. Therefore, r_m values measured at 20 ± 1 °C have been selected. When possible, information was categorised according to, e.g. degree of mobility: weak (aphids do not leave their feeding site when disturbed), medium (aphids leave their feeding site by walking away only when repeatedly disturbed) or highly mobile (aphids walk away, drop off their host plant or fly away when disturbed); degree of ant-tending: never, facultative, obligate.

Table 14.3 is a summary of the strength of the aphid-ant associations and degree of mobility when disturbed shown by aphids of a range of families and subfamilies. Because these results are collected from different authors they may be biased.

Of all aphid species, 39.5% are never tended by ants, 30.6% are facultative myrmecophiles and 29.9% are obligate myrmecophiles. Aphids which differ in their degree of ant attendance also significantly differ in their degree of mobility ($Chi^2 = 112.499$, $df = 4$, $p < 0.0001$). Thus, mobility and association with ants are not independent; 75% of non-tended aphid species are highly mobile and all obligate myrmecophiles show poor degrees of mobility. Means of r_m values significantly differ for mobile and non-mobile species [r_m(mobile): 0.208, r_m(non-mobile): 0.288; M-W-test: $U = 25.0$, $W = 80.0$, $p = 0.035$, $n = 21$]. The tendency to form dense colonies, like *Aphis fabae* Scopoli or *Brachycaudus*

Table 14.3. Literature survey on the number of genera and species of aphids with different degrees of ant associations (never, facultative, obligate) and mobility (high, medium, low). Only those species are included for which sufficient information is available. Numbers in brackets indicate percentages for that family/subfamily. The classification follows Heie (1980)

Family/subfamily	Genera	Species	Ant-tending			Mobility		
			Never	Fac.	Obl.	High	Medium	Low
Pemphigidae	1	2	0 (0)	0 (0)	2 (100)	0 (0)	0 (0)	2 (100)
Lachnidae	6	32	7 (21.9)	14 (43.7)	11 (34.4)	7 (21.9)	9 (28.1)	16 (50.0)
Lachninac	3	9	0 (0.0)	3 (33.3)	6 (66.7)	0 (0)	2 (22.2)	7 (77.8)
Cinarinae	3	23	7 (30.5)	11 (47.8)	5 (21.7)	7 (30.4)	7 (30.4)	9 (39.2)
Thelaxidae	1	1	0 (0.0)	1 (100)	0 (0.0)	0 (0)	1 (0)	0 (0)
Drepanosiphidae	19	54	33 (62.3)	12 (22.6)	8 (15.1)	25 (49.0)	12 (23.5)	14 (27.5)
Chaitophorinae	2	14	3 (21.4)	5 (35.7)	6 (42.9)	1 (7.1)	5 (35.7)	8 (57.2)
Drepanosiphinae	2	6	6 (100)	0 (0)	0 (0)	6 (0)	0 (0)	0 (0)
Phyllaphidinae	15	34	24 (72.7)	7 (21.2)	2 (6.1)	18 (58.1)	7 (22.6)	6 (19.3)
Aphididae	28	73	22 (31.9)	21 (30.4)	26 (37.7)	17 (23.3)	7 (9.6)	49 (67.1)
Pterocommatinae	2	10	3 (30.0)	0 (0)	7 (70.0)	0 (0)	0 (0)	10 (100)
Aphidini	8	19	0 (0)	8 (44.4)	10 (55.6)	2 (10.5)	2 (10.5)	15 (79.0)
Macrosiphini	18	44	19 (46.3)	13 (31.7)	9 (22.0)	15 (34.1)	5 (11.4)	24 (54.5)
Total	55	162	62 (39.5)	48 (30.6)	47 (29.9)	49 (30.8)	29 (18.2)	81 (51.0)

Data from Addicott (1978b); Blackman and Eastop (1984, 1994); Dixon (1958); Kunkel and Kloft (1985); Nault et al. (1976); Way (1963); Zwölfer (1958).

cardui (L.), may enhance ant-tending. The literature supports this view, as dense colonies are characteristic of all (100%, n = 47) species that are closely associated with ants, by 89% (n = 43) of the facultatively tended species but only by 27% (n = 16) of the non-tended species.

An interesting group of aphids are the facultative myrmecophiles, which are not always ant-attended. They may respond to the presence or absence of ants indirectly via the mortality pressures imposed by predators or parasitoids.

14.5 Dynamics in the Aphid-Ant Relationships and Possible Life-History Effects for Tended Aphids

Aphid-ant interactions fall into two extreme classes. Either ants prey on aphids or they tend them and collect honeydew (Buckley 1987). If aphids are tended, they may be sheltered or defended from natural enemies by the ants. However, the significance of ant attendance in terms of aphid population increase differs between species (Sudd 1987). In many cases, tending seems to benefit aphids as it has been demonstrated that the aphid colonies achieve a larger size (Skinner and Whittaker 1981; Fowler and Macgarvin 1985), are protected against predators (see, e.g. Buckley 1987) and show increased rates of feeding and excretion (Herzig 1937); aphid colonies are cleaned by the ants removing honeydew (Nixon 1951) or they produce fewer winged offspring (Johnson 1959). Ants, in contrast, may benefit from the easy access to a carbon-rich food source. However, the aphid-ant association can vary during the course of a growing season (Addicott 1979), indicating some form of cost for tending ants or effects on the life history of the aphids.

To analyse the seasonal aspects of aphid-ant relationships, the degree of ant-tending of *C. pilicornis* feeding on 3-year-old seedlings of Norway spruce was studied. Although *C. pilicornis* is classified as non-tended according to Scheurer (1964), fundatrices of this species and colonies of second-generation individuals were frequently tended by *Formica polyctena* (Först.) and *Formica fusca* L. at sampling sites in northern Bavaria, which indicates that this aphid is at least a facultative myrmecophile. The spruce seedlings were grown in pots at three locations (A–C) in close vicinity to each other. The three locations allowed the ants different access to alternative honeydew sources. In plot A ants were excluded from tending by placing the ten potted plants in vessels filled with water. In plot B the ten plants were placed in the vicinity of *L. niger* colonies. Plot A and B were in a meadow close to the institute. Plot C again consisted of ten plants placed on a sun terrace which was accessible to *L. niger*. Except for the experimental plants, no other plants were present on the sun terrace, so the experimental aphids were the only honeydew source at this location. In early May, each plant was infested with a single adult aphid of *C. pilicornis*, which was allowed to multiply freely. Every 2 weeks the number of tending ants was counted at midday. At the end of the season the number of eggs at each location was counted.

Figure 14.2 shows the seasonal change in the numbers of tending ants on the plants in the meadow (plot B) and terrace (plot C) locations. In May and June, aphids at both plots were readily visited by *L. niger*. From mid-June to mid-

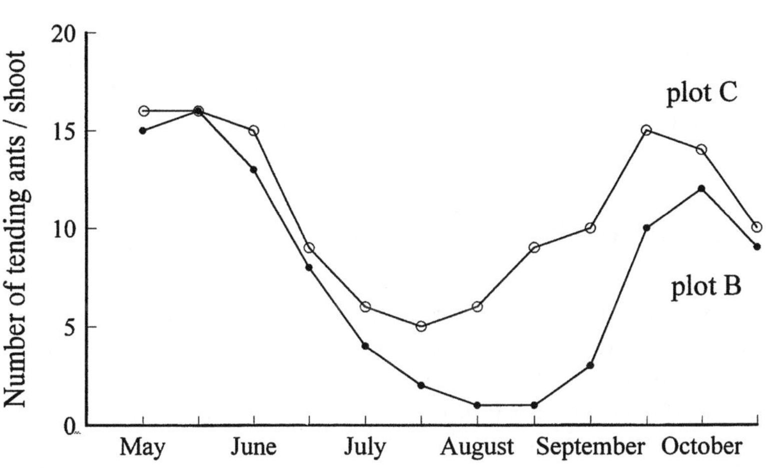

Fig. 14.2. Number of *L. niger* workers on shoots of 3-year-old Norway spruce seedlings infested with *C. pilicornis*. *Plot B* (●) consisted of ten potted plants located in the field (alternative sugar sources available); *plot C* (○) comprised ten plants on the sun terrace of the institute with the honeydew of *C. pilicornis* being the sole sugar source

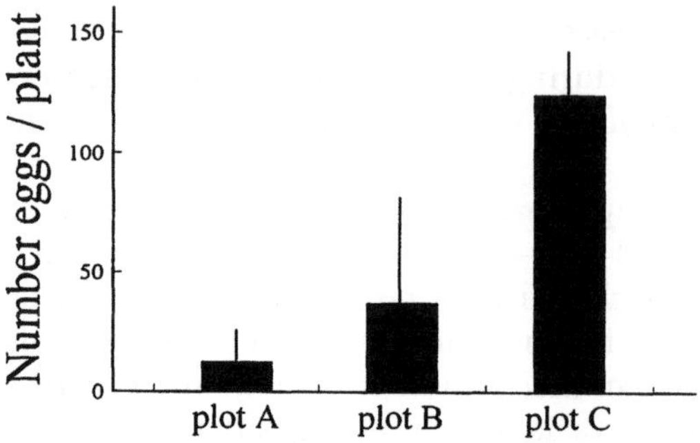

Fig. 14.3. Number of eggs of *C. pilicornis* found on 3-year-old Norway spruce seedlings at the end of October. *Plot A* was kept free of tending aphids; *plots B* and *C* as above; *vertical lines* standard deviation

July, tending intensity decreased at both plots. From July to the beginning of September, those aphids feeding on plants in plot B were rarely visited by ants but they were increasingly visited from mid-September, when sexuparae and oviparae were present. In plot C, where *C. pilicornis* was the only sugar source, tending increased again in mid-August after a short summer decrease. The decline in ant-tending at the end of October is probably due to the death of the oviparae after depositing their eggs on current-year needles.

Figure 14.3 shows the mean number of eggs on the plants at the end of October. At the location where ants were excluded (plot A), a mean of 12.3 eggs was found per plant. However, these eggs did not originate from the descendants of the original aphid colonist but from the descendants of alates of *C. pilicornis* that recolonised the plants from mid-June to July. The original colonies were completely destroyed, mainly by chrysopids and spiders. As a result, the original clone left no offspring (accumulated zero fitness) when ant-tending was prevented. In the plot with alternative sources of sugar (plot B), there was a mean of 37.1 eggs per plant, which represents 66.9% of all eggs left by the descendants of the clone which infested the plants in early May. Aphids in plot C laid an average of 124.4 eggs per plant (90.1% of all eggs might be attributed to the original clone) and this treatment differed significantly from both others K-W test: $Chi^2 = 10.685$, $p = 0.004$, $n = 30$).

14.6 Synthesis and Conclusions

Most aphids can have prodigious rates of increase, are host-specific and show marked phenotypic plasticity. They face selection pressures from different trophic levels, e.g. plant chemical defence substances (Dorschner 1990; Whitham et al. 1991), competition for resources, tending ants, (Addicott 1978b), or attack by natural enemies. To rank these in order of their importance as selective forces of the biological environment is difficult, as single optimal solutions are not expected to be found in a heterogeneous environment (Sibly 1995) and historical constraints are likely to be involved. Parthenogenetic reproduction and telescoping of generations (Dixon 1985) might be an adaptation to feeding on climax hosts, with reproduction early in adult life as a means of outcompeting conspecific or heterospecific species of aphids. Parthenogenetic reproduction possibly evolved 200 million years ago in the Triassic (Dixon 1985; Heie 1987) and sap-feeding probably is an early form of phytophagy (e.g. Zwölfer 1978; Chaloner et al. 1991). The long evolutionary association with particular groups of plants (especially woody hosts: Heie 1994) makes it likely that the most important biotic factor in the evolution of the life-history characters of aphids is their hosts, especially their nutritional quality, biochemistry (Dreyer and Campbell 1987; Montllor 1991; White 1993) or distribution (Dixon and Kindlmann 1990).

If an aphid has to leave its host plant, it is more likely to be successful in locating a new host if it can direct more energy to locomotive structures (e.g. wing muscles or soma) because during migration it cannot feed. Therefore, the ability to manipulate resource allocation might have evolved very early in aphids. However, what factors influence the willingness to migrate and ultimately the population dynamics of aphids is unclear. Dixon (1985, 1990) suggested that intraspecific competition for phloem sap may be important at least in deciduous tree-dwelling aphids and there is some evidence, e.g. from studies on the sycamore aphid, to support this idea (Chambers et al. 1985). Consequently, migratory behaviour might be a response to periodic changes in host-plant quality. However, the degree to which competition for phloem sap is a prevailing force for evolutionary change relative to predator pressure is still an open question; migratory forms invest less in reproduction (Dixon et al. 1993) plus showing a delayed onset to reproduction (Newton and Dixon 1990). In addition to adaptations shown by *U. jaceae* to seasonal changes in host-plant quality (Stadler 1992), this aphid induces changes in the growth and amino acid composition of its host plant (Tables 14.1, 14.2), which make the habitat even more unpredictable. This uncertainty is thought to be a major reason for the evolution of clonal phenotypic plasticity (Hoffmann and Parsons 1994).

With the major diversification of the angiosperm flora in the Early Cretaceous, changes in terrestrial ecosystems occurred (Heie 1987; Crane et al. 1995). The diversification of phytophagous insects induced a concomitant diversification of entomophagous insects (Zwölfer 1978) and natural enemies probably have become a more important force in the evolution of life-history characters and the behaviour of aphids. The evolution of soldiers, which appears to have happened at least four times in aphids (Aoki 1987), or successful biocontrol programs, is visible evidence of the selection pressures exerted by natural enemies. Figure 14.1A,B illustrates the different behavioural responses shown by *U. jaceae* to attacks by predators and parasitoids, with the dropping reaction shown only in response to predators. These differences could be due to the different attack strategies of predators and parasitoids or simply be a consequence of differences in size/speed of approach of predators and parasitoids. However, a controlled or induced usage of different escape and defence behaviours dependent on the nature of the predator can also demonstrate the existence of differential selection pressures.

If predators are a significant selective force, individuals should be favoured which maximise the ratio of reproductive potential, r, to attack rate of predators or parasitoids, a (Oksanen 1992). The ratio can be maximised either by increasing r of by reducing a. However, r and a might not be optimised independently. For aphids, a reduced investment in gonads should have fitness advantages, as it has been demonstrated that a high reproductive investment incurs costs in terms of survival if food uptake is interrupted (Ward et al. 1983; Stadler 1995). In this case, selection for early reproduction should be reversed. Consequently, an index such as potential fecundity cannot be equated with fitness if few or no aphids ever achieve this potential or achieve it only when host-plant quality is optimal. Houston and McNamara (1992) provide a solid foundation for evolutionary analysis by showing how the rate of population increase depends on the full range of habitats encountered by an asexually reproducing population. Their approach gives less importance to current reproduction as opposed to survival. In good habitats it is important to produce offspring quickly because of high growth rates there, but emphasis on rapid growth is less severe when offspring find themselves in different habitats than their parents. Producing ovipara and laying some eggs early in the season (as *C. pilicornis* does) might be another bet-hedging strategy reducing predator pressure not intra- or interspecific competition.

The literature indicates that roughly one third of the surveyed aphid species are not ant-tended, another third are facultative myrmecophiles and the last third obligate myrmecophiles. Ant-tending seems to be associated with lower degrees of mobility, and non-mobile aphids have higher intrinsic rates of increase than highly mobile species. Although three categories of ant-tending may capture the essence of this relationship and its evolutionary significance, Fig. 14.2 shows that the process is a dynamic one, influenced by season (e.g. low tending during summer) and habitat (availability of alternative sugar sources). Aphids feeding in habitats with few alternative sugar sources (plot C) are more likely to be tended by ants during summer, and the ants also seem to be more aggressive towards intruding natural enemies. The most important consequence is a more than 90% increase in egg number deposited by aphids on ant-attended trees (plot C) compared to the non-tended aphids (plot A) (Fig. 14.3). The eggs found on plants of plot A most likely originated from immigrating alates, recolonizing plot A in June/July. That is, the original clone left no offspring. Similar results were reported by Tilles and Wood (1982) showing that *Cinara occidentalis* (Davidson) was more likely to survive and produce oviparae when tended by *Camponotus modoc* Wheeler. However, tending ants do not necessarily provide absolute defence for aphids, especially as natural enemies can circumvent ant attacks behaviourally or chemically (Völkl 1992; Völkl and Mackauer 1993).

If environmental change is unpredictable, then aphids are unlikely to be selected to maximise r. The genotype most likely to survive is the one that can endure an array of environments, even if its performance is suboptimal in some of these environments. Successful aphid genotypes or phenotypes need not be optimal in an absolute sense but must be able to reproduce and leave surviving offspring under a variety of circumstances. Therefore, more emphasis should be laid on determining ranges of phenotypic responses to environmental variations (e.g. r_m values, abilities to survive, degrees of mobility, fluctuations in ant attendance) and analyse clonal performance along gradients of different environmental configurations. For aphids, the overal fitness of a clone is central when evaluating the evolution of life-history characters or the strength of different selection pressures.

Acknowledgements. I am very grateful to Tony Dixon, Gerhard Hübner and Wolfgang Völkl for constructive criticisms on earlier drafts of the manuscript. To Helmut Zwölfer I express my deepest gratitude for the enjoyable and rewarding years spent in his lab. His support and helpful advice over all the years provided the setting for a successful development. Financial support was given by the German Ministry for Research and Technology (Fördernummer: BMBF No. PT BEO 51-0339476B).

References

Addicott JF (1978a) The population dynamics of aphids on fireweed: a comparison of local populations and metapopulations. Can J Zool 56:2554–2564

Addicott JF (1978b) Competition for mutualists: aphids and ants. Can J Zool 56:2093–2096

Addicott JF (1979) A multispecies aphid-ant association: density dependence and species-specific effects. Can J Zool 57:558–569

Aoki S (1987) Evolution of sterile soldiers in aphids. In: Itô Y, Brown JL, Kikkawa J (eds) Animal societies: theories and facts. Japan Scientific Society Press, Tokyo, pp 53–65

Banks CJ (1957) The behaviour of individual coccinellid larvae on plants. Anim Behav 5:12–24

Blackman RL, Eastop VF (1984) Aphids on the world's crops. Wiley, Chichester

Blackman RL, Eastop VF (1994) Aphids on the world's trees. CABI, Wallingford

Buckley R (1987) Ant-plant-homopteran interactions. Adv Ecol Res 16:53–85

Cappuccino N (1988) Spatial patterns of goldenrod aphids and the response of enemies to patch density. Oecologia 76:607–610

Carter MC, Dixon AFG (1982) Habitat quality and the foraging behaviour of coccinellid larvae. J Anim Ecol 51:865–878

Carter MC, Sutherland D, Dixon AFG (1984) Plant structure and the searching efficiency of coccinellid larvae. Oecologia 63:394–397

Chaloner WG, Scott AC, Stephenson J (1991) Fossil evidence for plant-arthropod interactions in the Palaeozoic and Mesozoic. Philos Trans R Soc Lond 333:177–186

Chambers RJ, Wellings PW, Dixon AFG (1985) Sycamore aphid numbers and population density. II. Some Processes. J Anim Ecol 54:425–442

Crane PR, Friis EM, Pedersen KR (1995) The origin and early diversification of angiosperms. Nature 374:27–33

Dennis P, Sotherton NW (1994) Behavioural aspects of the staphylinid beetles that limit their aphid feeding potential in cereal crops. Pedobiologia 38:222–237

Dixon AFG (1958) The escape responses shown by certain aphids to the presence of the conccinellid *Adalia decempunctata* (L.). Trans R Entomol Soc Lond 110:319–334

Dixon AFG (1959) An experimental study of the searching behaviour of the predatory cocinellid beetle *Adalia decempunctata* (L.). J Anim Ecol 28:259–281

Dixon AFG (1971a) The role of aphids in wood formation. I. The effect of the sycamore aphid, *Drepanosiphum plantanoides* (Schr.) (Aphididae), on the growth of sycamore, *Acer pseudoplantanus* (L.). J Appl Ecol 8:165–179
Dixon AFG (1971b) The role of aphids in wood formation. II. The effect of the lime aphid, *Eucallipterus tiliae* L. (Aphididae), on the growth of lime, *Tilia vulgaris* Hayne. J Appl Ecol 8:393–399
Dixon AFG (1985) Aphid ecology. Blackie, Glasgow
Dixon AFG (1990) Population dynamics and abundance of deciduous tree-dwelling aphids. In: Watt AD, Leather SH, Hunter MA, Kidd NAC (eds) Population dynamics of forest insects. Intercept, Hampshire, pp 11–23
Dixon AFG, Kindlmann P (1990) Role of plant abundance in determining the abundance of herbivorous insects. Oecologia 83:281–283
Dixon AFG, Kindlmann P, Lepš J, Holman J (1987) Why there are so few species of aphids, especially in the tropics. Am Nat 129:580–592
Dixon AFG, Horth S, Kindlmann P (1993) Migration in insects: costs and strategies. J Anim Ecol 62:182–190
Dorschner K (1990) Aphid induced alternation of the availability and form of nitrogenous compounds in plants. In: Campbell RK, Eikenbayry RD (eds) Aphid-plant genotype interactions. Elsevier, Amsterdam, pp 225–235
Douglas AE (1993) The nutritional quality of phloem sap utilized by natural aphid populations. Ecol Entomol 18:31–38
Dreyer DL, Campbell US (1987) Chemical basis of host-plant resistance to aphids. Plant Cell Environ 10:353–361
Fowler SV, MacGarvin M (1985) The impact of hairy wood ants, *Formica lugubris*, on the guild structure of herbivorous insects on birch, *Betula pubescens*. J Anim Ecol 54:847–855
Furuta K (1988) Annual alternating population size of the thuja aphid, *Cinara tujafilina* (Del Guercio), and the impacts of syrphids and disease. J Appl Entomol 105:344–354
Hance T (1987) Predation impact of carabids at different population densities on *Aphis fabae* development in sugar beet. Pedobiologia 30:251–262
Harborne JB (1993) Introduction to ecological biochemistry. Academic Press, Oxford
Heie OE (1980) The Aphidoidea (Hemiptera) of Fennoscandia and Denmark. Fauna Entomol Scand 9:1–236
Heie OE (1987) Palaeontology and phylogeny. In: Minks AK, Harrewijn P (eds) Aphids; their biology, natural enemies and control, vol 2A. Elsevier, Amsterdam, pp 367–391
Heie OE (1994) Aphid ecology in the past and a new view on the evolution of Macrosiphini. In: Leather SR (ed) Individuals, populations and patterns in ecology, Intercept, Hampshire, pp 409–418
Herzig J (1937) Ameisen und Blattläuse. Z Angew Entomol 24:367–435
Hoffmann AA, Parsons PA (1994) Evolutionary genetics and environmental stress. Oxford Univ Press, Oxford
Houston AI, McNamara JM (1992) Phenotypic plasticity as a state-dependent life-history decision. Evol Ecol 6:243–253
Ives AR, Kareiva P, Perry R (1993) Response of a predator to variation in prey density at three hierarchical scales: lady beetles feeding on aphids. Ecology 74:1929–1938
Johnson B (1959) Ants and form reversal in aphids. Nature 184:740–741
Johnson N (1965) Reduced growth associated with infestations of Douglas-fir seedlings by *Cinara* species (Homoptera: Aphidae). Can Entomol 97:113–119
Kareiva P (1987) Habitat fragmentation and the stability of predator-prey interactions. Nature 326:388–391
Kareiva P, Odell G (1987) Swarms of predators exhibit "preytaxis" if individual predators use area-restricted search. Am Nat 130:233–270
Kidd NAC (1985) The role of the host plant in the population dynamics of the large pine aphid, *Cinara pinea*. Oikos 44:114–122

Kidd NAC, Smith SDJ, Lewis GB, Carter CI (1990) Interactions between host-plant chemistry and the population dynamics of conifer aphids. In: Watt AD, Leather SH, Hunter MA, Kidd NAC (eds) Population dynamics of forest insects. Intercept Hampshire, pp 183–193

Kindlmann P, Dixon AFG (1989) Developmental constraints in the evolution of reproductive strategies: telescoping of generations in the parthenogenetic aphids. Funct Ecol 3:531–537

Kindlmann P, Dixon AFG (1993) Optimal foraging in ladybird bettles (Coleoptera: Coccinellidae) and its consequences for their use in biological control. Eur J Entomol 90:443–450

Klingauf F (1967) Abwehr- und Meidereaktionen von Blattläusen (Aphididae) bei Bedrohung durch Räuber und Parasiten. Z Angew Entomol 59/60:277–317

Kunkel H, Kloft WJ (1985) Die Honigtau-Erzeuger des Waldes. In: Kloft WJ, Kunkel H (eds) Waldtracht und Waldhonig in der Imkerei. Ehrenwirth, München, pp 48–265

Leather SR (1994) Life-history traits of insect herbivores in relation to host quality. In: Bernays E. (ed) Insect-plant interactions. CRC Press, Boston, pp 175–207

Liebig G (1987) Der Massenwechsel der grünen Tannenhoniglaus *Cinara pectinatae* in verschieden stark erkrankten Tannenbeständen 1977–1985. Apidologie 18:147–162

Major EJ (1990) Water stress in Sitka spruce and its effect on green spruce aphid *Elatobium abietinum*. In: Watt AD, Leather SH, Hunter MA, Kidd NAC (eds) Population dynamics of forest insects. Intercept, Hampshire, pp 85–93

McEwen PK, Clow S, Jervis MA, Kidd NAC (1993) Alternation in searching behaviour of adult female green lacewings *Chrysoperla carnea* (Neur.: Chrysopidae) following contact with honeydew of the black scale *Saisseta oleae* (Hom.: Coccidae) and solutions containing acid hydrolysed L-tryptophan. Entomophaga 38:347–354

Mittler TE (1958) The excretion of honeydew *Tuberolachnus salignus* Gmelin. Proc R Entomol Soc Lond 33:49–55

Montllor CB (1991) The influence of plant chemistry on aphid feeding behaviour. In: Bernays E (ed) Insect-plant interactions, vol III. CRC Press, Boston, pp 125–173

Morris WF (1992) The effects of natural enemies, competition, and host-plant water availability on an aphid population. Oecologia 90:359–365

Nault LR, Montgomery ME, Bowers WS (1976) Ant-aphid association: role of alarm pheromone. Science 192:1349–1351

Newton C, Dixon AFG (1990) Pattern of growth in weight of alate and apterous nymphs of the English grain aphid, *Sitobion avenae*. Entomol Exp Appl 55:231–238

Nixon GEJ (1951) The association of ants with aphids and coccids. London Comm Inst Entomol, London, pp 1–36

Oksanen L (1992) Evolution of exploitation ecosystems I. Predation, foraging ecology and population dynamics in herbivores. Evol Ecol 6:15–33

Raven JA (1983) Phytophages of xylem and phloem: a comparison of animal and plant sapfeeders. Adv Ecol Res 13:136–234

Redfearn A, Pimm SL (1988) Population variability and polyphagy in herbivorous insect communities. Ecol Monogr 58:39–55

Sanders W (1979) Das Eiablageverhalten der Schwebfliege *Syrphus corollae* Fabr. in Abhängigkeit von der Größe der Blattlauskolonie. Z Angew Entomol 66:217–232

Scheurer S (1964) Zur Biologie einiger Fichten bewohender Lachnidenarten (Homoptera, Aphidina). Z Angew Entomol 53:153–178

Sibly RM (1995) Life-history evolution in spatially heterogeneous environments, with and without phenotypic plasticity. Evol Ecol 9:242–257

Skinner GJ, Whittacker JB (1981) An experimental investigation of interrelationships between the wood ant (*Formica rufa*) and some tree-canopy herbivores. J Anim Ecol 50:313–326

Smith BD (1966) Effects of parasites and predators on a natural population of the aphid *Acyrthosiphon spartii* (Koch) on broom (*Sarothamnus scoparius* L.). J Anim Ecol 35:255–267

Stadler B (1989) Untersuchungen zur Populationsökologie von *Uroleucon jaceae* (L.) (Homoptera, Aphididae) an Centaureen in Oberfranken. Diploma Thesis, University of Bayreuth, Bayreuth

Stadler B (1992) Physiological responses of *Uroleucon jaceae* (L.) to seasonal changes in the quality of its host plant *Centaurea jacea* L.: multilevel control of adaptations to the life cycle of the host. Oecologia 91:273–280
Stadler B (1995) Adaptive allocation of resources and life-history trade-offs in aphids relative to plant quality. Oecologia 102:246–254
Stadler B, Weisser WW, Houston AI (1994) Defence reactions in aphids: the influence of state and future reproductive success. J Anim Ecol 63:419–430
Stubbs M (1980) Another look at prey detection by coccinellids. Ecol Entomol 5:179–182
Sudd JH (1987) Ant-aphid mutualism. In: Minks AK, Harrewijn P (eds) Aphids: their biology, natural enemies and control, vol A. Elsevier, Amsterdam, pp 355–356
Tilles DA, Wood DL (1982) The influence of carpenter ant (*Camponotus modoc*) (Hymenoptera: Formicidae) attendance on the development and survival of aphids (*Cinara* spp.) (Homoptera: Aphididae) in a giant Sequoia forest. Can Entomol 114:1133–1142
van Emden HF (1972) Aphids as phytochemists. In: Harborne JB (ed) Phytochemical ecology. Academic Press, London, pp 25–43
Völkl W (1992) Aphids or their parasitoids: who actually benefits from ant attendance? J Anim Ecol 61:273–281
Völkl W, Mackauer M (1993) Interactions between ants attending *Aphis fabae* ssp. *cirsiiacanthoidis* on thistles and foraging parasitoid wasps. J Insect Behav 6:301–312
Völkl W, Stadler B (1996) Colony orientation and successful defence behaviour in the conifer aphid, *Schizolachnus pineti*. Entomol Exp Appl 78:197–200
Ward SA, Wellings PW, Dixon AFG (1983) The effect of reproductive investment on pre-reproductive mortality in aphids. J Anim Ecol 52:305–313
Way MJ (1963) Mutualism between ants and honeydew producing Homoptera. Annu Rev Entomol 8:307–344
Weisser WW (1994) Age-dependent foraging behaviour and host-instar preference of the aphid parasitoid *Lysiphlebus cardui*. Entomol Exp Appl 70:1–10
Wellings PW, Leather SR, Dixon AFG (1980) Seasonal variation in reproductive potential: a programmed feature of aphid life cycles. J Anim Ecol 49:975–985
White TCR (1993) The inadequate environment. Nitrogen and the abundance of animals. Springer, Berlin Heidelberg New York
Whitham TG, Maschinski J, Larson KC, Paige KN (1991) Plant responses to herbivory: The continuum from negative to positive and underlying physiological mechanisms. In: Price PW, Lewinsohn TM, Fernandes GW, Benson WW (eds) Plant-animal interactions: evolutionary ecology in tropical and temperate regions. Wiley, New York, pp 227–256
Wilson EO (1990) Success and dominance in ecosystems: The case of the social insects. In: Kinne OE (ed) Excellence in ecology. Ecology Institute, Oldendorf/Luhe
Zwölfer H (1958) Zur Systematik, Biologie und Ökologie unterirdisch lebender Aphiden. Z Angew Entomol 42:129–172
Zwölfer H (1978) Mechanismen und Ergebnisse der Co-Evolution von phytophagen und entomophagen Insekten und höherer Pflanzen. Sonderbd Naturwiss Ver Hamb 2:7–50

Part E

Community Organization and Diversity in Multitrophic Terrestrial Systems

15 Diversities of Aphidopha in Relationship to Local Dynamics of Some Host Alternating Aphid Species

D.-H. STECHMANN

15.1 Introduction

Aphids are about the oldest group of strictly phytophagous insects, yet they have established only a comparatively small number of species. Regional aphid faunas comprise a few hundred species at the most, with exceptionally low species richnesses in the tropics (Wolda 1979; Blackman and Eastop 1985, 1994; Dixon et al. 1987). On the other hand, aphids on plants are vertically related to a very large number of other insect species.

There are specialized Aphidophaga which reproduce and develop on aphids as food only. Members of the insect orders of Heteroptera, Dermaptera, Neuroptera, Coleoptera and Diptera are aphid predators (Frazer 1988) or aphid-specific parasitoids (Hymenoptera: Aphidiidae and Aphelinidae) (Starý 1988a,b). Furthermore, there are many other insect taxa that can be associated with aphids via honeydew (liquid faeces of Homoptera containing mainly water, sugars and some amino acids), e.g. adults of some Aphidophaga, as well as some Hymenoptera, Lepidoptera and Diptera in particular. Basically, honeydew-feeding is to be treated as a non-interactive food linkage, yet many of the honeydew-taking insects, like some Aphidophaga, as well as aphid-attending ants, can have a strong impact on insect assemblages associated with aphid host plants.

Although only a small fraction of the aphid species are rated as pests, they are among the most difficult insects to control in agriculture and forestry. From applied research on aphids, substantial knowledge has been gathered on aphid biology, host-plant relationship, impact of natural enemies, and population dynamics (see Minks and Harrewijn 1987, 1988 for a review). Little attention, however, has been given to the more general research approaches of the underlying community ecology and its relevance for biological control.

During investigations carried out at Bayreuth University, northern Bavaria, Germany, I had the opportunity to study various homopteran-based food webs in central Europe. I was particularly interested in the role that local processes play in the structuring of aphid-based vertical webs in both crops and adjacent habitats. The present chapter will use data gathered during studies in the vicinity of Bayreuth on the insect fauna of hedges, with special emphasis on cereal aphids, e.g. the relevance of hedge-field interactions for aphid

Ecological Studies, Vol. 130
Dettner et al. (eds.) Vertical Food Web Interactions

control. It can be shown that these vertical webs are predictable at the level of functional groups within regional species pools, yet the diversities are locally unpredictable due to the structuring forces acting within and between local food sources.

15.2 Sampling and Patterns of Local Cereal Aphid Populations

A detailed report on sites studied, methods employed, and results obtained for a wide range of spiders and insects associated with hedges and fields is available in Zwölfer et al. (1984). Details on landscape diversity of the region are given by Schulze and Gerstberger (1993). Some information, however, will be mentioned here for making the sampling methods in the cereal aphid-Aphidophaga associations available.

The investigations were carried out at two sites near Bayreuth differing significantly in hedge densities: (1) Hummelgau, including a river border at Thurnau: a hillsite covering at least 1000 ha of meadows and cereal fields in the black Jurassic loams, repeatedly cleared from hedges, with low field hedge density of <5 m of hedges per hectare, and (2) Stadtsteinach, including a river border at Neuenmarkt: an area of similar size and land use in the limestone, particularly rich in field hedges that grow on stone walls, with high field hedge density of >30 m of hedges per hectare. Field size is generally small (<2 ha per field), the cropping system is of low intensity due to the mountainous environment and the smallholder community of farmers. Cereal production is low (<40 dt/ha), insecticides for cereal aphid control are not used at all.

Aphids and Aphidophaga were assessed by two complementary methods, e.g. shoot or tiller counts, and beating or sweep-netting, respectively. For comparability, 1 m^2 of both shrub canopy and cereal fields were sampled, with 10–20 subsamples each taken randomly. Overwintering eggs were counted in February, spring morphs and summer morphs from the beginning of April to the end of October at weekly intervals, for three to four seasons (1978–1981).

The four cereal aphid species investigated are common and widely distributed species (Dixon 1987), with a holocyclic development in the region studied. The host-plant ranges are extremely narrow, as far as winter host plants are concerned, and extremely wide in the summer host plants. The heteroecious species *Metopolophium dirhodum* and *Sitobion fragariae* hibernate in the egg stage on wild roses, the *Rosa canina* agg. in particular. The third heteroecious species, *Rhopalosiphum padi*, hibernates exclusively on *Prunus padus*. The fourth aphid species, *Sitobion avenae*, is monoecious and hibernates on Poaceae.

In the life cycles of holocyclically developing aphids, three major types of population patterns can be distinguished:

1. Stationary populations: overwintering eggs on winter host plants; there is no immigration, emigration and growth, yet due to predation on winter eggs there is substantial mortality as a function of time.
2. Non-equilibrium populations in the sense of Price (1980): populations on growing plant parts, established either by stem mothers hatched from winter eggs, or from alatae females immigrated from other plants and/or other habitats, with exponential growth during a period of three to five aphid generations, ending with complete emigration from the host plant by winged aphid morphs. Non-equilibrium populations in the system studied are established for no longer than 6–8 weeks, depending on host-plant quality and population growth (intraspecific competition and morph determination). There is, however, seasonal overlap between local population phases of the same aphid on different host-plant species on Poaceae.
3. Aerial populations: winged aphids dispersing in the air obligatorily migrate for some hours up to some days, before settling again on a suitable host plant (e.g. Klingauf 1987). Aphids in the air are randomly distributed (Taylor 1974, 1984), and local populations on plants are re-assembled from various source populations according to prevailing air movement and distribution of suitable host plants.

In the context of aphid-based food webs, investigations are restricted to stationary and non-equilibrium populations.

15.3 Dynamics of Aphid Populations and Associated Aphidophagous Species

15.3.1 Sationary Populations on Wild Roses and Bird Cherry

Stationary populations of the cereal aphid species studied showed marked differences in abundance between wild roses and bird cherry (Table 15.1). On roses, two third of the branches are free of aphid winter eggs, with an average density of 1.4 eggs/m of branch (observed minimum 0.2, maximum 3.7 per sample). In contrast, bird cherry branches always carried *R. padi* eggs, with an average density of 67.9 eggs/m of branch (observed minimum 3.3, maximum 199.9). Studies in the UK and in Finland by Leather (1980), Leather and Lehti (1981) and Hand and Williams (1981) and in Germany by Gruppe (1985) revealed similar egg densities in *M. dirhodum* and *R. padi*.

The only aphidophagous species observed in association with stationary aphid populations was *Anthocoris nemorum* on bird cherry, which is active even during the winter. Numbers of *A. nemorum* caught prior to April (hatching of *R. padi* fundatrices) were low, amounting to less than one bug per ten beating samples on the average. It is during the establishment of non-equilibrium populations in April and May that *Anthocoris* appears in large

Table 15.1. Winter egg abundance of cereal aphids at a site with low hedge density (roses at Hummelgau, bird cherry at Thurnau), and at a site with high hedge density (roses at Stadtsteinach, bird cherry at Neuenmarkt)

Host plant	Aphid species	Location	Sample size (length of branches) (m)	No. of eggs	Branches with eggs (%)
Wild roses	*M. dirhodum*, *S. fragariae*	Hummelgau	962.70	1191	36.2
	"	Stadtsteinach	642.26	994	12.9
Bird cherry	*R. padi*	Thurnau	338.01	19119	100
		Neuenmarkt	324.68	25896	100

Table 15.2. Species occurrence matrix in the aphid-Aphidophaga complex associated with host plants of cereal aphids at two sites from 1977–1981

No.	Species	Cereals	Roses	Bird cherry	Black thorn and hawthorn
1	*Metopolophium dirhodum*	+	+	–	–
2	*Rhopalosiphum padi*	+	–	+	–
3	*Sitobion avenae*	+	+	–	–
4	*Sitobion fragariae*	+	+	–	–
5	*Adalia 2-punctata*	–	+	+	+
6	*Adalia 10-punctata*	–	+	+	+
7	*Anatis ocellata*	–	(+)	(+)	(+)
8	*Calvia 14-guttata*	–	+	+	+
9	*Coccinella 7-punctata*	+	(+)	(+)	(+)
10	*Propylaea 14-punctata*	+	(+)	(+)	(+)
11	*Dasysyrphus lunulatus*	+	–	–	(+)
12	*Episyrphus balteatus*	+	+	+	(+)
13	*Meliscaeva cinctella*	–	–	+	(+)
14	*Melanostoma mellinum*	+	–	–	(+)
15	*Eupeodes corollae*	+	–	–	–
16	*Eupeodes latifasciatus*	+	–	–	–
17	*Platycheirus clypeatus*	+	–	–	–
18	*Platycheirus cyaneus*	+	–	–	–
19	*Platycheirus manicatus*	+	–	–	–
20	*Platycheirus peltatus*	+	–	–	–
21	*Platycheirus scambus*	+	–	–	–
22	*Scaeva pyrastri*	+	+	–	–
23	*Scaeva selenitica*	–	–	–	+
24	*Sphaerophoria menthastri*	+	+	–	–
25	*Sphaerophoria scripta*	+	–	–	–
26	*Syrphus ribesii*	+	+	+	(+)
27	*Syrphus torvus*	+	+	+	(+)
28	*Syrphus vitripennis*	+	+	+	(+)
29	*Xanthogramma pedisequum*	+	–	–	–
30	*Chrysoperla carnea*	+	+	+	+
31	*Chrysopa septempunctata*	–	–	–	+

Table 15.2. *Continued*

No.	Species	Cereals	Roses	Bird cherry	Black thorn and hawthorn
32	*Nineta flava*	–	–	–	+
33	*Hemerobius humulinus*	–	+	+	+
34	*Hemerobius* sp.	–	–	–	+
35	*Anthocoris nemorum*	+	+	+	+
36	*Anthocoris nemoralis*	–	+	+	+
37	*Aphidius ervi*	+	+	–	–
38	*Aphidius urticae*	+	–	–	–
39	*Aphidius hieraciorum*[a]	+	–	–	–
40	*Aphidius picipes*	+	–	–	–
41	*Aphidius rhopalosiphi*	+	–	–	(+)
42	*Aphidius rosae*	–	+	–	–
43	*Aphidius uzbekistanicus*	+	–	–	–
44	*Ephedrus niger*[a]	+	–	–	–
45	*Ephedrus lacertosus*	–	+	–	–
46	*Ephedrus minor*	+	+	–	–
47	*Ephedrus plagiator*	+	+	–	–
48	*Praon abjectum*[a]	+	+	–	–
49	*Praon dorsale*[a]	–	+	–	–
50	*Praon exsoletum*	+	–	–	–
51	*Praon flavinode*	+	–	–	–
52	*Praon gallicum*	+	–	–	–
53	*Praon volucre*	+	+	+	(+)
54	*Trioxys angelicae*[a]	–	+	–	–

[a] Atypical species record for this assemblage (P. Starý, pers. comm.).
+ Juveniles and adults. (+) Adults only. – Not recorded.

numbers (see below). However, winter egg mortality in *R. padi* on bird cherry is substantial, ranging between 30 and 70% of the eggs being sucked out (own observ.; Leather 1980; Leather and Lehti 1981). Other aphidophagous insects were found only in the egg (hemerobiid eggs on wild rose) or pupal stage (aphid parasitoid mummies on wild rose; Table 15.2).

In conclusion, stationary winter populations support a very small aphidophagous complex, despite long exposure time, moderate to high frequencies of host-plant infestation, and moderate to very high aphid egg population densities. Both egg predation by *Anthocoris* (river borders as hibernation sites) and hibernation of Aphidiidae in the pupal stage (hibernation on wild roses in hedges) are site-specific determinants, independent of aphid egg abundance, or Aphidophaga diversity in the web.

15.3.2 Non-Equilibrium Populations on Wild Roses and Bird Cherry

On roses, the two aphid species establish non-equilibrium populations in spring for a time of about 6–8 weeks, with substantial variation between years.

The hatching of fundatrices in *M. dirhodum* and *S. fragariae* on wild roses commenced from the end of April to early May, the first larvae of the second generation appeared from mid-May onwards. By the middle to end of June, both species had left by emigration (Table 15.3).

Infestation rates of branches were low; of a total of 781 shoots sampled a fraction of 74% was free of aphids. Colonies of all sizes are constantly on the move, as could be shown by counting all aphid colonies on all leaves on three bushes at three intervals (April and May, 1979, 1980). While aphid abundance was increasing, new colonies were founded and old colonies became extinct on the same bushes, following the relation: $y = 98 - 25 (\log x)$ (y = Arcsin $\sqrt{\text{proportion}}$ of colonies still present, x = time in days after first record; $n = 8$, $r = 0.95$, $p \leq 0.001$). Thus, within 2 to 3 weeks, 80% of the colonies established in the beginning had disappeared from their founding location on the shoot.

Aphid colony size on wild rose averaged 10.5 aphids/colony (n = 357). In most cases there were a few exceptionally large colonies established on each bush (maximum observed is 155 aphids/colony). Aphid abundance was low even at peak levels prior to the decline of the non-equilibrium populations, following similar density patterns between years at all sites studied. In 1979 peak densities ranged between 2.2 and 18.7 aphids/shoot, in 1980 between 37.3 and 60.6 aphids/shoot, and between 2.4 to 13.9 aphids/shoot in 1981.

The complex of Aphidophaga associated with aphids on roses in spring is large and composed of 21 species (Table 15.2). Abundance of adult Aphidophaga was low, and beating samples (n = 68) yielded per m^2: Coccinellidae: 1.5; Chrysopidae: 0.6; Anthocoridae: 1.5.

Table 15.3. A non-equilibrium aphid population on wild roses in the Hummelgau area 1980, data after beating samples

Date (1980)	Total	Larvae	Nymphs[a]	Adults *M. dirhodum* Apter	Alat	*S. fragariae* Apter	Alat
16 April	41	41	–	–	–	–	–
25 April	10	10	–	–	–	–	–
30 April	6	6	–	–	–	–	–
7 May	3	2	–	1	–	–	–
14 May	66	56	–	3	–	7	–
21 May	30	25	17	3	–	2	–
28 May	37	13	13	2	–	21	1
4 June	112	67	54	1	2	37	5
12 June	112	67	54	1	2	29	7
19 June	11	4	3	–	1	5	1
25 June	33	27	24	–	3	1	2
4 July	5	3	–	–	–	–	2

[a] Larvae with wing buds, numbers included in number of larvae.

Immature stages appeared late in relation to aphid populations, shortly before peak abundances of aphids were reached. Eggs layed on shoots (Coccinellidae: *Adalia 2-punctata, A. 10-punctata*; Syrphidae: *Scaeva pyrastri, Episyrphus balteatus, Syrphus ribesii, S. vitripennis*) were found in mid-May, larvae after the end of May. Syrphid larvae abundance (y) correlated with aphid density (x) following the function: $\log y = 0.3 + 0.18 (\log x)$ ($n = 64$, $r = 0.35$, $p = \leq 0.01$). On average, less than 25 aphids were present on shoots with syrphid larvae, so that it is very likely that there was strong competition for food between predatory larval Aphidophaga, if daily food intake rates of syrphid larvae are considered (Bastian 1986; Medvey 1988).

On bird cherry, *R. padi* established non-equilibrium populations in spring from early April to the end of May. By early June, it has emigrated from the host completely. Spring populations on bird cherry occur 4 weeks earlier than populations on wild roses.

Both host-plant infestation rates and aphid abundance were high in bird cherry. Of a total of 462 shoots sampled, a fraction of only 15% was free of aphids. *R. padi* colonies can grow to extremely large size, even covering whole branches by amalgamation with previously separated colonies. Stem mother aggregates, however, are generally small, with lowest mean sizes of 1.8 to 4.2 stem mother per colony in April. Prior to hatching of winged emigrants in the second half of May, average colony size was 28.2 aphids/colony (n = 1099). At peak aphid densities in May, maximum colony sizes were 372 aphids/colony in 1978, 386 in 1979, and 504 in 1980.

The complex of Aphidophaga associated with *R. padi* on bird cherry in spring consisted of 16 species (Table 15.2).

The difference between aphid-Aphidophaga species complexes on wild rose and bird cherry arose from the lower species number of aphid parasitoids on the latter host plant (Table 15.2).

Abundance of adult Aphidophaga on bird cherry was higher than on roses, and beating samples (n = 27) yielded per m^2: Coccinellidae: 2.4; Chrysopidae: 0.6; Anthocoridae: 4.8. Aphidophaga adults appeared fairly early in relation to aphid population buildup (mid-April onwards). Eggs and larvae of Syrphidae, in particular, were the first Aphidophaga to appear, whereas adults of Coccinellidae were present at the same time, yet eggs and larvae were not found earlier than mid-May. On 452 shoots sampled, 135 larvae of *A. 2-puncata* and 319 syrphid larvae were found, 84% of the latter belonging to *E. balteatus* (identified after rearing to adults). Syrphid larvae abundance on shoots (y) correlated with aphid abundance (x) in the relationship: $\log(y + 1) = 0.16 + 0.17 \log(x + 1)$ ($n = 84$, $r = 0.35$, $p \leq 0.005$). Syrphid larvae were particularly common in 1980, when 71% of the shoots sampled carried syrphid larval instars. Peak syrphid larvae densities observed were 7.6 larvae/shoot at Neuenmarkt 1980, and 3 larvae/shoot at Thurnau 1980. Thus, despite high density of syrphid larvae, the food supply was good for concluding development, and syrphid larvae did not alter peak aphid densities significantly on this plant.

In conclusion, bird cherry as a cereal aphid winter host carried larger aphid populations, and larger Aphidophaga populations as well. However, the Aphidophaga species complex was slightly larger on wild roses due to the occurrence of more parasitoid species.

15.3.3 Non-Equilibrium Aphid Populations on Cereal Crops

In cereals, four aphid species were found in all fields investigated (n = 12 fields and 94 samples): *M. dirhodum*, *Sitobion avenae*, *S. fragariae* and *R. padi*. The distribution of individuals among species is given in Table 15.4. Since in the *Sitobion* species larvae cannot be separated to species level, total number of individuals is recorded at the generic level only. As far as adults of *Sitobion* are concerned, in most counts both species were about equally common, with *S. fragariae* being more common in some cases. The distribution of species between sites showed similar patterns during the whole study period. *Sitobion* had a clear preference for winter cereals, 82% of the specimens being found on this crop. *R. padi* was the least common species in all instances.

Cereal aphid non-equilibrium population establishment on cereal crops is strongly related to host-plant development. There was substantial variability in the time of immigration up to the disappearance of aphids, ranging in winter cereals for a period of up to 8 weeks (early June to late July). If plant

Table 15.4. Distribution of cereal aphid species on summer cereals (sc, sown in spring) and winter cereals (wc, sown in autumn) at two sites with low (Hummelgau) and high (Stadtsteinach) hedge densities

Year	Site	Crop	Total number and % fraction within each field of three aphid genera		
			Metopolophium	*Sitobion*	*Rhopalosiphum*
1979	Hummelgau	wc	2459	4276	19
		sc	2419	1055	109
	Stadtsteinach	wc	166	3361	25
		sc	995	456	18
1980	Hummelgau	wc	118	166	19
		sc	148	24	9
	Stadtsteinach	wc	217	263	12
		sc	550	72	60
1981	Hummelgau	wc	943	88	30
		sc	765	164	78
	Stadtsteinach	wc	1195	73	50
		sc	523	197	55
Total		wc	5098	8827	155
		sc	5400	1968	329

growth stages are considered for comparison, however, peak aphid abundances were almost always reached at milky ripeness. This growth stage occurred between the end of June in winter cereals on limestone in the Stadtsteinach area, and around mid-July in the other crops.

In the population development prior to peak densities, both infestation rates (of tillers) and density (mean number of aphids/tiller) were very variable, particularly in the establishment phases of the aphid populations. Maximum infestation rates reached in non-equilibrium populations were between 10 and 40% in 7 of the 12 fields investigated, and between 50 and 80% in 5 of the fields.

Peak aphid abundance remained low in all cases (Table 15.5), if compared to various published cereal aphid density data from outbreak areas in Europe available from the literature (e.g. Vickerman and Wratten 1979; Carter et al. 1980, 1982; Dixon 1987). Density patterns had similar trends within years and between sites. Differences in hedge density between sites do not always explain differences in aphid abundance, particularly in *M. dirhodum* and *R. padi*. In *Sitobion* spp., however, the low hedge density area (Hummelgau) had significantly higher population peaks in all cases.

Aphid infestation rates (x, proportion of infested tillers) and aphid density (y, mean number of aphids/tiller) showed highly significant correlations in all cases ($y = 0.19–0.09 (\arcsin \sqrt{x}) + 0.002 (\arcsin \sqrt{x^2})$; $r = 0.93$, $n = 87$, $p < 0.001$).

Table 15.5. Peak cereal aphid density observed at weekly counts in 12 crop fields in an area of high hedge density (ST = Stadtsteinach) and an area of low hedge density (HG = Hummelgau)

Site	Date	Crop	Mean no. of aphids per tiller at peak cereal aphid density			
			Mean total ± SE	*Sitobion*	*Metopolophium*	*Rhopalosiphum*
HG	12.7.1979	wc	9.41 ± 0.71	6.34	3.01	0.06
	20.7.1979	sc	5.59 ± 0.52	1.79	3.75	0.05
ST	18.7.1979	wc	3.54 ± 0.33	3.20	0.34	0
	18.7.1979	sc	1.14 ± 0.12	0.49	0.65	0
HG	17.7.1980	wc	0.26 ± 0.09	0.15	0.11	0
	4.7.1980	sc	0.39 ± 0.10	0.39	0	0
ST	4.7.1980	wc	0.71 ± 0.11	0.02	0.67	0.02
	4.7.1980	sc	0.71 ± 0.11	0.02	0.67	0.02
HG	11.7.1981	wc	6.96 ± 0.83	0.39	6.31	0.26
	17.7.1981	sc	2.48 ± 0.41	0.17	2.23	0.08
ST	9.7.1981	wc	5.69 ± 0.47	0.11	5.35	0.23
	9.7.1981	sc	1.87 ± 0.30	0.12	1.56	0.19

Aphid species: *Sitobion* = *S. avenae* and *S. fragariae*; *Metopolophium* = *M. dirhodum*; *Rhopalosiphum* = *R. padi*; cereals: wc = winter cereals (autumn-sown), sc = summer cereals (spring sown).

Colony size was generally small in all species, with the following averages (± SE): *M. dirhodum* 4.3 ± 11 (n = 2430), maximum observed 61 aphids/colony. *Sitobion* spp.: 3.7 ± 0.14 (n = 2933), maximum observed 230 aphids/colony. *R. padi*: 2.6 ± 0.23, maximum observed 28 aphids/colony.

The Aphidophaga complex associated with aphids on cereals comprised 35 species (Table 15.2). Abundance of adult Aphidophaga was high, and sweep netting (n = 96) yielded per sample (equivalent to roughly 1 m^2 crop area): Coccinellidae: 8.3; Chrysopidae: 1.7; Syrphidae: 13.6; Anthocoridae: 2.7; and Aphidiidae: 9.

C. 7-puncata and *Propylaea 14-punctata* eggs and larvae appeared late in relation to aphid non-equilibrium populations, the numbers increasing after aphids had surpassed peak densities. Honek (1982) could show that both species prefer high aphid densities for reproduction.

Syrphidae were the most abundant group of Aphidophaga, the dominating species (according to both sampling of larvae from tillers and breeding to the adult stage, and sweep-netting) were in decreasing sequence of abundance: *E. balteatus*, *Melanostoma mellinum*, *Sphaerophoria menthastri*, *Platycheirus clypeatus*, *Sphaerophoria scripta* and *Eupeodes corollae.*

There were remarkable differences in species composition between years. During the whole period covered, the species *E. balteatus*, *M. mellinum*, *S. menthastri*, *S. scripta* and *P. clypeatus* were always present in large numbers, whereas *M. corollae* was abundant in 1979 but rare in 1980 and 1981. The three *Syrphus* species were reproducing in the crops only in 1980, a year in which they could establish well on aphid winter host plants in spring.

Syrphidae appeared early in the crop (beginning of June), the grass pollen feeders *Platycheirus* and *Melanostoma* species being the first, *E. balteatus* following about 2 weeks later. Syrphid larvae density (y, mean number of larvae per 100 tillers) was significantly related to aphid tiller infestation rates (x, proportion of tillers with aphids) (1979: $y = 0.18 - 0.09 \arcsin \sqrt{x} + 0.002 \arcsin \sqrt{x}^2$ (n = 35, r = 0.796, $p \leq 0.001$); 1980: $y = 1.55 - 0.22 \arcsin \sqrt{x} + 0.009 \arcsin \sqrt{x}^2$ (n = 34, r = 0.56, $p \leq 0.001$)).

Syrphid larvae density had a significant influence on cereal aphid growth rates. Using path analysis in a multiple regression, 12 data sets from 2 regions were analysed. Criterion variable was peak aphid density reached in each non-equilibrium population. The predictor values of aphid instantaneous coefficient of growth $[r = /\ln N_t - \ln N_o)/t]$ were compared between flowering (N_o) and milky ripeness (N_t), together with syrphid density at flowering. Syrphid larva density was the stronger predictor ($p \leq 0.01$) of peak aphid abundance. Low aphid abundance in the crops in 1980 could thus be explained by mortality caused by Syrphidae.

The dominating aphid parasitoid species in all years were *P. volucre*, *Aphidius uzbekistanicus*, *A. picipes*, *A. rhopalosiphi* and *Ephedrus plagiator*. These species are known to parasitise all four recorded cereal aphid species (Starý 1981). The aphid parasitoids appeared late in relation to aphid population buildup, yet the mean number of aphid mummies the week after aphid counts (y) correlated positively with the mean number of aphids per tiller (x) ($y = 1.16 + 2.7x$; $n = 58$, $r = 0.75$, $p \leq 0.001$). This is the steepest slope found in all aphid-Aphidophaga abundance correlates in this investigation.

In conclusion, cereal aphid distribution and abundance was very variable between years, but consistent within years between areas of low and high hedge densities. As compared to non-equilibrium populations on winter host plants, in cereals both infestation rates and peak abundances reached were variable as well, alternating between years with low and high values, respectively. On the other hand, Aphidophaga were rich in species, with about twice as many species present in the crop, as compared to the unmanaged winter host associations in hedges. Syrphidae and Aphidiidae, however, were present in exceptionally larger numbers of species.

The result of a significantly larger Aphidophaga species complex associated with the non-apparent and low plant species diversity cereal crop sites requires some detailed comparisons between the two subsystems studied.

15.4 Differentiation of Diversities

The last comparison is within and between sample diversities in the web. The dimensions of species diversity are related to both area and time scales over which species have been recorded. According to Whittaker's (1972) classification (see Giller 1984), a limited number of species is expected to coexist at any locality. Species records from such local assemblages are called inventory or α diversities. Due to horizontal species replacements, local situations are expected to differ in species composition. By numerical comparison between habitats, we can differentiate components of α diversities. These measurements are called β diversity. If the area of reference is further expanded (over space and time), we record species accumulations that are called species richnesses or γ diversities. Natural species assemblages are governed by vertical interactions, thus α and β diversities are measurements based on ecological processes, whilst γ diversity (e.g. regional species lists, herbivorous species recorded from particular host plants) are a complex of potentially coexisting species, which are not necessarily interactive at a particular site. Thus, γ diversity covers species pools of definable faunas, from which α and β are a site-specific fraction (see Zwölfer 1987; Schaefer 1995).

Local α diversities have been treated in Section 15.3. The total of species recorded is given in Table 15.2. These species records, though compilations of records from two sites, are treated here as α diversities, since there have been no significant differences in species found between the two areas studied, and differences in species composition between the years occurred simultaneously at the two study sites.

In the original investigations of northern Bavarian hedges (Zwölfer et al. 1984), hawthorn and black thorn bushes at the two sites studied for cereal aphids had also been included in beating samples at the same level of intensity as in wild roses and bird cherry (Stechmann 1984). Since these four hedge plant species cover the most common Rosaceae in hedges in the region (Achtziger 1996), the Aphidophaga on hawthorn and black thorn will be included in an analysis of species diversities.

On black thorn and hawthorn, 19 species of Aphidophaga were found in the spring (Table 15.2). Thus, this complex contains fewer Aphidiidae and more species in the Syrphidae and Neuroptera (Chrysopidae and Hemerobiidae) than were found on both cereals and wild roses.

In summary, on the various hedge plants and cereal crops studied, the lowest number of aphidophagous species was (not surprisingly) recorded during stationary winter populations on aphid winter host plants (1–3 species). Medium species diversity was recorded on non-equilibrium spring populations on these plants (16–19 species), and highest species number occurred in the crops (35 species).

The distribution of species within samples (α diversity distribution) is compared for functional groups with large numbers of species, e.g. Syrphidae and Aphidiidae (Table 15.6). Potentially, 17 and 14 species, respectively, can be expected to coexist in each sample. Yet, within-sample distributions showed a realised maximum of six and five co-occurring species, with 70–75% of the assemblages being composed of one or two species only. In both Syrphidae and Aphidiidae, about 50% of all samples contained specimens of the one dominating species in each group, e.g. *E. balteatus* and *P. volucre*. Thus, at the α diversity level a very limited number of species actually contributes to the diversity of local assemblages at a particular time of population development.

Further reducing the scale of assessment, e.g. looking at Aphidophaga species and single aphid colonies, is impossible due to limitations in accuracy, since the fraction of specimens not identifiable at the species level, as well as failure in breeding field-collected instars to the adult stage due to the occurrence of parasitoids and hyperparasitoids, becomes numerically

Table 15.6. Local (within sample) species diversity of Syrphidae and Aphidiidae in cereal fields in two areas at the Bayreuth region (1978–1981)

Group	Number of co-occurring Aphidophaga species and distribution of samples within classes							
	No. of co-occurring species	1	2	3	4	5	6	Total samples
Syrphidae	No. of samples in class	36	37	19	9	5	2	108
	Proportion (%)	33.3	34.3	17.6	8.3	4.6	1.9	100
Aphidiidae	No. of samples in class	20	38	14	6	1	0	79
	Proportion (%)	25.3	48.1	17.7	7.6	1.3	0	100

more important the smaller the total number of host aphids per unit of relation.

To assess β diversities in order to compare species similarity between Aphidophaga associated with cereal aphids on different host plants, we use the SØRENSEN index, which is calculated as $QS = 2\ j/(a + b)$. QS = quotient of similarity, j = number of species common in both systems; a = species associated on plant a; b = species associated with plant b. According to Müller-Dombois and Ellenberg (1974) and Wolda (1981), this index is very robust if samples of similar size are investigated. Great similarity (SQ approaches 1) will result in a slow increase in β diversity only, since adding new habitats along a transect will increase α diversities only slightly due to recruitment of new species. Thus, there is a relationship between faunistical similarity SQ and β diversity according to $\beta = (1 - SQ)$.

In the Aphididae, only species living on aboveground plant parts (shoot, leaf, flower) will be considered. In bird cherry γ diversity is 1, since only *R. padi* has been recorded on this plant for central Europe. *Myzus padellus* HRL, reported by Hille Ris Lambers (1947), has not been recorded from bird cherry ever since (Eastop and Hille Ris Lambers 1976). According to Börner (1952), on hawthorn four species are known, if glasshouse-dwelling aphids in Europe are excluded, e.g. *Aphis pomi*, *Eriosoma lanigerum*, *Ovatus crataegarius* and *Rhopalosiphum insertum*. On wild roses, six species are known, if again glasshouse dwellers are excluded. These species are *Chaetosiphon tetrarhodum*, *Longicaudus trirhodus*, *Macrosiphum rosae*, *Metopolophium dirhodum*, *Myzaphis rosarum* and *Sitobion fragariae*. On blackthorn we have records for seven species, *Brachycaudus cardui*, *B. helichrysi*, *B. persicae*, *B. prunicola*, *Hyalopterus pruni*, *Phorodon humuli* and *Rhopalosiphum nympheae* (Börner 1952).

In cereals (barley, oat, rye, wheat), six species are recorded for central Europe, the four species found during this study, as well as *Metopolophium festucae* and *Rhopalosiphum maidis*. The latter two are locally common but lacking in southern Germany (Müller 1964; Blackman and Eastop 1985; Dent and Wratten 1986).

Thus aphid species on hedge plants differ completely ($SQ = 0$, β diversity = 1), and species similarity between hedges and cereal crops accounted only for wild roses (β diversity = 0.33) and bird cherry (β diversity = 0.6). Although γ diversity in hedge plants and cereals is of similar magnitude, the species-host-plant relationship is totally different. In hedges, host-plant specificity is extremely high, whereas in cereals it is extremely wide.

The greatest similarity between aphidophagous species complexes were found within the hedge plants. Aphidophaga complexes in cereals had the highest similarity with those found on wild roses (Table 15.7).

In Coccinellidae $\gamma = 6$, on aphid winter hosts the species *Adonia variegata*, *Aphidecta obliterata* and *Coccinella 5-puncta* were also found in samples from hedges in the area taken some 10 years later (Geyer 1988). *C. 11-punctata*, recorded in cereal crops for the Czech Republic (Honek 1982), is absent in

Table 15.7. Faunistic similarity of aphidophagous insects associated with cereal aphids (cereals, wild rose, bird cherry) and other homopterans on black thorn, hawthorn and in fallow land in the Bayreuth region

Quotient of similarity according to Sørensen

Plant	Cereal	Wild rose	Bird cherry	Black thorn and hawthorn
A) All Aphidophaga species recorded				
Cereal	–	0.48	0.35	0.44
Wild rose	–	–	0.72	0.71
Bird cherry	–	–	–	0.91
B) Syrphidae recorded in the Hummelgau area				
Fallow land	0.56	0.30	0.16	0.20
Cereals	–	0.58	0.36	0.42
Wild rose	–	–	0.67	0.71
Bird cherry	–	–	–	0.83

Germany (e.g. Basedow 1982). The species found on hedge plants were similar ($b = 0$), and between cereals and hedges $\beta = 0.5$. Aphids on shrubs develop too early for most of the ladybird beetles to reproduce on them, yet feeding of adults on homopterans on hedges allows for early immigration of the species into the cropping area (Stechmann 1982).

In Syrphidae, most investigations are based on trap catches; up to 23 species were recorded (Grosser and Klapperstück 1977; Kröber and Carl 1991). In the Hummelgau area, Nakott (1983) intensively sampled fallow land located within the site investigated in this study. Thus, α diversities found in this study approach values of γ diversities of the region. The greatest similarity was found within hedge plant associations, and the closest similarity of associations in the crop occurred towards fallow land and wild roses (Table 15.7). According to Bankovska's classification (Bankovska 1980), the hoverfly assemblages in the crop are composed equally of forest (*Scaeva*, *Syrphus*, *Episyrphus*) and meadow species (*Melanostoma*, *Platycheirus*, *Sphaerophoria*). Some of the species in the former group develop on hedges in the early season, and visit these plants during flowering (Korman 1975; Clausen 1980; Röder 1992). Therefore, it seems to be obvious that high α diversities in the crop are supported by high β diversities generated by hedge plants. The hedges bring forest components into the agricultural landscape.

In Aphidiidae, there are few data on species records from hedges (Bode 1980; Starý 1981), with very few species on hedge plants ($\gamma = 5$), and a large complex associated with crops ($\gamma = 15$). Thus α diversities revealed in this study are comparatively great (Table 15.2). Similarity between hedges and fields gave a medium $\beta = 0.58$ (wild roses versus crops). According to the habitat classification of central European Aphidiidae (Starý 1970), the complex in cereals is composed of species of both deciduous forests (*P. volucre*) and steppe-type meadows (the numerous *Aphidius* spp.).

15.5 Conclusions

The following can be deduced from the analysis of inventory and differential diversities of cereal aphids and Aphidophaga in this study:

1. Diversities of primary consumers (aphids) were of equal size in both successionally young, low-diversity cereal crops, and successionally intermediate, medium-diversity hedge plant communities. γ Diversities of grasslands, as well as rosaceous shrubs in other parts of the world, support equally low numbers of aphid species (see Blackman and Eastop 1985). Since aphids are restricted to growing plant parts for food intake, their feeding niche is restricted to only a fraction of the feeding niches offered by vascular plants. Thus, aphids cannot radiate to high diversities on structurally rich, apparent plant species, and there is no significant difference in γ diversities between trees, shrubs and herbaceous plants.
2. Differences regarding host-plant specificity of aphids between hedge plants and gramineous crops are important determinants of γ diversities. High specificity in aphids on Rosaceae results in cumulative effects on aphid species within hedge-plant communities. High β diversities between different hedge plants create high α and γ diversities for hedge habitats. Each additional hedge-plant species is likely to contribute another three to five aphid species to the local pool.

   In contrast, low host-plant specificity of aphids associated with cereal crops results in low β diversities between habitats, and low β diversities do not allow for increases in γ diversities. For natural, species-rich grasslands, however, high β diversities can be expected in central Europe, since a number of rather host-specific aphid species are likely to occur, e.g. *Chaetosiphella tshernavini*, *Atheroides serrulatus*, *Cryptaphis poae* on *Festuca ovina* or *Diuraphis muehlei* on *Phleum pratense* (Müller 1964). On the average, however, the cumulative effect on grasses is expected to be rather low, in the order of only 0.2 aphid species per grass species. Thus, the potential contribution of high plant species diversity in grasslands in increasing α and γ diversities seems to be significantly smaller than in hedges.
3. The vast majority of the Aphidophaga species recorded here and elsewhere are rather non-specific, as far as their host/prey species relationship is concerned. Therefore low to intermediate β diversities occur if assemblages on both particular plant species, or different habitats are compared. This is well documented by monographs on particular Aphidophaga functional groups (Hodek 1973; Hill 1977; Starý and Rejmanek 1981; Canard et al. 1984; Bastian 1986). Additionally, the low habitat specificity and high dispersal capacity of most aphidophagous insect species (Starý 1970; Hodek 1973; Bankowska 1980; Stechmann 1982, 1988) can explain the low β diversities observed between many habitats.

4. The result of the highest α and γ diversities of Aphidophaga in the crop can be explained simply by the population size in primary consumers in the crop (species-area effects). The actual composition of the Aphidophaga assemblages in these attractive habitats, however, largely depends on the processes within the web occurring outside particular definable populations, e.g. at another space unit (adjacent areas such as hedges and field margins), and in another time unit (previous development and current events of subunits of non-equilibrium populations).

Acknowledgements. I am grateful to Helmut Zwölfer for introducing me to the study of aphids and aphidophagous insects, and for the substantial support that I received during the research on hedgerow ecology and biological control. Technical assistance was provided by Annick Servant and Marion Preiß, and by the students K. Benisch, H. Eschenbacher, F. Foeckler, M. Komma, E. Kreuzer, H. Möller, J. Nakott, M. Romstöck, E. Sauer-Gareis, and W. Wolf. Dr. P. Starý (Aphidiidae) and Dr. S. Vidal (Syrphidae) kindly helped to identify insect material.

References

Bankowska R (1980) Fly communities of the family Syrphidae in natural and anthropogenic habitats of Poland. Mem Zool 33:3–93

Bastian O (1986) Schwebfliegen (Syrphidae). Neue Brehm-Bücherei, Wittenberg-Lutterstadt Bd 576, 168 pp

Blackman RL, Eastop VF (1985) Aphids on the world's crops: an identification guide. Wiley, Chichester, 466 pp

Blackman RL, Eastop VF (1994) Aphids on the world's trees. An identification and information guide. CAB International, Wallingford; Natural History Museum, London, 987 pp + 16 plates

Bode E (1980) Untersuchungen zum Auftreten der Haferblattlaus *Rhopalosiphum padi* (L.) (Homoptera: Aphididae) an ihrem Winterwirt *Prunus padus* L. I. Biologie der Haferblattlaus *Rhopalosiphum padi* (L.) am Winterwirt. Z Angew Entomol 89:363–377

Börner C (1952) Europae centralis Aphides (Die Blattläuse Mitteleuropas). Mitt Thür Bot Ges Beih 3:1–484

Canard M, Semeria Y, New TR (eds) (1984) Biology of Chrysopidae. Junk, The Hague, 297 pp

Carter N, McLean IFG, Watt AD, Dixon AFG (1980) Cereal aphids: a case study and review. Appl Biol 5:271–349

Carter N, Dixon AFG, Rabbinge R (1982) Cereal aphid populations: biology, simulation and prediction. Pudoc, Wageningen, 97 pp

Clausen L (1980) Die Schwebfliegen des Landesteiles Schleswig-Holstein (Diptera, Syrphidae). Faun Ökol Mitt Kiel Suppl Bd 1:3–79

Dent DR, Wratten SD (1986) The host-plant relationships of apterous Virginoparae of the grass aphid *Metopolophium dirhodum cerealium.* Ann Appl Biol 108:567–576

Dixon AFG (1987) Cereal aphids as an applied problem. Agric Zool Rev 2:1–57

Dixon AFG, Kindlmann P, Leps J, Holman J (1987) Why are there so few aphid species, especially in the tropics? Am Nat 129:580–592

Eastop VF, Hille Ris Lambers R (1976) Survey of the world's aphids. Junk, The Hague, 573 pp

Frazer B (1988) Predators. In: Minks AK, Harrewijn P (eds) Aphids 2B. Elsevier, Amsterdam, pp 217–230

Geyer A (1988) Verinselungseffekte an der Entomofauna der Heckenrose (*Rosa canina* L.). Diploma Thesis, Bayreuth University, Bayreuth, 74 pp

Giller PS (1984) Community structure and the niche. Chapman & Hall, London, 176 pp

Grosser N, Klapperstück J (1977) Ökologische Untersuchungen an Syrphiden zweier Agrarbiozönosen, Hercynia NF 14:124–144
Gruppe A (1985) Beobachtungen zur Entwicklung der Frühjahrspopulationen von *Metopolophium dirhodum* (Wlk.) auf dem Winterwirt. Anz Schädlingskunde 58:51–55
Hand SC, Williams CT (1981) The overwintering of the rose-grain aphid *Metopolophium dirhodum* on wild roses. In: Thresh JM (ed) Pests, pathogens and vegetation. The role of weeds and wild plants in the ecology of crop pests and diseases. Pitman, London, pp 307–314
Hill AR (1977) The seasonal distribution of *Anthocoris* spp. (Hem., Cimicidae) in a deciduous wood in west central Scotland. Entomol Mon Mag 113:139–146
Hille Ris Lambers D (1947) Contributions to a monogrpah of the Aphididae of Europe. II. The genera *Dactynotus* Rafinesque, 1818; *Straticobium* Mordvilko, 1914; *Macrosiphum* Passerini, 1860; *Masonaphis* nov. Gen.; *Pharalis* Leach, 1826. Teminckia 14:1–134
Hodek I (1973) Biology of the Coccinellidae with keys for the identification of larvae by co-authors. Junk, The Hague, 260 pp
Honek A (1982) Factors that determine the composition of field communities of adult aphidophagous Coccinellidae (Coleoptera). Z Angew Entomol 94:157–168
Klingauf F (1987) Host plant finding and acceptance. In: Minks AK, Harrewijn P (eds) Aphids 2A. Elsevier, Amsterdam, pp 209–223
Kormann K (1975) Schwebfliegen als Blütenbesucher an frühblühenden Sträuchern und Blumen (Diptera, Syrphidae). Nachrichtenbl Bayer Entomol 24:9–13
Kröber T, Carl K (1991) Cereal aphids and their natural enemies in Europe – a literature review. Biocontrol News Inf 12:357–371
Leather SR (1980) Egg survival in the bird cherry-oat aphid, *Rhopalosiphum padi*. Entomol Exp Appl 27:96–97
Leather SR (1986) Insect species richness of the British Rosaceae: the importance of host range, plant architecture, age of establishment, taxonomic isolation and species-area relationships. J Anim Ecol 55:841–860
Leather SR, Lehti JP (1981) Abundance and survival of eggs of the bird cherry-oat aphid, *Rhopalosiphum padi* in southern Finland. Ann Entomol Fenn 47:125–130
Medvey M (1988) On the rearing of *Episyrphus balteatus* (Deg.) (Diptera: Syrphidae) in the laboratory. In: Nemzcyk E, Dixon AFG (eds) Ecology and effectiveness of Aphidophaga. SPB Academic Publishing, The Hague, pp 61–63
Minks AK, Harrewijn P (eds) (1987) Aphids. Their biology, natural enemies and control. World Crop Pests 2A. Elsevier, Amsterdam, 450 pp
Minks AK, Harrewijn P (eds) (1988) Aphids. Their biology, natural enemies and control. World Crop Pests 2B. Elsevier, Amsterdam, 364 pp
Müller FP (1964) Merkmale der in Mitteleuropa an Gramineen lebenden Blattläuse (Homoptera: Aphididae). Wiss Z Univ Rostock 18: Math Naturwiss Reihe 23:269–278
Müller-Dombois D, Ellenberg H (sen) (1974) Aims and methods of vegetation ecology. Wiley, New York, 547 pp
Nakott J (1983) Untersuchungen über Ansprüche der Imagines von Syrphinae (Syrphidae, Diptera) bezüglich Klima und Nahrung (Pollen). Diploma Thesis, Bayreuth University, Bayreuth
Price, PW (1980) Evolutionary biology of parasites. Princeton University Press, Princeton, 237 pp
Röder G (1992) Biologie der Schwebfliegen Deutschlands (Diptera: Syrphidae). Bauer, Keltern-Weiler, 575 pp
Schaefer M (1995) Die Artenzahl von Waldinsekten: Muster und mögliche Ursachen der Diversität. Mitt Dtsch Ges Allg Angew Entomol 10:387–395
Schulze ED, Gerstberger P (1993) Functional aspects of landscape diversity: a Bavarian example. Ecological studies, vol 99. Springer, Berlin Heidelberg New York, pp 453–466
Starý P (1970) Biology of aphid parasites (Hymenoptera: Aphidiidae) with respect to integrated control. Junk, The Hague, 643 pp
Starý P (1981) Biosystematic synopsis of parasitoids on cereal aphids in the western Palaearctic (Hymenoptera, Aphidiidae; Homoptera, Aphidoidea). Acta Entomol Bohemoslov 78:382–396

Starý P (1988a) Aphidiidae. In: Minks AK, Harrewijn P (eds) Aphids 2B. Elsevier, Amsterdam, pp 171–184

Starý P (1988b) Aphelinidae. In: Minks AK, Harrwijn P (eds) Aphids 2B. Elsevier, Amsterdam, pp 185–188

Starý P, Rejmanek M (1981) Number of parasitoids per host in different systematic groups: the implications for introduction strategy in biological control (Homoptera, Aphididae; Hymenoptera, Aphidiidae). Entomol Scand (Suppl) 15:341–351

Stechmann D-H (1982) Zur Ökologie aphidophager Insekten in Hecken und Feldern Oberfrankens: Beobachtungen an Coccinelliden in den Jahren 1978/79. Jahres ber Naturwiss Verein Wuppertal 35:38–42

Stechmann D-H (1984) Ergebnisse des Klopfprobenprogrammes. In: Zwölfer H, Bauer G, Heusinger G, Stechmann D (eds) Beiheft 3/2 Berichte Akademie Naturschutz Landschaftspflege, Laufen/Salzach, pp 38–48

Stechmann D-H (1988) Aktionsräume bedeutender Prädatoren der Agrarbiozönose. VDLUFA-Schriftenreihe, Kongreßband 1988 Teil II. Verein Deutscher Landwirt-Schaftlicher Untersuchungs- und Forschungsanstalten, Darmstadt, pp 1187–1197

Vickerman GP, Wratten SD (1979) The biology and pest status of cereal aphids (Hemiptera: Aphididae) in Europe: a review. Bull Entomol Res 69:1–32

Taylor LR (1974) Monitoring change in the distribution and abundance of insects. Rothamsted Exp St Annu Rep 1973 part 2, pp 202–239

Taylor LR (1984) Assessing and interpreting the spatial distributions of insect populations. Annu Rev Entomol 29:321–357

Whittaker RH (1972) Evolution and measurement of species diversity. Taxon 21:213–251

Wolda H (1979) Abundance and diversity of Homoptera in the canopy of a tropical forest. Ecol Entomol 4:181–190

Wolda H (1981) Similarity indices, sample size and diversity. Oecologia 50:296–302

Zwölfer H (1987) Species richness, species packing, and evolution in insect-plant systems. In: Schulze ED, Zwölfer H (eds) Potentials and limitations of ecosystem analysis. Ecological studies, vol 61. Springer, Berlin Heidelberg New York, pp 301–319

Zwölfer H, Bauer G, Heusinger G, Stechmann D-H (eds) (1984) Die tierökologische Bedeutung und Bewertung von Hecken. Beih 3/2 Berichte Akademie Naturschutz Landschaftspflege, Laufen, 155 pp

16 Organization Patterns in a Tritrophic Plant-Insect System: Hemipteran Communities in Hedges and Forest Margins

R. ACHTZIGER

16.1 Introduction

A central question of community ecology is "whether ecosystems are organized in a predictable way and which processes have formed them" (Zwölfer 1987, p 301). The term organization of ecological systems in this context means that one can identify patterns and structures that are ordered in some spatial and temporal way; there should be driving forces and adaptations of species that result in a system which can be predicted in – at least – basic structures (Zwölfer 1986). Plant-insect systems and their associated food webs build suitable models for analysing both the selective forces that underlie specific life-history traits of organisms and the vertical interactions between life histories on different trophic levels (Zwölfer 1988, 1994).

All organisms live and reproduce under extrinsic constraints (deriving from the biotic or abiotic environment) and intrinsic constraints (physiological, morphological, phylogenetic factors deriving from the evolutionary history of the organism) that limit their performance (Barbault and Stearns 1991). The responses of an organism to environmental conditions in ecological time, which is the basis of actual community structure, will be determined by its evolutionary adaptations to the spectrum of constraints in the habitat (see Mattson 1980; Barbault 1991; Brown 1995).

Analyzing data sets of hemipteran communities on woody plants in hedge and forest edge ecosystems, this chapter presents some patterns of vertical interactions between traits on plant and consumer levels and shows some implications for community organization patterns.

16.2 The Tritrophic Plant-Insect System: Hemiptera on Woody Plants of Hedges and Forest Margins

16.2.1 Data Sets and System Components

Hedges and forest margins are linear arrangements of shrubs and trees which established naturally on boundaries between arable fields, pastures and forests (Pollard et al. 1974; Zwölfer et al. 1984; Schulze and Gerstberger 1993). In

Ecological Studies, Vol. 130
Dettner et al. (eds.) Vertical Food Web Interactions

different regions of northern Bavaria (Germany) local hemipteran communities associated with the dominant shrub species of these biotopes were studied, i.e. blackthorn (*Prunus spinosa* L.), wild rose (mainly *Rosa canina* L., *R. rubiginosa* L) and hawthorn trees (mainly *Crataegus laevigata* (Poiret) DC., *C. monogyna* Jacq.). Using standardized beating samples (cf. Zwölfer et al. 1984), data on the distribution and abundance of 82 shrub-dwelling hemipteran species (41 Heteroptera, 28 Auchenorrhyncha, 9 Aphidoidea, 4 Psylloidea; 55 on rose, 61 on blackthorn, 55 on hawthorn) were collected from 1989 to 1995 in hedgerows (vicinity of Bayreuth, northern Bavaria, Germany) and at forest margins (vicinity of Feuchtwangen, northern Bavaria) varying in habitat conditions and age (Achtziger 1995).

The woody plants (producers), phytophagous (plant consumers) and predatory hemipteran species form a complex tritrophic plant-insect system (see Zwölfer 1994) with vertical interactions concerning both trophic and structural aspects: life-history traits and adaptations of a plant (e.g. phenology, physiology, growth and differentiation modes, "plant architecture"; see Küppers 1992) determine food resource properties (e.g. temporal and spatial availability of plant sap, fruits, buds, leaves) and habitat structure (e.g. quantity and diversity of microhabitats, oviposition sites, hibernation localities, foraging space) for consumer organisms on higher trophic levels. According to the main feeding styles and plant tissues used, two main guilds and six subguilds can be identified among the 82 shrub-dwelling Hemiptera (cf. Brown and Southwood 1983; Dolling 1991):

1. Phytophage guild (48 spp.): (a1) phloem suckers (23 spp., Aphidoidea, Psylloidea, Auchenorrhyncha: Cicadellidae, Cixiidae, Issidae, etc.), (a2) xylem suckers (2 spp., Auchenorrrhyncha: Membracidae, Cicadidae), (a3) mesophyll feeders (17 spp., Auchenorrhyncha: Typhlocybinae; Heteroptera: Tingidae) and (a4) species which feed on shoots, fruits, buds and other non-vascular plant tissues (6 spp., Heteroptera: Miridae, Pentatomidae, Lygaeidae).
2. Predator guild (34 spp.): (b1) mainly entomophagous species (17 spp., Heteroptera: Anthocoridae, Miridae, Pentatomidae etc.) and (b2) species that feed on a combination of plant and animal material (= entomophytophagous, 17 spp., Heteroptera: mainly Miridae).

A first step to describe the vertical interactions within the plant-insect system is a characterization of food resources and habitat structure in hedges and forest margins from the view of a hemipteran insect.

16.2.2 Traits on the Plant Level: Resource Availability

1. Food resources. Although overall food resources in hedges may be regarded as highly predictable from year to year because of the longevity of shrubs, the temporal predictability of resource abundance (e.g. quantity of certain plant

organs and tissues) and resource availability in terms of food quality (e.g. the content of nutrients) differs in time and space (Wiens 1984; Kareiva 1986): quantity and quality of resources may be affected by various biotic (e.g. insect infestation) and abiotic factors (e.g. temperature) modifying both plant physiology (e.g. Flückinger and Oertli 1978; White 1993) and plant performance (e.g. time of bud burst), and thus may influence synchronization of phytophagous insects. For example, variation of rose hip abundance in hedges of the Bayreuth study region was mainly determined by the degree of bud infestation by the geometrid winter moth, *Operophtera brumata* L. (Bauer 1986). As another example, lower spring temperatures in 1991 resulted in a delayed bud burst in woody plants, affecting the time of resource availability for herbivorous insects (Achtziger 1995). Furthermore, food availability can underlie endogeneous rhythms: hawthorn trees, for instance, tend to have biennal flowering patterns with food quality differing between flowering and non-flowering trees (Sutton 1984).

Note that resource availability is not the same as resource abundance (see Wiens 1984): whereas resource abundance is the quantity of a resource in the environment, independent of the consumer, resource availability refers to the portion of resources that is directly accessible to the consumer (Johnson 1980). For sap-feeding insects like Hemiptera, food availability is mainly determined by the nutritional quality of plant tissues, i.e. the content of soluble and hence digestable organic nitrogen compounds (e.g. amino acids), an essential limiting factor for the development of organisms (Mattson 1980; White 1993). Several studies suggest that patterns of seasonal variation and overall nutritional quality differ between plant tissues used by hemipteran guilds (McNeill and Southwood 1978; Mattson 1980; Strong et al. 1984). As a general pattern, phloem sap, young (= meristematic) leaves and flowers can be regarded as a seasonal pulsing resource type (Price 1984), because periods of high food availability are limited to the short flush period in spring (allocation of stored nutrients into meristematic tissues) and, in phloem sap, again before leaf fall. In contrast, the sap of mesophyll cells in full-developed leaves (pallisade parenchym) consumed by some hemipteran taxa (Dolling 1991, see below), may be considered as a resource type of relatively low overall food availability, because most nitrogen of leaf cells is bound in complex, indigestible plant proteins (Prestidge 1982). However, abundance (green foliage) and durational stability of food during the season are higher in mesophyll cells than in phloem sap. Higher levels of soluble nitrogen in leaf cells may be expected shortly after leaf flush, when photosynthesis reaches a maximum, and during the senescense and allocation phase. To complete the list of resource types, ripe fruits that are available in summer and autumn may be regarded as slowly increasing resource type (cf. Price 1984) with relatively low overall abundance.

2. Habitat structure. For shrub-dwelling insects, shrubs also provide the physical structure for feeding, development, reproduction and foraging (see

also Price et al., Chap. 18, this Vol.). Analogous to landscape on larger spatial scales, Samways (1994: 71) proposed the term plantscape for the surface of the plants and vegetation. The structural constitution of this plantscape, its area and its complexity, determine the availability and diversity of essential structural resources for the development of organisms, e.g. hiding/hibernation localities, feeding niches and oviposition places. Therefore, resource diversity and surface area of a plant were proposed to be important determinants for the number of insect species living on it, in both evolutionary and ecological times (e.g. Lawton 1986; Leather 1986; Zwölfer 1987 and literature herein).

Recent studies suggest that complex plant surfaces and their outlines cannot be described by means of euclidean geometry using integer dimensions as 1 (line) or 2 (area), but must be regarded as fractals with dimensions lying between 1 and 2 (Morse et al. 1985; Williamson and Lawton 1991). As measure for the complexity of the plantscape in hedges, the fractal dimensions D were determined for the outlines of distal parts of branches differing in age taken from rose, blackthorn and hawthorn bushes, by using a slightly modified version of the box-counting method of Morse et al. (1985; see Achtziger 1995). Fractal dimensions D of hedgerow plants varied from 1.30 to 1.67 across all branches; the mean, 1.51 ± 0.10, is identical to the average value of 1.5 found in another analysis of plant species with different complexity (Morse et al. 1985). The structural complexity of the plantscape and its fractal dimension is mainly determined by the foliage. After the removal of all leaves by hand, mean fractal dimension of branches consisting of twigs, thorns and shoots was only 1.17 ± 0.10. Nevertheless, fractal dimensions of both conditions were highly correlated ($r_S = 0.85$, $p < 0.0001$, $n = 23$).

D did not differ significantly between the three shrub species (Kruskal-Wallis-ANOVA: $Chi^2 = 0.34$, $p = 0.84$, $n = 26$), but was highly correlated to the age of the branch. Fractal dimension D increased from 1-year-old to 5-year-old branches ($r_S = 0.54$, $p < 0.001$, $n = 26$). Similar results were obtained in a study of branches without leaves, taken in winter ($r_S = 0.89$, $p < 0.0001$, $n = 15$, mean = 1.35 ± 0.17; Achtziger 1995). In conclusion, these results indicate that the structural complexity and the surface area of a branch, a shrub, and hence of the whole plantscape of a hedge, generally increase as a result of time-dependent scrub differentiation and growth (see Sect. 16.4). For example, the width of the hedge body is determined by the growth of the shrubs and the lateral spread of hedges through time (see Küppers 1992). In fact, results of a study of 20 hedges showed that hedge width significantly increased with age ($r_S = 0.79$, $p < 0.001$, $n = 20$; Achtziger 1995).

16.2.3 Traits on Consumer Levels

The characteristics of nutritional and structural resources constitute important constraints and driving forces for consumer levels of the plant-insect system. Vertical food web interactions may determine the evolution of specific resource exploitation strategies and morphological adaptations (see Sects. 16.3, 16.5), and may affect abundance and species composition via population responses in ecological time. In the hemipteran fauna of hedges and at forest margins considered here, a variety of resource exploitation strategies are evolved. To track the patchy distribution of resources and to achieve a high temporal synchronization with high-quality food, feeding styles were combined with specific life cycles, life-history traits (e.g. generation time, reproductive rate) and different physiological, ecological, behavioural and morphological adaptations including host-plant or prey specifity, migratory behaviour (resource tracking), foraging strategies and body sizes (see Sect. 16.5). The next sections give some examples of interactions between resource availability and resource exploitation strategies in dominant hemipteran species and their implications for community structure.

16.3 Food Web Interactions: Resource Dynamics, Resource Exploitation, Population Variability and Community Dynamics

Results of a study on the population ecology and dynamics of the dominant hemipteran species living on rose, blackthorn and hawthorn at several forest margins under comparable habitat conditions in the Feuchtwangen region (1990–1994) showed that total annual abundances and year-to-year variability of abundances differed considerably between species and guilds (Table 16.1): In the phytophagous guild, the dominant aphid species on rose (*Metopolophium dirhodum*, *Macrosiphum rosae*) and on blackthorn (*Hyalopterus pruni*, *Phorodon humuli*) and two psyllid species on hawthorn (*Cacospylla peregrina*, *C. melanoneura*) reached the highest abundances. Highest year-to-year variability was observed in aphids, which showed extremely high population densities in 1991 and 1994, whereas psyllid species and dominant mesophyll-feeding leafhoppers (*Edwardsiana* spp., *Zygina* spp.) showed lesser variability (Table 16.1). In the entomophagous guild the dominant anthocorid species (*Orius minutus*, *Anthocoris nemorum*) showed higher fluctuations than dominant entomophytophagous Heteroptera (*Phytocoris ulmi*, *Heterocordylus tumidicornis*) (Table 16.1).

Dominant aphids mainly feed on phloem sap of meristematic leaves and shoots. Therefore, they are highly adapted to exploit a pulsing resource type. Because of high reproductive power and short generation times (plurivoltism, telescoping of generations, parthenogenesis), host-plant specifity and nutrient-mediated host-plant alternation (Minks and Harrewijn 1987, 1988, see also Stechmann, Chap. 15, this Vol.), aphid populations respond rapidly to resource availability. Therefore, aphid densities in general are high under favourable nutritional conditions (high synchronization with plant phenology) or weather conditions. The tremendous year-to-year fluctuations

Table 16.1. Biological and ecological characteristics of the dominant hemipteran species found on rose, blackthorn and hawthorn in hedges and forest margins of the Feuchtwangen region (1990–1994). Data from the literature and personal observations (Achtziger 1995)

Guild/species name	T	HPP			Mean local annual abundance (all sites, 1990 to 1994)			Population variability (CV)			Biology/ecology				BM	BL
		Ro	Ps	Cr	Ro (n = 23)	Ps (n = 29)	Cr (n = 13)	Ro	Ps	Cr	FS	PVL	GEN	HIB	(mg)	(mm)
Phloem feeders																
Macrosiphum rosae (L.)	Ap	●			107.2 ± 166.7			1.38			Pp	HPA	>2	E	0.49	3.45
Metopolophium dirhodum (Walk.)	Ap	●			161.5 ± 261.9			1.44			Pp	HPA	>2	E	0.38	2.50
Hyalopterus pruni (Geoffr.)	Ap		●	○		349.2 ± 814.1			1.64		Pp	HPA	>2	E	0.27	2.20
Phorodon humuli (Walk.)	Ap		●			69.2 ± 186.3			1.55		Pp	HPA	>2	E	0.27	2.55
Psylla pruni (Scop.)	P		●			43.2 ± 38.8			0.56		Pp	S	1	A	0.24	2.50
Cacopsylla melanoneura (Först.)	P			●			1127.0 ± 1753.7			0.53	Pp	S	1	E	0.29	2.60
Cacopsylla peregrina (Först.)	P			●			302.7 ± 257.5			0.23	Pp	S	1	A	0.47	2.80
Empoasca vitis (Goethe)	Au	□	□	□	31.8 ± 27.0	22.1 ± 15.3	9.8 ± 8.8	0.45	0.30	0.35	Pp,m	S	1–2	A	0.24	3.45
Mesophyll feeders																
Physatocheila dumetorum (H.-Sch.)	H		●			23.5 ± 46.5			0.73		Pm	S	1	A	0.33	2.90
Edwardsiana rosae (L.)	Au	●			67.0 ± 32.9			0.37			Pm	S	2	E	0.31	3.60
Edwardsiana prunicola (Edw.)	Au		●			21.4 ± 12.0			0.48		Pm	S	2	E	0.31	3.70
Edwardsiana crataegi (Dougl.)	Au			●			13.9 ± 11.4			0.83	Pm	S	2	E	0.21	3.35
Zygina angusta (Leth.)	Au	●	○	●	6.0 ± 5.3		11.1 ± 7.5	0.63		0.26	Pm	S	1	A	0.15	2.95
Zygina flammigera (Fourcr.)	Au	□	□	□		47.8 ± 25.2			0.32		Pm	S	1	A	0.17	3.30
Feeders of fruits etc.																
Palomena prasina (L.)	H	□	□	□	5.1 ± 3.7	0.9 ± 0.9	1.2 ± 1.3	0.75	0.73	0.71	Pn,Pf	F + S	1	A	38.75	13.00
Entomophagous																
Anthocoris nemorum (L.)	H	□	□	□	2.2 ± 3.6	5.5 ± 4.9	3.6 ± 4.5	1.03	0.71		Ep	F + S	2–3	A	0.61	3.95
Anthocoris nemoralis (Fabr.)	H	○	○	●			20.2 ± 25.7			0.76	Eo	S	1–2	A	0.78	3.70
Himacerus apterus (Fabr.)	H	□	□	□	5.9 ± 5.6	7.3 ± 7.1		0.95	0.74		Ep	S	1	E	8.92	10.00
Orius minutus (L.)	H	□	□	□	4.7 ± 7.5	11.7 ± 9.7	3.0 ± 3.4	0.81	0.81	1.21	Eo	F + S	2	A	0.22	2.30

Entomophytophagous																
Heterocordylus tumidicornis (H.-S.)	H		●			31.3 ± 24.1			0.47		EpPn	S	1	E	1.27	4.45
Atractomus mali (M.-D.)	H	○	○	●			13.0 ± 14.4			0.58	EpPn	S	1	E	0.69	3.30
Phytocoris ulmi (L.)	H	□	□	□	5.7 ± 4.2	4.62 ± 2.5	2.4 ± 2.1	0.73	0.37	1.06	EpPn	S	1	E	3.20	6.80

T = Taxon: H = Heteroptera, Au = Auchenorrhyncha, Ap = Aphidoidea, P = Psylloidea.
HPP = host plant preference: Ro = *Rosa* spp., Ps = *Prunus spinosa*, Cr = *Crataegus* spp.
● = species has a high association with the shrub because it is monophagous or oligophagous (phytophagous species) or has a preference for a specific prey on the plant (entomophagous species).
□ = polyphagous and eurytopic species without specific plant preference.
○ = species has its main preference on another plant species or is polyphagous and is found only in small numbers.
mean population abundance: n = number of sites/years sampled for the plant species.
CV = coefficient of variance between years (mean CV calculated over sites that were sampled in at least three years in the Feuchtwangen region).
FS = feeding strategy: Ep = entomophagous, polyphagous; Eo = entomophagous, oligophagous (prey preference); Ep-Eo = entomophagous oligo- bis polyphagous; EpPn, EpPf = entomophytophagous, Pn = phytophagous on non-vascular plant tissues (buds, shoots, leafs); Pf = phytophagous on fruits; Pm = mesophyll feeder; Px = xylem feeder; Pp = phloem feeder.
PVL = preferred vegetation layer: S = species lives in the shrub and tree layer only (shrub dwelling species), F + S = species lives on woody plants and in the field layer as well; F > S = species changes from herbs to woody plants during its life cycle; HPA = host plant alternation in aphids.
GEN = numbers of generations per year.
HIB = stage of hibernation: E = egg, L = larva, A = adult.
BL = body length (in mm; geometric mean of the range given in the literature).
BM = mean body mass (in mg dry weight): mean body mass determinded by weighting of adult individuals (up to ten per specimens, if available) after drying them for 48 h at 60 °C; if no individual of species was available, the value of a related, similarly sized species was used.

of dominant aphid species observed at all sites during the study period (with factors of 200 between total annual abundances) may be explained by specific weather conditions (temperature) that altered the synchronization of aphids with resource quality, aphid developmental time and activity of predators, especially coccinellids.

The dominant psyllid species on hawthorn also reach high total annual densities (Table 16.1), probably due to the high overall food quality of phloem sap. Because of their ability to synchronize with the flush and flowering period and to track feeding sites of high food quality (e.g. inflorescences, leaf clusters and apical shoots; Sutton 1984; Novak 1994), psyllid population abundances in general are highly correlated with resource availability (nitrogen content, Sutton 1984). In contrast to most aphids, they have only one generation each year and year-to-year variability is considerably lower (Table 16.1; Novak 1994). As in aphids, population control derives from a mixture of density-dependent (predation, intraspecific competition) and density-independent factors (resource availability, temperature) (Novak 1994; Novak and Achtziger 1995). With the exception of *Balcanocerus larvatus* (H.-S.), all other phloem feeders show habitat alternation during their life cycles. In these cases, larvae develop in the ground (e.g. *Cixius* spp., *Tachycixius pilosus* (Oliv.) or in the field layer [e.g. *Allygus mixtus* (Fabr.)].

In the mesophyll-feeding guild (mainly leafhoppers of the subfamily Typhlocybinae), even dominant species showed lower overall densities than psyllids in all years and than aphids in gradation years (Table 16.1). That may be due to the lower level of food quality (see Sect. 16.2.2). As was shown by Prestidge (1982), mesophyll feeders probably compensate for this shortage of nitrogen by higher consumption rates. Perhaps as a consequence of high overall abundance and durational stability of full-developed leaves, various life cycles (univoltine, bivoltine, egg and adult hibernation) and resource exploitation strategies, including host-plant specialization, migration and partial host plant alternation (see Claridge and Wilson 1978), are evolved in this guild. Within the guild, population response and population variability differs between species according to their ability to track plant parts, localities and periods of higher soluble nitrogen levels and the impact of larval predators and parasitoids (e.g. Dryinidae, Pipunculidae, see Dolling 1991).

Within the entomophagous guild, most predatory species feed on aphids, psyllids and other small insects (most mirid and anthocorid bugs), some also feed on insect eggs, e.g. anthocorid bugs (see Stechmann, Chap. 15, this Vol.) or on lepidopteran larvae, e.g. larger bugs like *Himacerus apterus* (Fabr.) or *Pentatoma rufipes* (L.). In dominant anthocorid species, data suggest that local population abundances were mainly determined by prey availability. (1) Mean local annual abundances of the anthocorid *A. nemorum* on blackthorn were significantly higher in 1991, a year of high aphid infestation, than in 1992 with low aphid densities (U-test: $z = -2.19$, $p < 0.05$, $n = 17$). (2) In 1991, *A. nemorum* built a second larval generation on shrubs, whereas in 1992, only a small second generation could be observed (Achtziger 1995). (3) Local abundances of the anthocorids *A. nemorum* and *O. minutus* on blackthorn ($\log N + 1$) were correlated with aphid abundances per site ($\log N + 1$) (*O. minutus*: $r = 0.74$, $p < 0.001$, $n = 17$; *A. nemorum*: $r = 0.43$, $p < 0.1$, $n = 17$). Additionally, on rose there was a significant correlation between *O. minutus* abundances and the abundances of mites per site ($r = 0.54$, $p < 0.05$, $n = 16$). These results are consistent with other studies that found correlations between anthocorid densities and prey availability (e.g. Dixon and Russel 1972; Novak and Achtziger 1995). Thus, the observed population fluctuations (Table 16.1) in these predatory species may be explained by rapid numerical responses to prey availability and resource tracking as a consequence of their flexible and opportunistic exploitation strategy and mobile foraging behaviour (Anderson 1962; Novak and Achtziger 1995). For example, because *Anthocoris* females hibernate in the adult stage, are active early in the season (see Stechmann, Chap. 15, this Vol.) and can reproduce in a variety of habitats, they may be able to adjust the locality and the reproduction rate (numbers of eggs layed) to prey availability. In contrast, in most entomophytophagous heteropteran species (e.g. *Heterocordylus tumudicornis* on blackthorn) oviposition occurs in the preceding season. Therefore, egg-laying females cannot respond directly to prey availability and cannot predict reproductive success. In such cases, selection should favour a relatively fixed clutch size (Wilbur et al. 1974). This may explain the observed lower population variability. Additionally, facultative phytophagous species can "buffer" periods of low prey densities to some degree by switching to plant food. Studies by Jonsson (1987) on *Atractotomus mali*, an entomophytophagous mirid on apple and hawthorn, showed that mortality of nymphs decreased with increasing portions of animal material (psyllid larvae) in the food. This observation is consistent with the positive relationship between *A. mali* densities and maximal annual abundances of psyllid nymphs found by Novak and Achtziger (1995).

In conclusion, total abundance and overall population variability can be explained by the population response of a species (which is constrained by its life-history traits like generation time or lifetime reproduction) to the temporal and spatial dynamics of resource availability (which is influenced by biotic and abiotic conditions, see Sect. 16.2.2). As Blackburn et al. (1996) showed for British birds, species with faster development generally have higher abundances. They suggest that "some taxa have access to a larger resource base and that this is reflected in the abundance a species can attain, and the rate at which it can provision offspring" (Blackburn et al. 1996, p. 60). Comparisons of dominant guild members presented here suggest similar patterns: Higher levels of overall resource availability may be reflected in higher overall abundances of aphids or psyllids feeding on phloem compared with numbers of dominant mesophyll feeding species or fruit feeding species (e.g. *Palomena prasina*; Table 16.1). Since species feeding on pulsing resource types also have traits that enable them to respond rapidly to increasing food availability ("flush feeding type", White 1993), they often show higher population variability than species feeding on resources that are available more constantly throughout the season, e.g. mesophyll feeders ("senescence feeding type", White 1993). As indicated by correlations presented above, the relationship between resource availability and abundance also holds true on local scales (e.g. sites). To some extent, the local abundance of a species and thus species-abundance patterns and guild structure of a local community may reflect the availability of resources at that site in a season (cf. Kolasa 1989).

Of course, the tentative resource-based (bottom-up) explanations given here cannot describe population dynamics completely and have to be verified in detail. Furthermore, factors such as dispersal, migration and exchange of individuals between biotopes (patch dynamics, see Wu and Loucks 1995) as well as density-dependent processes may influence population density and hence community structure at a given site or in a season. However, the study on the spatiotemporal dynamics of Hemiptera communities showed that among factors potentially influencing or controlling individual population densities, none could be identified as playing a dominant role in community structure ("diffusive density control"). For example, there were no signs of density compensation within one guild, niche space saturation or negative relationships between population densities of potential competitors (Achtziger 1995). Therefore, hemipteran communities may be regarded as non-interactive communities with horizontal, interspecific species interactions playing a more minor role in community structure than vertical interactions (see Lawton 1990).

16.4 Resource and Habitat Diversity, Surface Area and Local Species Richness

This section will consider local species richness or α diversity, i.e. the number of hemipteran species at a given site (on a single shrub species in a hedge or at a forest edge) and its dependence on resource diversity and surface area of the plantscape in a hedge. According to the hypothesis mentioned in Section 16.2.2, hedges with higher habitat heterogeneity, resource diversity and surface area should provide more living conditions and microhabitats for species with different traits (niches) and thus should contain more insect species than hedges with a lower supply of microhabitats and resources (Huston 1994). To test this habitat-heterogeneity/surface-area hypothesis, data sets were analyzed on Hemiptera communities (Heteroptera and Auchenorrhyncha) on rose, blackthorn and hawthorn of 20 hedges that differed in age class (ranging from younger than 10 years to older than 50 years) and hedges' parameters (hedge width, hedge length, number of woody plant species, Table 16.2; 1989, vicinity of Bayreuth). Resource diversity and surface area of a hedge can hardly be measured directly. As indirect measures for these parameters were used: hedge width, the lateral extension of the hedge body (see Sect. 16.2.2) and a parameter of overall structural diversity that was originally derived for

Table 16.2. Correlations between hemipteran species numbers on rose, blackthorn and hawthorn per hedge with hedge parameters (Bayreuth region, 1989)

Parameter	Number of species		
	On rose n = 15	On sloe n = 14	On hawthorn n = 16
Hedge width (m) (all age classes)	$r = 0.72^{**}$	$r = 0.70^{**}$	$r = 0.65^{**}$
Factor for overall structural diversity ("hedge evaluation factor")	$r = 0.75^{**}$	$r = 0.79^{***}$	$r = 0.60^{**}$
Length of hedge (m)	$r = 0.13^{ns}$	$r = 0.15^{ns}$	$r = 0.11^{ns}$
Number of woody plant species	$r = -0.07^{ns}$	$r = -0.14^{ns}$	$r = -0.21^{ns}$
Age class	$r_s = 0.77^{***}$	$r_s = 0.65^{*}$	$r_s = 0.71^{**}$

ns = not significant, * = $p < 0.05$, ** = $p < 0.01$, *** = $p < 0.001$.
r = Pearson correlation, r_s = Spearman rank correlation.

hedgerow evaluation in nature conservation ("hedge evaluation factor", see Zwölfer et al. 1984) and which integrates (1) age class diversity of woody plants, (2) identity and number of woody plant species and (3) regional hedge density.

Consistente with the hypothesis, local hemipteran species richness on rose, blackthorn and hawthorn showed a highly positive correlation with hedge width and the hedge evaluation factor (Table 16.2). No significant correlations were obtained, however, between species diversity on a plant species and the numbers of woody plant species in the hedge or the length of the hedge. Of course, if one considers the total hemipteran species numbers in a hedge, summarized over all woody plant species, a significant increase in insect species richness with plant diversity is to be expected (Zwölfer and Stechmann 1989; Achtziger 1995, see also contributions of Stechmann, Chap. 15, and Boller et al., Chap. 17, this Vol.). Besides other factors like land use activities, shrub species diversity in a hedge is determined by hedge length as a result of species-area relationships ($r = 0.63$, $p = 0.004$, $n = 20$) (see also MacDonald and Johnson 1995).

Furthermore, species richness did increase with the age of the hedge, measured as age classes (Table 16.2). On average, species numbers were significantly higher in older hedges (older than 20 years) than in younger ones (younger than 20 years) (U-tests, $p < 0.05$). This pattern may be explained by two time-dependent processes: (1) the increase in resource diversity, habitat heterogeneity and surface area with time as indicated by the age-dependent increase of fractal dimension and lateral hedge extension (= hedge width) in Section 16.2.2, and (2) pure colonization time effects, i.e. the fact that species need some time for finding and establishing in a new hedge. Therefore, time plays a key role for the development of both structural and taxonomic diversity in hedges and forest margins (cf. Pollard et al. 1974; Cameron et al. 1980).

In an attempt to separate the effects of time-dependent increase in resource diversity/quantity and of time per se, data were analyzed of hemipteran species accumulation during colonization of young, 2-year-old blackthorn and hawthorn shrubs that were planted in experimental plots at forest margins in the Feuchtwangen region (see Kögel et al. 1993). As a measure of structural resource diversity and surface area at a site, the mean height of ten individual shrubs was determined at each site (Achtziger 1995). According to local climatic and edaphic conditions at a site, shrub growth and hence mean shrub height differed even between sites of the same age. In Fig. 16.1a the number of hemipteran species per site on blackthorn is plotted with the mean shrub height per site. The age of the planting (= time for colonization) is

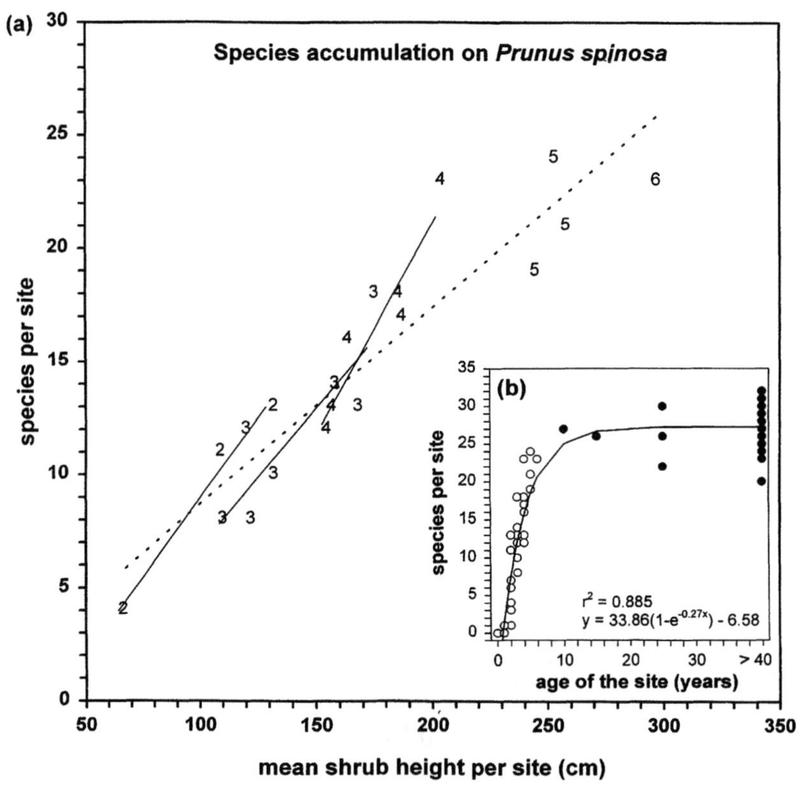

Fig. 1.1a,b. Hemipteran species accumulation during colonization of blackthorn shrubs planted at experimental forest margins (Feuchtwangen region). **a** Number of hemipteran species per site and year plotted with mean height of shrubs at the site (see text); the age of each site is indicated by *numbers* (2 to 6 years); *dotted line* overall regression: $y = 0.02 + 0.09x$, $r = 0.92$, $p < 0.001$, $n = 20$ regressions for each age class: 2 years: $r = 0.99$, $p < 0.1$, $n = 3$, 3 years: $r = 0.86$, $p < 0.05$, $n = 7$, 4 years: $r = 0.95$, $p < 0.01$, $n = 6$; 5 years: $r = 0.53$, not significant. **b** Number of hemipteran species on blackthorn plotted with the age of the site; ○ = on young shrubs planted at forest margins, ● = shrubs growing at older forest margins in the Feuchtwangen region

indicated as number of years. According to their migratory ability and habitat requirements, hemipteran species originating from the regional species pool colonized the new biotopes (Achtziger 1995): local species richness increased linearly up to 5 years, when bushes reached a height of about 2.50 m (dotted line in Fig. 16.1a, open circles in Fig. 16.1b). Comparisons with 10-year-old planted hedges, middle-aged (about 25 years) and old (more than 40 years) forest margins suggest an asymtotical course of hemipteran species accumulation on shrubs (Fig. 16.1b). Interestingly, the correlation between species diversity and mean shrub height was also obtained when sites of the same age were analyzed separately (Fig. 16.1a, regression lines for 2-, 3- and 4-year-old sites). Forest margins with larger shrubs and thus with a higher supply of resources, resource types and microhabitats can support more species than those with a poorly developed shrub layer, even if the time for colonization is equal. These results suggest that the number of species establishing in a new scrub habitat during colonization is primarily determined by the quality and quantity of resources (microhabitats, surface area) available at a site. Potential colonizers can establish successfully at a site only when their habitat requirements and traits ("niches", see Brown 1995) match the environmental conditions (resource types, microhabitats, foraging area, oviposition sites etc.). Therefore, species accumulation rate during colonization is constrained by the time-dependent development of structural and trophic resources.

16.5 Linking Ecology and Evolution: Body Size Relationships in Hemipteran Communities

The evolution of the body size of a species underlies both intrinsic constraints (e.g. tradeoffs, phylogenetic constraints) and extrinsic constraints (derived from the environment, Barbault 1991). Therefore, body size is correlated with most of a species life-history traits by allometric relations or trade-offs (e.g. Peters 1983) and thus may provide a convenient surrogate for life-history variables that are hard to estimate or are not known in detail (Blackburn and Gaston 1994). In the shrub-dwelling hemipteran species, for instance, species with two and more generations per year (with shorter generation times) had significantly lower body weights than univoltine species (U-test: $z = -3.69$, $p < 0.001$, $n = 78$). This is consistent with the well-known relationship between body size and generation time which was also found in other insect taxa (e.g. Gaston and Reavey 1989). Within phytophages, mono- and oligophagous species were smaller on average than polyphagous ones (U-test: $z = -2.87$, $p < 0.05$, $n = 35$), which appears to be a general rule in many insect assemblages (e.g. Gaston and Lawton 1988; Lindström et al. 1994).

Furthermore, there are relationships between body size and population and community parameters, e.g. species abundance (see Blackburn and Lawton

1994). Figure 16.2 shows the abundance-body size relationship of hemipteran species on blackthorn at forest margins (pooled data over all sites in the Feuchtwangen study region from 1990 to 1994). Log abundance was negatively correlated with log body size (Fig. 16.2), but with a great scatter around the regression line ($r^2 = 0.13$): The data show a roughly triangular pattern with low abundances in large species (e.g. fruit feeders) and both high and low abundances in small species (aphids, Typhlocybinae). In contrast to the energy equivalence rule, stating a slope of −0.75 because of energetic requirements (Damuth 1991), the slope was only −0.51. Similar results were obtained in an analysis of 44 data sets including local hemipteran communities and data on assemblages pooled over years and localities on rose, blackthorn and hawthorn (Achtziger 1995). Since these patterns are consistent with results of many other insect assemblages that do not follow the predictions of the energy equivalence rule, non-metabolic explanations for abundance-body size relationships were proposed (see reviews in Lawton 1990; Blackburn and Lawton 1994).

In hemipteran assemblages, the abundance-size relationships differ between guilds: whereas abundances of phytophages show a highly significant relationship with body size ($r = -0.53$, $p < 0.005$, $n = 27$), in the entomophagous

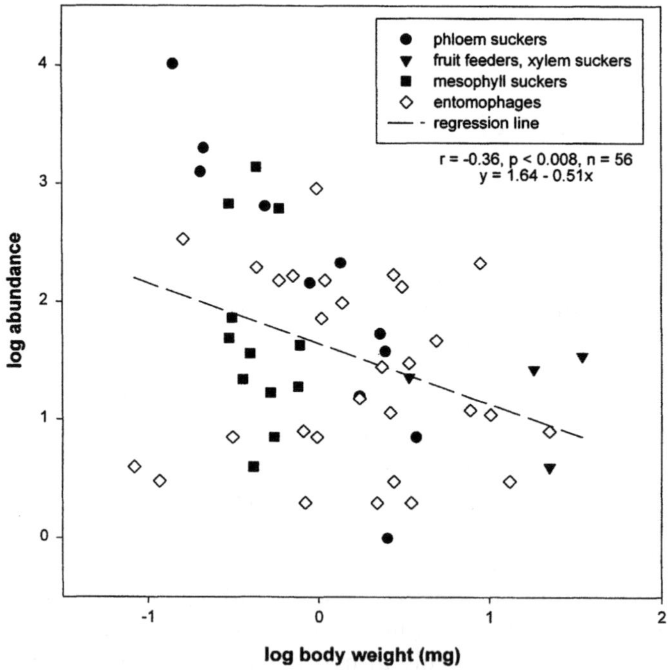

Fig. 16.2. Abundance-body size relationship for hemipteran species and guilds on blackthorn at old forest margins; abundances are pooled over all study years (1990–1994) and all sites in the Feuchtwangen region

guild no significant correlation could be obtained (r = –0.1, p = 0.61, n = 29). If the regression is calculated without the two smallest species (points lying in the left lower corner of Fig. 16.2), a marginally significant trend appears (r = –0.34, p = 0.08, n = 27). An interesting observation is the different abundance-size relationship in mesophyll suckers (Typhlocybinae, Tingidae) and phloem suckers (aphids, psyllids, some Auchenorrhyncha; Fig. 16.2): In phloem feeders log abundance is highly correlated with log body mass (r = –0.9, p = 0.0001, n = 12), whereas in the mesophyll feeding guild no correlation exists (r = –0.2, p = 0.54, n = 12). A similar result is obtained for hemipteran species on hawthorn (phloem suckers: r = –0.65, p = 0.03), n = 11; mesophyll suckers: r = –0.15, p = 0.71, n = 9).

These results correspond to the observation that, though the range of abundances are similar (Fig. 16.2), the ranges of body size differ between the species of the subguilds (Fig. 16.3): mesophyll feeders are restricted to a body size of about 0.30 mg/3.5 mm, whereas phloem feeders range from 0.25 mg/1.8 mm in aphids to 6 mg/8.5 mm in some Auchenorrhyncha.

Within plant suckers, species feeding on the content of phloem or mesophyll cells on leaves (e.g. aphids, psyllids, Typhlocybinae, lace bugs) are

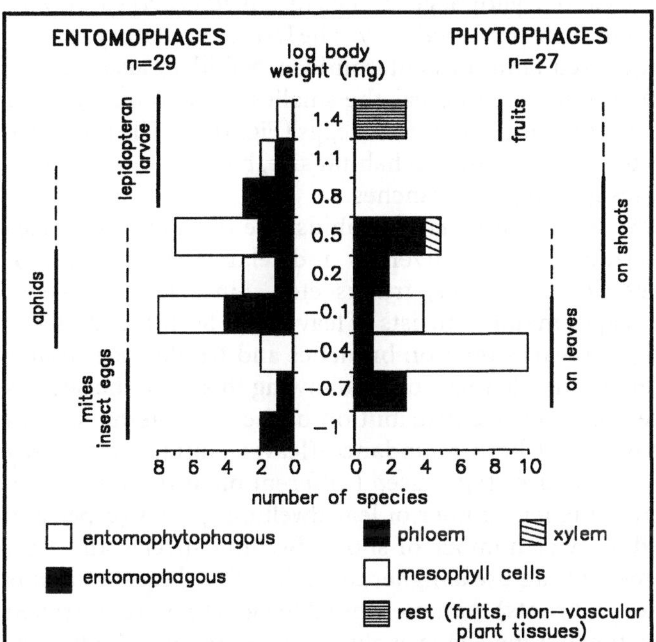

Fig. 16.3. Species-body size distribution of the hemipteran fauna on blackthorn. Frequency distribution of the logarithm of body weight (dry weight in mg), separated for entomophagous (*left*) and phytophagous subguilds (*right*). *Vertical lines* indicate range of prey type (entomophages) or plant organs (phytophages) used by species of the corresponding body sizes; *solid line* main resource type

smaller than plant bugs lacerating tissues of special plant organs like fruits, buds, shoots etc. (cf. Dixon and Kindlmann 1994). Therefore, body size is related to the type and size of the resource used and - besides phylogenetic constraints - functional constraints derived from the resource or the habitat may have affected the evolution of species body size (see Griffiths 1986; Holling 1992). For example, adult body size in aphids, which is highly correlated with proboscis length, mainly depends on the depth of the phloem elements in the plant organs on which the species is feeding (Dixon et al. 1995). As a consequence, aphid species feeding on branches, shoots or trunks of woody plants are larger than those feeding on leaves (Dixon and Kindlmann 1994).

Similar functional constraints may explain the larger body size of leafhoppers feeding on shoots (the second peak in phloem feeders in Fig. 16.3) and the larger size of xylem feeders [*Aphrophora alni* (L.)], because xylem elements lie deeper in the plant and can be reached only by long and robust probosces. Similar considerations may explain that predatory species were larger than phytophagous ones (U-test: $z = -1.75$, $p < 0.08$, $n = 56$). Predator-prey size relations with predators being larger than their preferred prey were found in many food webs (Cohen et al. 1993). In the data set considered here, the mainly aphidophagous and psyllophagous bugs (e.g. *Anthocoris* spp.) were about two times larger than their preferred prey. The largest predatory species (Heteroptera of the large-sized families Pentatomidae, Nabidae) also feed on larger prey like lepidopteran larvae, whereas the smallest species mainly feed on mites, small insects (aphid nymphs) or insect eggs (Fig. 16.3). The body size of the two smallest species is also related to habitat size, because these species live in lichens and mosses growing on branches.

As Dixon et al. (1995) showed for British aphids, the diversity in species body sizes may be determined by the diversity and relative surface area of specific plant structures (leaves, shoots, trunks etc.). Since the surface of woody plants (the plantscape) mainly consists of leaves (see Sect. 16.2.2), there are more species feeding on leaves than on branches and trunks (see similar arguments in Stechmann, Chap. 15, this Vol.). Following this idea, the species composition and hence the body size distribution of the hemipteran species pool on blackthorn may reflect both abundance (leaves, surface area, prey densities) and diversity of resource types used (different plant tissues or prey size classes), resulting in a higher number of leaf-dwelling species (especially mesophyll suckers) and a lower number of shoot-dwelling species and fruit suckers. Since phloem feeders can use a higher diversity of feeding sites (veins in leaves, twigs, branches and roots) than mesophyll feeders that are restricted to leaf cells of equal size, the first may have evolved a wider range of body sizes than the latter.

As pointed out by Blackburn et al. (1996) and in Section 16.3, abundance is often related to fast development. Because body size is also positively correlated with developmental time (small species have shorter generation times, see above), small phloem suckers may reach higher abundances than larger ones, resulting in an abundance-size relationship among species of this subguild (Fig. 16.2).

Since the species composition of a local community is mainly influenced by habitat features like resource diversity and surface area (Sect. 16.4) and species abundance in many cases is correlated with resource availability (Sect. 16.3), abundance-body size relationships may reflect both resource diversity (via species composition and associated traits, e.g. body size) and resource availability (via species abundances). Abundance-body size patterns can serve as comprehensive pictures of community structure and resource situation in a local ecosystem.

16.6 Conclusions

Communities are the result of processes acting on different spatial, temporal and organizational scales (Giller and Gee 1987). To understand their structure and dynamics, one has to consider both proximate (ecological) and ultimate (evolutionary) factors (Zwölfer 1987). The fauna living on woody plants of hedges and forest margins represents a good example for an ecological system whose structure, composition and dynamics is determined both by ecological/environmental and evolutionary/phylogenetic processes. As a consequence of their anthropogeneic origin (e.g. Schulze and Gerstberger 1993), hedges are, in evolutionary terms, very young ecosystems and therefore lack an autochthonous hedge fauna (Zwölfer and Stechmann 1989). Nevertheless, they inhabit a particular mixture of species with different life-history traits, resource exploitation patterns, and ecological, physiological, behavioural and morphological adaptations (see Sects. 16.2.3, 16.3, 16.5), resulting in specific community patterns.

The actual composition of the fauna, regional species pools and local communities can be regarded as the result of individual adaptations of each species to ecological constraints and driving forces in evolutionary and ecological time (e.g. dynamics and patchiness of nutritional and structural resources, see Sects. 16.2.2, 16.4, 16.5). During their individual evolutionary history, each species adapted to different environmental constraints by the evolution of specific resource exploitation strategies (Sect. 16.3), body size (Sect. 16.5) and other species traits. The taxonomic relatedness and common habitat requirements of the three rosacean plant species (original habitats are suspected to be the understorey of light forests and edges of natural clearings, Ellenberg 1986) may have promoted the colonization by a common species pool and the radiation of some genera (*Edwardsiana*, *Zygina*). Differences in number and identity of aphid and psyllid species, however, also indicate different and idiosyncratic colonization processes. In historical time (after the emergence of hedges and secondary forest margins as a result of human land use), especially oligophagous, polyphagous and many entomophagous species whose adaptations, life histories and exploitation strategies matched the specific environmental conditions (resource dynamics, habitat structure) established in the new biotopes. Since species respond differently to environmental variation according to their traits, and environmental conditions vary in space and time (e.g. within a region, during colonization and succession, e.g. Brown and Southwood 1983) these "matching processes" between organismal traits and environmental conditions find their expression in local community dynamics (Sects. 16.3, 16.4) and biogeographical range dynamics (Hengeveld 1994). The observation of some species that extend their biogeographical ranges and have in some regions established local hemipteran communities in the recent years (e.g. *Balcanocerus larvatus*, Achtziger 1995) may provide examples for biogeographical processes. In conclusion, vertical interactions of the plant-insect system concerning both trophic (Sect. 16.3) and structural aspects (Sects. 16.4, 16.5) are major driving forces for the evolution of species traits in evolutionary time and important determinants for community structure and organization (e.g. species-abundance patterns, species richness) in ecological time. Consequently, local hemipteran communities are not completely random assemblages of the total species pool, but exhibit distinct spatio-temporal patterns that can be predicted at least in basal structures. With the exception of aphid gradations, essential community organization features such as species ranks, identity of dominant species, guild structures, qualitative species composition and local species richness remained relatively stable and were reasonably predictable in both temporal (between years) and spatial terms (between sites within a region) (Achtziger 1995). Compared with other plant-insect systems (e.g. the fauna of thistle flower heads, Zwölfer 1994), the Hemiptera communities of hedges and forest margins take an intermediate position within a continuum from highly organized systems/predictable structure and chance-constructed systems/without predictable structure.

Acknowledgements. I thank Professor Dr. Helmut Zwölfer for many comments and much helpful advice during my work. I benefited greatly from his fundamental investigation on hedge ecosystems and his view of nature, emphasizing both details and major processes. W. Völkl and G. Bauer kindly corrected the manuscript. The study was funded by the German Federal Environmental Ministery (BMUNR) and the Federal Department of Nature Conservation (BfN) within the study project Development of Rich-Structured Forest Edges (N1-072(89)-89211-2/2).

References

Achtziger R (1995) Die Struktur von Insektengemeinschaften an Gehölzen. Die Hemipteren-Fauna als Beispiel für die Biodiversität von Hecken- und Waldmantelökosystemen. PhD Thesis, Bayreuth University, Bayreuth, Bayreuther Forum Ökologie (BFÖ) 20, 216 pp

Anderson NH (1962) Bionomics of six species of *Anthocoris* (Heteroptera: Anthocoridae) in England. Trans R Entomol Soc Lond 114:67–95

Barbault R (1991) Ecological constraints and community dynamics: linking community patterns to organismal ecology – the case of tropical herpetofaunas. Acta Oecol 12(1):139–163

Barbault R, Stearns S (1991) Towards an evolutionary ecology linking species interactions, life-history strategies and community dynamics: an introduction. Acta Oecol 12(1):3–10
Bauer G (1986) Life-history strategy of *Rhagoletis alternata* (Diptera: Trypetidae), a fruit fly operating in a "non-interactive" system. J Anim Ecol 55:785–794
Blackburn TM, Gaston KJ (1994) Animal body size distributions: patterns, mechanisms and implications. TREE 9:471–205
Blackburn TM, Lawton JH (1994) Population abundance and body size in animal assemblages. Philos Trans R Soc Lond B 343:33–39
Blackburn TM, Lawton JH, Gregory RD (1996) Relationships between abundances and life histories of British birds. J Anim Ecol 65:52–62
Brown JH (1995) Organisms and species as complex adaptive systems: linking the biology of populations with the physics of ecosystems. In: Jones CG, Lawton JH (eds) Linking species and ecosystems. Chapman & Hall, New York, pp 16–24
Brown VK, Southwood TRE (1983) Trophic diversity, niche breadth and generation times in exopterygote insects in a secondary succession. Oecologia 56:220–225
Cameron RAD, Down K, Pannett DJ (1980) Historical and environmental influences on hedgerow snail faunas. Biol J Linn Soc 13:75–87
Claridge MF, Wilson MR (1978) Seasonal changes and alternation of food plant preferences in some mesophyll-feeding leafhoppers. Oecologia 37:247–255
Cohen JE, Pimm SL, Yodzis P, Saldanas J (1993) Body sizes of animal predators and animal prey in food webs. J Anim Ecol 62:67–78
Damuth J (1991) Of size and abundance. Nature 351:268–269
Dixon AFG, Kindlmann P (1994) Optimum body size in aphids. Ecol Entomol 19:121–126
Dixon AFG, Russel RJ (1972) The effectiveness of *Anthocoris nemorum* and *Anthocoris confusus* (Hemiptera: Anthocoridae) as predators of the sycamore aphid, *Drepanosiphon platanoides*, II. Searching behaviour and the incidence of predation in the field. Ent Exp Appl 15:35–50
Dixon AFG, Kindlmann P, Jarosik V (1995) Body size distribution in aphids: relative surface area of specific plant structures. Ecol Entomol 20:111–117
Dolling WR (1991) The Hemiptera. Oxford University Press, Oxford, 274 pp
Ellenberg H (1986) Vegetation Mitteleuropas mit den Alpen in ökologischer Sicht. Ulmer, Stuttgart, 989 pp
Flückinger W, Oertli JJ (1978) Observations of an aphid infestation of hawthorn in the vicinity of a motorway. Naturwissenschaften 65:654–655
Gaston KJ, Lawton JH (1988) Patterns in the distribution and abundance of insect populations. Nature 331:709–712
Gaston KJ, Reavey D (1989) Patterns in life histories and feeding strategies of British Macrolepidoptera. Biol J Linn Soc 37:367–381
Giller GS, Gee JHR (1987) The analysis of community organization: the influence of equilibrium scale and terminology. In: Gee JHR, Giller GS (eds) Organization of communities: past and present. Symp Br Ecol Soc 27:519–542
Griffiths D (1986) Size-abundance relationships in communities. Am Nat 127:140–166
Hengeveld R (1994) Biogeographical ecology. J Biogeogr 21:341–351
Holling CS (1992) Cross-scale morphology, geometry, and dynamics of ecosystems. Ecol Monogr 62(4):447–502
Huston MA (1994) Biological diversity. Cambridge University Press, Cambridge
Johnson DH (1980) The comparison of usage and availability measurements for evaluating resource preference. Ecology 61:65–71
Jonsson N (1987) Nymphal development and food consumption of *Atractotomus mali* (Meyer-Dür) (Hemiptera: Miridae), reared on *Aphis pomi* (DeGeer) and *Psylla mali* Schmidberger. Fauna Norv Ser B 34:22–28
Kareiva P (1986) Patchiness, dispersal, and species interactions: consequences for communities of herbivorous insects. In: Diamond J, Case TJ (eds) Community ecology. Harper & Row, New York, pp 192–206

Kögel K, Achtziger R, Blick T, Geyer A, Reif A, Richert E (1993) Aufbau reichgegliederter Waldränder - ein E&E-Vorhaben. Natur Landschaft 68:386–394

Kolasa J (1989) Ecological systems in hierarchical perspective: breaks in community structure and other consequences. Ecology 70:36–47

Küppers M (1992) Changes in plant ecophysiology across European hedgerow ecotone. In: Hansen AJ, diCastri F (eds) Landscape boundaries. Ecological Studies, vol 92. Springer, Berlin Heidelberg New York, pp 285–303

Lawton JH (1986) Surface availability and insect community structure: the effects of architecture and fractal dimensions of plants. In: Juniper BD, Southwood TRE (eds) Insects and the plant surface. Edward Arnold, London, pp 317–331

Lawton JH (1990) Species richness and population dynamics of animal assemblages. Patterns in body size: abundance space. Philos Trans R Soc Lond B 330:283–291

Leather SR (1986) Insect species richness of the British Rosaceae: the importance of host range, plant architecture, age of establishment, taxonomic isolation and species-area relationships. J Anim Ecol 55:841–860

Lindström J, Kaila L, Niemelä P (1994) Polyphagy and adult body size in geometrid moths. Oecologia 98:130–132

MacDonald DW, Johnson PJ (1995) The relationship between bird distribution and the botanical and structural characteristics of hedges. J Appl Ecol 32:492–505

Mattson WJ (1980) Herbivory in relation to plant nitrogen content. Annu Rev Ecol Syst 11:119–161

McNeill S, Southwood TRE (1978) The role of nitrogen in the development of insect/plant relationships. In: Harborne JB (eds) Biochemical aspects of plant and animal coevolution. Academic Press, London, pp 77–98

Minks AK, Harrewijn P (eds) (1987) Aphids. Their biology, natural enemies and control. World Crop Pests 2 A. Elsevier, Amsterdam, 450 pp

Minks AK, Harrewijn P (eds) (1988) Aphids. Their biology, natural enemies and control. World Crop Pests 2 B. Elsevier, Amsterdam, 364 pp

Morse DR, Lawton JH, Dodson MM, Williamson MH (1985) Fractal dimension of vegetation and the distribution of arthropod body lengths. Nature 314:731–733

Novak H (1994) Wechselbeziehungen zwischen Weißdornpsylliden und ihren Gegenspielern und Trophobiosepartnern. PhD Thesis, Bayreuth University, Bayreuth, 139 pp

Novak H, Achtziger R (1995) Influence of heteropteran predators (Heteroptera. Anthocoridae, Miridae) on larval populations of hawthorn psyllids (Hom., Psyllidae). J Appl Entomol 119:479–486

Peters RH (1983) The ecological implications of body size. Cambridge University Press, Cambridge

Pollard E, Hooper M, Moore NW (1974) Hedges. Collins, London, 256 pp

Prestidge RA (1982) Instar duration, adult consumption, oviposition and nitrogen utilization efficiencies of leafhoppers feeding on different quality food (Auchenorrhyncha: Homoptera). Ecol Entomol 7:91–101

Price PW (1984) Alternative paradigms in community ecology. In: Price PW, Slobodchikoff CN, Gaud WS (eds) A new ecology - novel approaches to interactive systems. John Wiley, New York, pp 353–383

Samways MJ (1994) Insect conservation biology. Chapman & Hall, London, 358 pp

Schulze E-D, Gerstberger P (1993) Functional aspects of landscape diversity: a Bavarian example. In: Schulze E-D, Mooney HA (eds) Biodiversity and ecosystem function. Ecological studies, vol 99. Springer, Berlin Heidelberg New York, pp 453–466

Strong DR, Lawton JH, Southwood TRE (1984) Insects on plants - community patterns and mechanisms. Blackwell, Oxford, 313 pp

Sutton RD (1984) The effect of host plant flowering on the distribution and growth of hawthorn psyllids (Homoptera: Psylloidea). J Anim Ecol 49:209–224

White TCR (1993) The inadequate environment. Nitrogen and the abundance of animals. Springer, Berlin Heidelberg New York, 425 pp

Wiens JA (1984) Resource systems, populations, and communities. In: Price PW, Slobodchikoff CN, Gaud WS (eds) A new ecology: novel approaches to interactive systems. Wiley, New York, pp 397–406

Wilbur HM, Tinkle DW, Collins JP (1974) Environmental certainty, tropic level, and resource availability in life history evolution. Am Nat 108:805–817

Williamson MH, Lawton JH (1991) Fractal geometry of ecological habitats. In: Bell SS, McCoy ED, Mushinsky HR (eds) Habitat structure – the physical arrangement of objects in space. Chapman & Hall, London, pp 70–86

Wu J, Louks OL (1995) From balance of nature to hierarchical patch dynamics: a paradigm shift in biology. Q Rev Biol 70:439–465

Zwölfer H (1986) Insektenkomplexe an Disteln – ein Modell für die Selbstorganisation ökologischer Kleinsysteme. In: Dress A, Hendrichs H, Küppers G (eds) Selbstorganisation – zur Bedeutung eines neuen disziplinenübergreifenden Paradigmas für die Einzelwissenschaften. Piper, München, pp 181–217

Zwölfer H (1987) Species richness, species packing, and evolution in insect-plant systems. In: Schulze E-D, Zwölfer H (eds) Potentials and limitations of ecosystem analysis. Ecological studies, vol 61. Springer, Berlin Heidelberg New York, pp 301–319

Zwölfer H (1988) Evolutionary and ecological relationships of the insect fauna of thistles. Annu Rev Entomol 33:103–122

Zwölfer H (1994) Structure and biomass transfer in food webs: stability, fluctuations and network control. In: Schulze E-D (ed) Flux control in biological systems: From the cell to the ecosystem level. Academic Press, San Diego, pp 365–420

Zwölfer H, Stechmann D-H (1989) Struktur und Funktion von Hecken in tierökologischer Sicht. Verh Ges Ökol 17:643–655

Zwölfer H, Bauer G, Heusinger G, Stechmann D-H (1984) Die tierökologische Bedeutung und Bewertung von Hecken. Berichte der Akademie für Naturschutz und Landschaftspflege (ANL), Beih 3, Teil 2, Laufen/Salzach

17 Biodiversity in Three Trophic Levels of the Vineyard Agro-Ecosystem in Northern Switzerland

E.F. Boller, D. Gut, and U. Remund

17.1 The Vineyard Agro-Ecosystem in Northern Switzerland

17.1.1 Biodiversity as Important Element of Modern Sustainable Viticulture

Biodiversity and the important role it plays in agro-ecosystems have been studied in many agricultural crops, the subject being covered in an excellent review by Altieri (1994). In this chapter we present our own findings and conclusions reached after some 12 years of investigations conducted in vineyards of northern Switzerland.

Modern vineyards situated in the foothills of the Alps and managed according to the basic principles of sustainable agriculture already show even from a distance that traditional viticulture has been substantially transformed: there is green cover between and below the grapevines throughout the year with great diversity of flowering plants. Not so obvious at first sight, but detected on closer examination by an experienced observer, is an astonishing richness of arthropod species estimated to range between 2000 and 3000 species in our research vineyard at Walenstadt. Indeed, modern vineyards have successfully made the transition from monotonous monocultures to diversified agro-ecosystems of considerable complexity (Boller and Remund 1986; Boller 1988).

Vineyards have a special capacity to establish and maintain a high degree of biodiversity and to establish rather complex ecosystems as important elements of a sustainable agriculture. Unlike arable crops in a short-lived annual agro-ecosystem, the grapevine occupies the site for some 30 or more years and has the potential to establish a rich community of plant and animal species and hence agroecosystems of high complexity. As opposed to modern orchards with more intensive management procedures (e.g. heavy machinery, frequent pesticide applications), it is possible for vineyards situated on steep slopes and small-scale terraces to develop a considerable species richness within the Habitat (α diversity) as well as a great habitat diversity within the vineyard itself (β diversity). Stone walls, shrubs, hedges and trails largely unaffected by continuous human interference constitute important components of habitat diversity. Hence, ecological compensation areas can be integral parts of ecologically diversified vineyards and do not have to be situated outside the crop area as is the rule in most annual agro-ecosystems.

Ecological Studies, Vol. 130
Dettner et al. (eds.) Vertical Food Web Interactions

The biological diversity displayed by such a modern vineyard has not only high aesthetical and environmental value (diversification of landscapes) but is also of considerable economic importance. The concept of a permanent green cover in vineyards was originally developed in our region in the late 1960s to prevent erosion on steep slopes caused by annual precipitation ranging from 800–1300 mm (Perret and Koblet 1973). This objective triggered an intensive search for suitable cover plants and weed management systems that do not compete with the grapevine for nutrients and water (Stalder and Potter 1979). Later, entomologists recognised the potential of the undergrowth to increase the abundance of important natural enemies of major pest species by providing food, shelter and a supply of alternate prey and host species.

The considerable impact of natural enemies on major pest species observed over the years stimulated further research into the quality of the plants required to achieve the desired effects (Boller et al. 1988; Boller and Frey 1990; Boller and Wiedmer 1990; Remund et al. 1989, 1992; Björnsen 1995). Soon it became evident that two factors were of particular interest: (1) the presence of a permanent supply of (suitable) flowering plants, and (2) a wide range of plant species with a high proportion of perennial dicotyledones.

One simple measure was introduced in 1985 that met both requirements and found immediate acceptance by grape growers: the alternating mowing regime (Remund et al. 1989). The basic idea is to cut only every second alley or terraced slope when the green cover is in full bloom, and mow the rest when new flowering plants have regrown on the mowed surfaces. Hence, 50% of the vineyard surface exhibits flowering plants throughout spring and summer, providing nectar for flower-visiting arthropods and larval food for a wide array of herbivores developing on biomass of perennial forbs. In addition, the seeds necessary for the enhancement of a species-rich plant composition are produced.

17.1.2 Characteristics of the Study Sites

First investigations on the botanical and faunistic aspects of the agro-ecosystem vineyard were carried out between 1984 and 1986 in our 2-ha vineyard at Walenstadt (Remund et al. 1989). The conclusions drawn from these detailed studies were verified in a 4-year survey conducted from 1987–1990 in 21 vineyards scattered over the entire viticultural area of German-speaking Switzerland. All participating grape growers practised either integrated production (IP) or organic farming and started at the beginning of the project to apply the alternating mowing regime. All grapevines were trained on a wire frame. Five of the 21 vineyards under investigation were situated on steep slopes with small terraces, whereas 16 vineyards were located on relatively flat foothills directly accessible to mechanisation. Soil conditions varied greatly among the farms. Local climate was predominantly characterised by substantial differences in precipitation during the growing season (April-October) ranging from 500 to 900 mm. Both factors strongly influenced the type of plant species and the ensuing arthropod fauna.

17.2 First Trophic Level: Flora

17.2.1 The Development of Permanent Ground Cover

The traditional form of weed control, manual hoeing, is probably as ancient as viticulture itself. This cultivation system led to a typical flora, adapted to regular soil disturbances. Mechanical tillage was, where possible, general practice in the first half of the 20th century. The development of effective herbicides resulted in a decline of soil tillage. Damage caused by erosion and a degradation of the soil structure prompted the investigation of cover crops in the vineyards (Perret and Koblet 1973). For the past two decades, vineyard soils in the eastern parts of Switzerland have been mostly covered either by sown ground covers or, more frequently, by natural vegetation.

New management techniques changed the botanical composition and led to the appearance of other vegetation types (Wilmanns and Bogenrieder 1991). The traditional wild-herb community type of vineyard undergrowth, the *Geranio rotundifolii-Allietum vinealis*, has undergone dramatic changes due to the widespread establishment of permanent ground covers in the last decades. Intensive mulching, instead of the traditional soil tillage, hoeing or herbicide application, resulted in a partial impoverishment of the flora (bulb geophytes and some therophytes disappeared almost completely). This management often promoted the development of grass-dominated plant communities with a relatively small number of plant species.

Improving plant species richness is an important goal of today's undergrowth management in Swiss vineyards. This can be achieved to a limited extent by creating new habitats within the boundaries of the vineyard (stone walls, borders of stairs, etc. increase β diversity). Increasing species richness in the vineyard itself (α diversity) consists of a suitable management of the existing plant communities in the strip underneath the vines, in the alleys between rows, and in the banks of terraced vineyards.

17.2.2 Spatial Differentiation of Habitats in the Vineyard

In traditional vineyards with high-density vine plantings, soil cultivation and weed control were uniform over the whole surface. The resulting vegetation was therefore similar throughout the vineyard. Wider planting in rows in the direction of the slope allowed a more or less permanent ground cover to grow in the alleys between the rows, i.e. a second plant community complementary to the strip underneath the vines still with regular weed control. In terraced vineyards there is even a third community on the banks different to those in the alleys or underneath the vines.

The differences in the botanical composition of the two or three undergrowth communities are the result of different management and site conditions. The grower focuses in soil management on maintaining the vigour and yield of the vines (Perret et al. 1993). There is a gradient of potential influence from the banks, with little direct influence, to the strip underneath the vines, with the greatest potential influence (this ranking depends on the site conditions and on the proportion of the width of the strip underneath to the strip between the vine rows; Table 17.1). The possibilities for the grower to improve plant species richness are limited because the vigour of the vines has first priority in his decisions concerning undergrowth management.

17.2.3 Plant Communities in Relation to Habitat and Management Techniques

Underneath the vines is an area that is of particular interest to the grower in regulating the vitality of the vines as well as in influencing yield quality and quantity (Table 17.1). In years with poor growth due to high competition for water and nitrogen, the grower often has to eliminate the vegetation cover in a strip below the vines. The duration of the period with bare soil depends on the vitality of the vines and on the distribution of the precipitation. It is important to reduce the competition for nitrogen mainly during and after bloom of the vines, i.e. from June to July (Löhnertz et al. 1989). The vegetation is therefore destroyed in May to promote a reduction in N uptake by weeds (Gut et al. 1996b) and to increase mineralised N in the soil (Perret et al. 1993) in June.

Table 17.1. Estimation of potential interference between plant communities of vineyard habitats and grapevines, and possible influences of undergrowth management on plant species richness

Site/habitat	Relative influence of undergrowth on vines: competition for		Potential to increase plant species richness by management	Remarks
	Water	Nitrogen		
Banks of terraced vineyards	Low	Low	High	Spatial separation from vines
Alleys (between rows)	Medium	Medium	Medium	In areas without long dry periods
	Medium to high	(Medium to) high	Small to medium	In areas with longer-lasting summer drought
Strip below vines	High	High	Small	Wide strips: less competition than narrow ones

Vegetation is controlled by leaf-applied herbicides or by mechanical weed control (e.g. blade cultivator). Regular disturbances of this kind favour life-forms such as therophytes and/or some well-adapted rhizomatous or deep-rooting perennials (Table 17.2). If carried out regularly and intensively, this soil management corresponds to a certain extent to traditional management and generates agroforms of the *Geranio rotundifolii-Allietum vinealis* (Wilmanns 1993). Sometimes, sites below the vines are the last refuges for endangered bulb geophytes (e.g. *Muscari* spp.; Arn 1996).

The competitive effect of this regularly disturbed plant community on the deep-rooting vines may be small due to the reduced soil cover (often less than

Table 17.2. Plant communities with characteristic species in vineyards of eastern Switzerland in different habitats of the vineyard (typical number of species compiled from relevés of ca. 70 vineyards)

Site/habitat	Plant community	Characteristic species	No. of species
Alleys (between rows)	Intensively mulched *Poa trivialis-Lolio-Potentillion*	*Agrostis stolonifera* *Poa trivialis* *Lolium perenne* *Ranunculus repens* *Potentilla reptans* *Trifolium repens*	20–30
	Extensively mowed *Arrhenatherion*	*Arrhenaterum elatius* *Vicia sepium* *Galium mollugo* *Achillea millefolium* *Crepis biennis*	20–50
Strip below vines	With bare soil Species of the *Stellarietea mediae*	Therophytes such as: *Polygonum* spp. *Chenopodium* spp. *Stellaria media* *Veronica* spp. *Sonchus* spp. (Rhizomatous or deep-rooting) perennials such as: *Elymus repens* *Convolvulus arvensis* *Calystegia sepium*	20–30
	Without bare soil (only mulched) *Poa trivialis-Lolio-Potentillion*	See alleys	20–30
Banks of terraced vineyards	Species of the *Arrhenatherion* (and partly of the *Mesobrometum*)	*Arrhenaterum elatius* *Bromus erectus* *Brachypodium pinnatum* *Vicia sepium* *Salvia pratensis* *Galium mollugo* *Scabiosa columbaria* *Achillea millefolium*	30–100

50%) and to the concentration of the main root biomass of the dominating therophytes in the upper soil layers (Kutschera 1960); but rhizomatous and deep-rooting perennials compete strongly for below-ground resources due to coincidence of root characteristics with the vines. These often undesirable species are in part encouraged by mechanical weed control alone. Where appropriate, they are suppressed by herbicides before they become dominant.

When the strip underneath the vines is regularly mulched, i.e. mowed without removing the biomass, the plant community is similar to the one in mulched alleys.

In the alleys between the vine rows the interference of the plant community with the vines is often smaller than in the strip directly under the vines. Therefore, the grower can take care of the botanical composition of the undergrowth without detriment to the grapevine. The plant community in the alleys varies from site to site but it can also be influenced by soil management practices: a high mulching or mowing intensity promotes a dense cover of creeping perennials (Table 17.2; Wilmanns and Bogenrieder 1991; Gut et al. 1996c) of the *Poa trivialis-Lolio-Potentillion* community. A more extensive management of the undergrowth promotes more species characteristic for meadows of the *Arrhenatherion* (Table 17.2; Ellenberg 1996).

Intensively mulched grass species of the *Poa trivialis-Lolio-Potentillion* community reduce their root-shoot ratio (Kmoch 1952) compared to species of the *Arrhenatherion*. Besides an expected short-term reduction in transpiration due to frequent clipping of the leaves, the grower also anticipates less competition than from an extensively managed *Arrhenatherion* with more profound root mass (Kutschera and Lichtenegger 1982, 1992) and higher potential coincidence of the rooting system with the vines.

Occasional soil movements in the alleys to conserve water efficiently and to make N available for the vines (Perret et al. 1993) significantly alter the botanical composition: therophytic species germinate in the bare soil or gaps and establish themselves for a certain period before their dominance is reduced again by more competitive perennials (penetration, Wilmanns and Bogenrieder 1991). Annual as well as perennial species are enhanced by tillage or spading due to regeneration from roots, rhizomes or shootbases (e.g. *Taraxacum officinale*, *Elymus repens*, *Crepis capillaris*), whereas perennials without such regeneration organs are suppressed (Gut et al. 1996c). Soil movement is also important for the conservation of endangered bulb geophytes such as *Gagea arvensis* or *Muscari racemosum* (Arn et al. 1997).

Banks of terraced vineyards are the habitat with the most extensive management: they are normally unfertilised and mowed not more than once or twice per year. We expect only minor interference of the bank community with the vines (Table 17.1). Banks normally show a higher plant species richness than alleys or strips underneath the vines (Table 17.2). Most species belong to the *Arrhenateretum* or the *Mesobrometum* (Ellenberg 1996). The vegetation cover is normally not dense, moreover there are often gaps in the vegetation due to local erosion or animal activity (e.g. rodents). Such open microsites are important for botanical richness in offering habitats for ruderals and segetals, but also for animals depending on bare soil (e.g. reptiles, spiders).

17.2.4 Ways to Enhance Plant Species Richness in the Different Habitats

The potential to increase the species richness of a community depends on its location in the vineyard. Banks that interfere little with vines can be managed mainly with respect to the botanical composition. On the other hand, where competition of the undergrowth below the vines or in the alleys is higher, management is primarily related to the vigour and microclimate of the vines, and, of course, to practical considerations related to vineyard activities, such as canopy management or pesticide application.

Our present ability to increase plant species richness underneath the vines is still limited. Currently, we can only give general recommendations, e.g. to change weed control methods from time to time to reduce selection pressure caused mainly by herbicides.

In the alleys the potential for improving botanical composition by management is higher, because the measures do not have to be strictly limited to the requirements of the vines but can also be adjusted to the existing vegetation. The easiest way to increase species richness in the alleys is to mow the sod as extensively as possible (Gut et al. 1996c).

Designing a vineyard habitat to obtain increased biodiversity is discussed in the last part of our chapter.

17.3 The Second and Third Trophic Level: Herbivores and Entomophagous Species

17.3.1 Sampling Methods

The sampling techniques applied in our investigations have been described in detail by Remund et al. (1989, 1992, 1994). Two faunistic profiles were established annually for 4 consecutive years at two periods that provided optimum results: end of May (3 weeks before bloom) and mid-August (beginning of berry coloration). Larger arthropods on the vine leaves were collected with the beating/funnel technique (collection area of funnel 50 × 50 cm, 3 taps per vine with a rubber-coated stick, 25 vines per plot) and small arthropods with the washing method (Boller 1984; 4 × 25 leaves collected per plot, soaked in detergent, rinsed and filtered). Arthropods on the cover plants were collected with the net (4 × 20 catcher movements per plot) and by means of a mouth-operated aspirator (45 min per plot).

Important taxa (e.g. acari, thrips and hymenoptera) were identified at the species level, whereas economically or ecologically less important taxa were identified only at the family, subfamily or genus level, respectively. The individual groups of arthropods exhibiting different levels of taxonomic identification will be referred to as taxa.

17.3.2 Composition and Dynamics of the Faunistic Complex of 21 Vineyards in Northern Switzerland

A total of 41 309 individual arthropods was collected in this investigation and 987 taxa were identified. In view of a future use of the results, in practical recommendations to the growers we divided the collected taxa subjectively into three economically relevant classes: potential pest species, beneficials (entomophagous species) and indifferent species. At the taxa level, 76% of the collected arthropods were considered as economically indifferent, 22.5% as beneficial and only 1.5% as potential pests of the grapevine. The respective classification at the abundance level showed 54, 31 and 15%. The overall dynamics of the arthropod fauna during the 4-year investigation are presented in Fig. 17.1. The increase in indifferent and beneficial arthropods stimulated by the alternating mowing regime is evident.

A general overview of the taxa collected at the beginning of the investigation in 1987 and of the status of the faunistic complex after 4 years is given in Table 17.3.

17.3.3 Comments to Table 17.3

Araneida. Dominant families are Araneidae, Linyphiidae, Philodromidae and Theridiidae. Great species richness and abundance of net-forming species

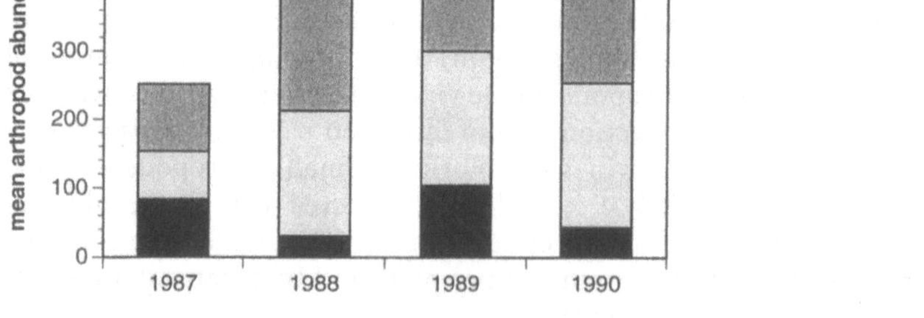

Fig. 17.1. Average arthropod abundance of 21 vineyards assessed by three collecting methods; the increase in indifferent and beneficial arthropods as a result of the alternating mowing system with permanent presence of flowering plants

Table 17.3. Faunistic composition of 21 vineyards in northern Switzerland at the start and termination of a 4-year survey conducted from 1987 till 1990

Arthropod order	Ecology	Abundance 1987	Abundance 1990	Dominance 1990 (%)	Abundance change 87–90 (%)	Observed taxa 87–90
Araneida	B	427	1256	9.14	194.15	73
	P	0	0	0.00	0.00	0
	I	0	0	0.00	0.00	0
	total	427	1256	9.14	194.15	
Acari	B	337	1918	13.96	469.14	9
	P	700	237	1.72	−66.14	4
	I	27	88	0.64	225.93	15
	total	1064	2243	16.32	110.81	28
Coleoptera	B	68	250	1.82	267.65	33
	P	5	6	0.04	20.00	2
	I	278	550	4.00	97.84	145
	total	351	806	5.87	129.63	180
Diptera	B	27	73	0.53	170.37	14
	P	13	15	0.11	15.38	1
	I	541	1736	12.63	220.89	149
	total	581	1824	13.27	213.94	164
Heteroptera	B	75	406	2.95	441.33	27
	P	6	4	0.03	−33.33	1
	I	163	1797	13.08	1002.45	116
	Total	244	2207	16.06	804.51	144
Homoptera	B	0	0	0.00	0.00	0
	P	15	12	0.09	−20.00	1
	I	235	1467	10.68	524.26	53
	Total	250	1479	10.76	491.60	54
Hymenoptera	B	412	285	2.07	−30.83	54
	P	28	46	0.33	64.29	2
	I	103	858	6.24	733.01	118
	Total	543	1189	8.65	118.97	174
Lepidoptera	B	0	0	0.00	0.00	0
	P	43	42	0.31	−2.33	2
	I	226	573	4.17	153.54	84
	Total	269	615	4.48	128.62	86
Thysanoptera	B	3	9	0.07	200.00	2
	P	561	560	4.08	−0.18	1
	I	195	342	2.49	75.38	19
	Total	759	911	6.63	20.03	22
Others	B	102	235	1.71	130.39	10
	P	19	1	0.01	−94.74	1
	I	256	976	7.10	281.25	51
	Total	377	1212	8.82	221.49	62
Total beneficials (B)		1451	4432	32.25	205.44	222
Total pests (P)		1339	923	6.72	−31.07	15
Total idifferents (I)		2084	8387	61.03	302.45	750
Total arthropods		4874	13742	100		

favoured by highly structured habitats and plentiful prey. An interesting species is *Argyope bruennichi* (zebra-spider) found in all organic vineyards and less frequently in integrated production (IP) plots, occurring in marginal parts of the vineyard where there is no or very little human interference.

Acari. Beneficial families: *Phytoseiidae, Anystidae and Trombididae.* Given the economic importance of this group we present the relevant details in Table 17.4.

The data on the spider-mite species *Panonychus ulmi* and *Tetranychus urticae* demonstrate that their abundance becomes negligible in the presence of the phytoseiid predators. A recent analysis of 230 grape farms practising IP confirmed that 91% of the growers no longer apply acaricides since predatory mites have solved the spider mite problem (Boller et al. 1994). Other pest species such as the eriophyid mites *Calepitrimerus vitis* and *Colomerus vitis* could not be collected with the methods applied. Both species provide an important food source for the predatory mites, of which *Typhlodromus pyri* is the most important species, exhibiting by far the greatest abundance and constancy.

Coleoptera. Dominant indifferent families are Chrysomelidae, Curculionidae, Elateridae and Malachiidae. Beneficial families include Coccinellidae, Staphylinidae, Carabidae and Cantharidae. Of ecological interest is the rose chafer *Cetonia aurata* (Scarabeaeidae) as visitor of the flowers *Daucus carota* and *Aegopodium podagraria*, as well as the bioindicator species *Trichodes apiarius* (Cleridae) indicating the presence of (wild) bees. Potential pests are the black vine weevil *Otiorhynchus sulcatus* (Curculionidae) and *Melolontha melolontha.*

Table 17.4. Average abundance, dominance and constancy of Acari collected in 21 vineyards in northern Switzerland in 1990

Taxa	Ecology	Abundance per leaf	Dominance (%)	Constancy (%)
Typhlodromus pyri	Beneficial	0.39	73.38	90.48
Amblyseius andersoni	Beneficial	0.03	6.42	28.57
Amblyseius finlandicus	Beneficial	0.004	78	23.81
Paraseiulus subsoleiger	Beneficial	0.003	0.6	4.76
Anystidae	Beneficial	0.02	4	100
Thrombididae	Beneficial	0.01	0.2	47.62
Panonychus ulmi	Pest	0.01	2	42.86
Tetranychus urticae	Pest	0.03	6	38.1
Others[a]	Indifferent	0.02	4	71.43
Total		0.5	100	

[a] Acaridae, Ameroseiidae, Ascidae, Dermanyssidae, Eugamasidae, Eviphidae, Macrochelidae, Oribatidae, Parasitidae, Tarsonemidae, Tenuipalpidae, Tydeidae.

Diptera. Syrphidae, with 50 taxa identified mostly at the species level, is the dominant family classified as agronomically indifferent for viticulture (no aphids as pests). Most abundant representatives are *Melanostoma mellinum* and *Sphaerophoria scripta*. Beneficial families consist of Asilidae as predators and Tachinidae as potential larval parasitoids of the grape moths. Potential pests: Drosophilidae as vectors of bacterial (vinegar) and fungal diseases (e.g. *Botrytis*, *Penicillium*) of mature grapes.

Heteroptera. Indifferent families: Lygaeidae, Miridae, Pentatomidae and Rhopalidae. Beneficial families are Anthocoridae with *Orius minutus* and *Anthocoris nemorum* occurring in large numbers in the foliage and especially in the grape cluster, where they prey on spider mites and larvae of the grape moths, respectively; Nabidae with *Himacerus apterus* and *Nabis pseudoferus*. Other potential predators belong to the families of Miridae, Reduviidae and Saldidae. A potential pest species of local importance is *Lygus spinolaii*.

Homoptera. Most dominant families are Aphididae, Cicadellidae, Cixidae, Dephacidae, Issidae and Membracidae.

Potential pest species is *Empoasca vitis* (Typhlocybinae), which can be efficiently controlled by the egg parasitoids *Anagrus atomus* and *Stethynium triclavatum* (see Hymenoptera). Minor pests are scale insects, e.g. *Eulecanium corni* (Coccidae). An occasional curiosity as to size of insect and type of damage is *Stictocephala bisonia* (Membracidae).

Hymenoptera. The superfamily Apoidea is represented by 39 of 174 taxa including numerous and in part rare wild bee species. Most abundant families are Formicidae, lchneumonidae, Pompilidae, Sphecidae, Tenthredinidae and Braconidae.

Important parasitoids of pest species:

- *Trichogramma cacoeciae* (Trichogrammatidae) as most important egg parasitoid of both species of grape moths. Alternate hosts are eggs of indifferent lepidoptera developing on plants of the green cover and in hedges in the proximity of the vineyard. Can be enhanced by honeydew of aphids in the green cover.
- *Anagrus atomus* and *Stethynium triclavatum* (Mymaridae). Egg parasitoids of the grape leafhopper *Empoasca vitis*. Hibernation and development of the first spring generation in eggs of indifferent leafhopper species occurring in hedges, especially on *Rosa canina* and *Rubus fructicosus*.
- *Agrypon minutum* (lchneumonidae). Larval parasitoid of the grape moths with parasitation rate up to 30% in the first generation of the pests. As in *Pimpla* the abundance of this parasitoid is enhanced by the presence of indifferent lepidopteran species in the green cover and hence by a great plant species richness.

- *Pimpla* (= *Coccigomimus*) *turionellae* (Ichneumonidae). Pupal parasitoid of the grape moths and many indifferent lepidoptera species of the green cover. Adults find nectar especially on flowers of *Daucus carota* and *Aegopodium podagraria*.

Potential pest species. *Paravespula vulgaris* and *P. germanica* can cause heavy damage to early ripening grape varieties.

Lepidoptera. Dominating among the 86 taxa observed are Geometridae (16), Nymphalidae (10), Pyralidae (9).

Olethreutes lacunana (Tortricidae) showed highest abundance. Important key pests are the two tortricid grape moths *Eupoecilia ambiguella* and *Lobesia botrana*. Of ecological interest is *Euplagia quadripunctata* (Arctideae). Indifferent lepidopteran species play an important role as alternate hosts for egg, larval and pupal parasitoids with impact on the grape moths (e.g. the noctuid moth *Autographa gamma* developing on *Lamium maculatum*).

Thysanoptera. *Drepanothrips reuteri* is the dominant thrips species, followed by *Thrips tabaci*, *T. fuscipennis* and 17 other (indifferent) thrips species. Predacious thrips of the family Aeolothripdidae occur only sporadically. Potential pest species do not constitute a serious problem where predatory mites occur at adequate population densities.

Various other insect orders. Indifferent taxa: Blattodea, Collembola, Crustacea, Diplopoda, Ephemeroptera, Saltatoria, Mecoptera, Megaloptera, Odonata, Neuroptera, Plectoptera, Psocoptera, Myriopoda, Trichoptera.

Interesting species. *Gryllus campestris* (Saltatoria) exhibiting often high population densities in terraced vineyards; *Tettigonia viridissima* (Saltatoria) formerly very abundant in cereal fields; *Libelloides coccaius* (Neuroptera: Ascalaphidae), rare species occurring in only 1 of the 21 collection sites (Walenstadt) and reported from vineyards in the Kaiserstuhl area in southern Germany. Potentially beneficial species: *Forficula auricularia* (Dermaptera) with high abundance in funnel samples collected from old vines and vines with compact grape clusters. Exceedingly high numbers can have a negative influence on must quality. *Chrysoperla carnea* (Chrysopidae): general predator classified presently as being of little importance in the regulation of pests in Swiss vineyards.

17.4 Interactions Between the Three Trophic Levels

17.4.1 General Influence of Plant Species Richness on Arthropod Diversity

Whilst Fig. 17.1 shows an overall picture of the development of the faunistic complex in all 21 vineyards under investigation, we can now include the botanical elements in the interpretation of the increasing faunistic diversity. Figure 17.2 shows the number of plant species found in each of the 21 vineyards plotted against the number of arthropod taxa found at the same site. This figure was included to visualise the general trends observed in our investigations. However, the correlations shown have to be interpreted with caution because the identification of the collected arthropods was not carried out down to the species level. Hence the taxa differ in the number of individual species they represent in reality. This affects especially the correlation for the class designated agronomically indifferent herbivores.

A considerable increase in indifferent herbivores and a significant but more modest increase in the beneficial entomophagous species with increasing number of plant species were to be expected, and confirm numerous reports in the literature (see review of Altieri 1994). The analysis of the data as to the quality of the plant species (i.e. number of perennial dicot species) revealed a correlation almost identical to that shown in Fig. 17.2. This confirmed our observation that especially the proportion of (flowering) perennial dicot plants in the plant association is responsible for the increase in the arthropod taxa (r = 0.72 for indifferent herbivores, $p \leq 0.01$; r = 0.55 for beneficial entomophages, $p \leq 0.05$). On the other hand, evidently the number of potential pest species does not change with increasing botanical species richness. Furthermore, we observed that in vineyards exhibiting flora with a large number of plant species, the tendency in most pest species (grape moths, spider mites, eriophyid mites, thrips and noctuid larvae) was to fluctuate much less at

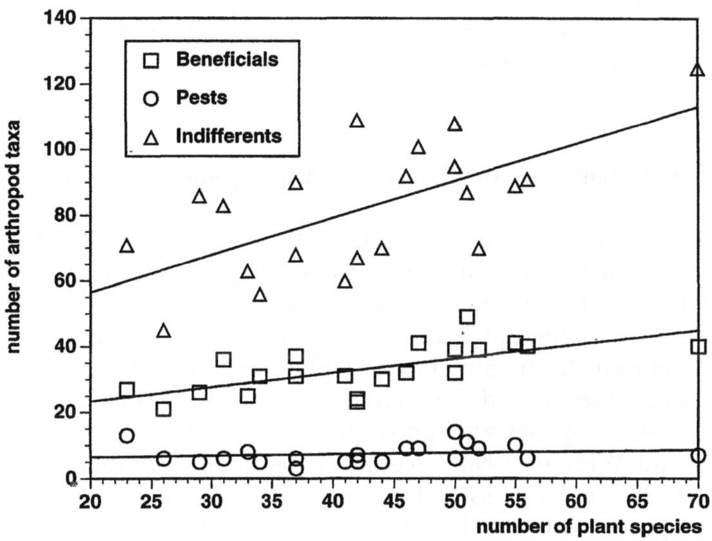

Fig. 17.2. Relation between botanical species richness and number of arthropod taxa observed on 21 vineyards of northern Switzerland in 1990. (Indifferents r = 0.65; Beneficials r = 0.67). Explanations in text

significantly lower levels than pest populations in botanically poorer situations. Three examples are given to demonstrate this point.

17.4.2 Plant Species Richness and the Grape Moth *Eupoecilia ambiguella*

Wide variations in plant species richness occur even within the same vineyard in Walenstadt where the grape moth populations always remain below the economic threshold level (i.e. <2% of infested grapes) in the steep terraced part of the vineyard, which exhibits a very high diversity of plant species. In contrast, regular control measures are required to control to deal with this key pest (infestation rates of >15%) in the flat part of the vineyard, which has a more trivial botanical composition due to the greater impact of mechanisation.

Accumulating evidence indicates that this low population density of the grape moth in the terraces is intimately linked to the more favourable conditions for the establishment and augmentation of egg, larval, and pupal hymenopteran parasitoids as discussed in Section 17.3.2.

Experiments conducted in another vineyard in eastern Switzerland (Remund and Boller 1991) with two distinctly different levels of plant species richness showed that artificially exposed eggs of *E. ambiguella* reached a parasitation rate by *Trichogramma cacoeciae* of 85% (1st generation early June) in the botanically richer part of the vineyard, while only 64% were parasitised in the area showing a more trivial flora. Such differences were also observed during the second generation in July when parasitation reached 35 and 14%, respectively. It also became evident that the short-lived *Trichogramma* could only bridge the gap between the two grape moth generations by utilising eggs of indifferent lepidopteran species occurring on perennial herbaceous plants in the green cover.

17.4.3 Flowering Cover Plants and the Predatory Mite *Typhlodromus pyri*

Another positive influence of a permanent supply of flowering plants has been described in connection with the dominant predatory mite, *Typhlodromus pyri*. This species does not leave the grapevine in search of food when all animal prey (such as spider mites and thrips) has been devoured, but will shift its nutritional basis from animal prey to plant pollen. Pollen profiles established in two vineyards showed that tree pollen transported over large distances prevails in early spring and grass pollen from the green cover in summer (Boller and Widmer 1990). Feeding experiments conducted in the laboratory proved that most grass pollen can provide adequate nutrition for survival and reproduction of the summer generations (Boller and Frey 1990). In most vineyards of northern Switzerland, the spider mite and thrips populations are successfully controlled by this predator (Boller et al. 1994).

17.4.4 Host Plants Outside the Vineyard Essential for the Parasitoids of the Grape Leafhopper *Empoasca vitis*

A sequence of events involving all three trophic levels leads to the regulation of the grape leafhopper *Emposca vitis* by the egg parasitoid *Anagrus atomus* (Cerutti 1989; Remund and Boller 1995). In this case, the parasitoid depends on alternate leafhopper species developing especially on *Rosa canina* and *Rubus fructicosus agg.* in nearby hedges (Remund and Boller 1996). Two of the generations of *Anagrus* developing outside the vineyard coincide with the life cycle of *E. vitis*, which produces two generations in our region. Trap catches in the vineyard of the leafhopper and its parasitoid clearly indicate an increasing parasitoid density with increasing proximity of the wild host plants in the hedges. Leafhopper outbreaks disappeared in the vicinity of suitable hedges after insecticide treatments against the leafhopper had stopped. A similar case has been described in California for *Anagrus epos*, where the removal of blackberries as important host plant of the alternate homopteran host had disrupted the vital food chain (Doutt and Nakata 1973).

17.4.5 Present Knowledge of the Ecological Significance of Individual Plant Species in the Undercover and Adjacent Hedges

Our present evaluation of the significance of individual plant species for the enhancement of arthropods of economic interest is summarised in Table 17.5 (U. Remund, unpubl.).

17.5 Habitat Management in Vineyards: How to Improve Botanical Diversity?

17.5.1 Aims of the Vineyard Undergrowth Management

Besides increasing biodiversity per se and optimising the habitat for beneficial arthropods mainly by fostering the number of perennial forb species, there are other goals involved in vineyard soil management:

- Soil protection from erosion or compaction and creation of an optimal microclimate for the grapevines.
- Control of vigour and yield of the grapevines by regulating undergrowth competition for water and nutrients (in most cases reducing competition, sometimes increasing competition to slow down excessive growth of the vines).
- Reduction of labour related to plant protection measures, canopy management, harvest etc.

Table 17.5. Present knowledge as to the ecological significance of individual plant species in the vineyard undercover and in adjacent areas (e.g. hedges) in northern Switzerland

Ecological significance of undercover and adjacent marginal areas (e.g. hedges)		Nectar source for lepidopterans	Nectar source for parasitoids	Egg parasitoids	Larval parasitoids	Pupal parasitoids	Predatory mites	Pollen for predatory mites	Egg parasitoids for leafhoppers
Undercover									
Urtica dioica	Common nettle			×	×	×		×	
Lamium maculatum	Spotted dead-nettle	×		×	×	×			
Daucus carota	Wild carrot	×	×	×					
Aegopodium podagraria	Ground-elder	×	×	×					
Galium mollugo	Hedge bedstraw		×	×	×	×			
Origanum vulgare	Marjoram	×	×						
Hedges									
Rosa canina	Dog rose			×	×	×	×		×
Rubus fruticosus	Bramble			×	×	×	×		×
Lonicera xylosteum	Fly honeystuckle						×		×

Nevertheless, it may be possible to integrate the requirements of the economy, productivity of labour, and habitat management in the same vineyard.

17.5.2 Enhancing Habitat Diversity (β Diversity)

Traditional vineyards exhibit just one single plant community produced by the same soil management practice over the whole surface. Modern planting systems, together with at least partial undergrowth, establish two or more different plant communities. One of them develops underneath the vines. It can be similar to plant communities of the traditional viticulture with bare soil, but mostly it consists of degraded agroforms (Wilmanns and Bogenrieder 1991). The second plant community develops in alleys and can be similar to the one typical for meadows. If it is occasionally spaded, it also contains annuals of the community underneath the vines (penetration, Wilmanns and Bogenrieder 1991). Further improvements can be achieved by creating more habitats, realised, e.g. in terraced vineyards with banks rich in plant species (Gut 1996a).

An important option to improve habitat diversity is to now, mulch or cultivate the soil only in every second alley. This alternating management of undergrowth was originally developed to prolong the flowering period of forbs in order to foster beneficial arthropods and to protect arthropod populations from injury by mulching of the entire surface; but it also affects botanical diversity positively. Soil movement in the alleys carried out every 3r to 4th year has proved to be an effective method in reducing an undesirably dense cover of dominating grasses (with exception of *Elymus repens*). It favours also, in the first period of bare soil, annual, biennial and afterwards also perennial forbs (Gut et al. 1996c). With spading carried out in alternate rows, there simultaneously exist recently spaded alleys with dominating annuals and adjacent alleys spaded 3 or 4 years previously where perennials dominate. Hence we can promote on a small area plant populations of different successional stages composed of different life-forms and species.

17.5.3 Increasing Species Richness in the Plant Communities of the Vineyard

Creating additional habitats that have their own plant communities (β diversity) is one way to increase species richness. Another possibility is to manage the existing plant communities underneath the vines, in alleys and on banks aiming at high species richness, especially of perennial forbs (α diversity).

In alleys, the number of perennial forb species correlates with the undergrowth management. Investigations in the Grisons (eastern Switzerland) showed that species richness can be increased without negative effects on vine performance by soil spading every 3rd to 4th year, combined with a low mulching or mowing intensity (Gut et al. 1996c). For an optimal management system, site conditions and long-term vigour of the vines have to be taken into account.

The water and nutrient supply in the soil can be adjusted to the demands of the vines by appropriate management of the area underneath the vines. Short-term or minor water deficiencies can be corrected by eliminating the vegetation only below the vines. In dry regions it might be necessary to destroy also the vegetation in alternate alleys (Perret et al. 1993). Whether it is possible to protect relicts of the *Geranio rotundifolii-Allietum vinealis* including several endangered species by maintaining bare soil underneath the vines is under investigation (Arn 1996; Arn et al. 1997). To differentiate the habitat underneath the vines from the habitat of the alleys and to increase plant species richness, the area below vines should be disturbed once a year or at least every second year. This can be done either by soil movement or leaf-applied herbicides. These procedures ensure that annuals are not too strongly suppressed by undesirable competitive perennials invading from the alleys.

Banks should be managed as extensively as possible to allow the establishment of a plant community similar to that in meadows with low management intensity (*Arrhenatherion*). This community contains several perennial forbs which are valuable in fostering beneficial hymenoptera (especially parasitoids of the grape moth (Björnsen 1995; Remund et al. 1992, 1994).

Acknowledgements. We gratefully acknowledge the valuable suggestions of Peter Duelli and Otto Holzgang during manuscript preparation.

References

Altieri MA (1994) Biodiversity and pest management in agroecosystems. Food Products Press, Haworth Press New York, 185 pp

Arn D (1996) Frühjahrs-Zwiebelgeophyten in Rebbergen der Nordostschweiz. Diploma, ETH Zürich, 60 pp

Arn D, Gigon A, Gut D (1997) Bodenpflege-Massnahmen zur Erhaltung gefährdeter Zwiebelgeophyten in begrünten Rebbergen der Nordostschweiz. Schweiz Z Obst-Weinbau 133 (no 2) (in press)

Björnsen A (1995) The role of flowering plants as nectar sources for larval parasitoids of the grape berry moth. Diploma work, ETH Zürich, 78 pp

Boller EF (1984) Eine einfache Ausschwemm - Methode zur schnellen Erfassung von Raubmilben, Thrips und anderen Kleinarthropoden im Weinbau. Schweiz Z Obst-Weinbau 120:16–17

Boller EF (1988) Das mehrjährige Agro-Ökosystem Rebberg und seine praktische Bedeutung für den modernen Pflanzenschutz. Schweiz Landw Forsch 27:55–61

Boller E, Frey B (1990) Blühende Rebberge in der Ostschweiz: 1. Zur Bedeutung des Pollens für Raubmilben. Schweiz Z Obst-Weinbau 126:401–405

Boller EF, Remund U (1986) Der Rebberg als vielfältiges Agro-Ökosystem. Schweiz Z Obst-Weinbau 122:45–50

Boller E, Wiedmer U (1990) Blühende Rebberge in der Ostschweiz: 2. Zum Pollenangebot auf den Rebblättern. Schweiz Z Obst-Weinbau 126:426–431

Boller E, Basler P, Giezendanner U (1994) VINATURA Deutschschweiz: guter ökologischer Leistungsausweis im Jahre 1993. Schweiz Z Obst-Weinbau 130:417–419

Boller EF, Remund U, Candolfi MP (1988) Hedges as potential sources of *Typhlodromus pyri* – the most important predatory mite in vineyards of northern Switzerland. Entomophaga 33:249–255

Cerutti F (1989) Modellizzazione della dinamica delle popolazioni di *Empoasca vitis* Goethe (Hom., Cicadellidae) nei vigneti del cantone Ticino e influsso della flora circonstante sulla presenza del parasitoide *Anagrus atomus* Haliday (Hym., Mymaridae). PhD Thesis, ETH Zürich, no 9019, 117 pp

Doutt RL, Nakata J (1973) The *Rubus* leafhopper and its egg parasitoid: an endemic biotic system useful in grape-pest management. Environ Entomol 2:381–386

Ellenberg H (1996) Vegetation Mitteleuropas mit den Alpen. Eugen Ulmer, Stuttgart

Gut D (1996a) Habitat-Management in Ostschweizer Rebbergen. Obstbau-Weinbau 33:200–201

Gut D, Barben E, Riesen W (1996b) Critical period for weed competition in apple orchards: preliminary results. IOBC wprs Bull 19(4):273–277

Gut D, Holzgang O, Gigon A (1996d) Weed control methodes to improve plant species richness in vineyards. Proc Int Weed Contr Congr 2:987–992

Kmoch HG (1952) Ueber den Umfang und einige Gesetzmässigkeiten der Wurzelmassebildung unter Grasnarben. Z Acker-Pflanzenbau 95:363–380

Kutschera L (1960) Wurzelatlas mitteleuropäischer Ackerunkräuter und Kulturpflanzen. DLG, Frankfurt/M

Kutschera L, Lichtenegger E (1982) Wurzelatlas mitteleuropäischer Grünlandpflanzen, vol 1. Moncotyledonae. Gustav Fischer, Stuttgart

Kutschera L, Lichtenegger E (1992) Wurzelatlas mitteleuropäischer Grünlandpflanzen, vol 2. Pteridophyta und Dicotyledonae. Gustav Fischer, Stuttgart

Löhnertz O, Schaller K, Mengel K (1989) Nährstoffdynamik in Reben. III. Mitteilung: Stickstoffkon-zentration und Verlauf der Aufnahme in der Vegetation. Wein Wiss 44:192–204

Perret P, Koblet W (1973) Ergebnisse von Bodenpflegeversuchen im Weinbau. Schweiz Z ObstWeinbau 109:116–128, 151–161

Perret P, Weissenbach P, Schwager H, Heller WE, Koblet W (1993) "Adaptive nitrogen management" – a tool for the optimisation of N-fertilisation in vineyards. Vitic Enol Sci 48:124–126

Remund U, Boller E (1991) Möglichkeiten und Grenzen von Eiparasiten zur Traubenwicklerbekämpfung. Schweiz Z Obst-Weinbau 127:535–540

Remund U, Boller E (1995) Untersuchungen zur Grünen Rebzikade in der Ostschweiz. Schweiz Z Obst-Weinbau 131:200–203

Remund U, Boller EF (1996) Bedeutung von Heckenpflanzen für die Eiparasitoide der Grünen Rebzikade in der Ostschweiz. Schweiz Z Obst-Weinbau 132:238–241

Remund U, Niggli U, Boller E (1989) Faunistische und botanische Erhebungen in einem Rebberg der Ostschweiz. Einfluss der Unterwuchsbewirtschaftung auf das Ökosystem Rebberg. Landwirtsch Schweiz 2:393–408

Remund U, Gut D, Boller EF (1992) Rebbergflora, Rebbergfauna: Beziehungen zwischen Begleitflora und Arthropodenfauna in Ostschweizer Rebbergen. Schweiz Z Obst-Weinbau 128:527–540

Remund U, Boller EF, Gut D (1994) Nützlinge in Rebbergen mit natürlicher Begleitflora – wie kann man sie erfassen? Schweiz Z Obst-Weinbau 130:164–167

Stalder L, Potter CA (1979) Zur natürlichen Begrünung der Rebböden – eine Orientierung für die Praxis. Schweiz Z Obst-Weinbau 114:253–256,265

Wilmanns O (1993) Plant stragegy types and vegetation development reflecting different forms of vineyard management. J Veg Sci 4:235–240
Wilmanns O, Bogenrieder A (1991) Phytosociology in vineyards – results, problems, tasks. In: Esser G, Overdieck D (eds) Modern ecology – basic and applied aspects. Elsevier, Amsterdam, pp 399–441

18 Landscape Dynamics, Plant Architecture and Demography, and the Response of Herbivores

P.W. Price, H. Roininen, and T. Carr

18.1 Introduction

Helmut Zwölfer's global perspective of nature has always been enriching. His broad geographical approach to the ecology and evolution of phytophagous insects on thistles, coupled with detailed knowledge on individual interactions between plant and herbivore, provide a comprehensive view seldom achieved in the literature (e.g. Zwölfer 1986, 1987, 1988; Zwölfer and Romstöck-Völkl 1991; Zwölfer and Arnold-Rinehart 1993). Recognizing the larger picture by understanding the details of interactions is a long-standing feature of his contributions in evolutionary biology (e.g. Zwölfer 1978, 1982; Zwölfer and Bush 1984; Zwölfer and Romstöck-Völkl 1991). We all have much to learn from his example and much to appreciate in his extensive publications.

In a vein similar to the approaches of Helmut Zwölfer, we have attempted to generate a broad understanding of plant and insect interactions, although on a narrower taxonomic scale. Our studies have focused on one plant family, the Salicaceae, with only two genera, *Salix* and *Populus*, the associated herbivorous sawflies (Hymenoptera: Tenthredinidae), and the carnivorous parasitoids associated with them. For over 15 years we have concentrated on the details of the interactions among these taxa and their consequences for population dynamics, with emphasis on the gall-inducing genus *Euura* and related genera. Much of the general literature and our specific studies are reviewed by Price and Roininen (1993) and Price et al. (1995).

We now endeavor to continue broadening our perspective to larger scales in space and time. The dimensions of space will cover the landscape, the latitudinal gradient over which most of the species are distributed, and the Holarctic biogeographic realm in which the Salicaceae and Tenthredinidae radiated. The time dimension includes the aging process of host plants, plant succession following disturbance, and the dynamics of vegetational change since the Pleistocene glaciations.

Superimposed on these templates of space and time is the demography of the host-plant species, the particulars of architectural differences among species and their consequences for the ecology and evolution of the associated herbivores and carnivores. The willows (*Salix*) show a remarkable range in architecture from low creeping shrubs to large trees, while all poplars (*Populus*) are trees. When we describe how the sawflies relate so

Ecological Studies, Vol. 130
Dettner et al. (eds.) Vertical Food Web Interactions

closely to host-plant module size and quality, it will become clear how plant modular development and structure, which ultimately define the architecture of an individual and a species, impact the kinds of sawflies utilizing a species.

Our challenge has been to meld the elements of space, time, demography, and architecture to portray landscapes which are usually highly dynamic in relation to the requirements of sawflies. Given this habitat template, to use Southwood's (1977) terminology, we attempt to generate an understanding of sawfly population dynamics and the resulting commonness or rarity of various species on landscapes. We build from landscape dynamics to host-plant demography and architecture, to the dynamics of herbivores and their carnivores. With this construction, perhaps we can glimpse an evolutionary understanding of the number of insect species per host plant species. This sequence we use as a plan for the chapter.

18.2 The Biotic Components

18.2.1 The Salicaceae

The regeneration niche in willows (cf. Grubb 1977) is molded by the very small, wind-dispersed seeds with brief viability. On reaching moist mineral soil, germination is rapid, and establishment is successful only in the absence of competition from grasses and herbs. Several years of above-average precipitation are necessary for establishment of seedlings in the more southern latitudes such as Arizona (Sacchi and Price 1992).

In their architecture, willow species range from low shrubs to large trees (Fig. 18.1), while poplars are represented by trees only. Many species are clonal. The essential feature of a shrub is the lack of terminal dominance, so multiple stems develop with usually frequent initiation of basal shoots. Trees develop with strong dominance of a terminal shoot, usually with no development of basal shoots, and thus physiological aging as the tree grows is irreversible under natural conditions.

18.2.2 The Herbivores and Carnivores

Sawflies emerge as adults in the spring when host-plant growth is rapid. Oviposition with the saw-like ovipositor places an egg into plant tissue in the species we have studied (Price and Roininen 1993). In the genus *Euura*, oviposition stimulates a gall in a characteristic location for each species, such as on shoots, petioles, buds, or leaf midribs. In the genus *Pontania*, galls are produced on leaves, and in *Phyllocolpa* a leaf fold is developed. Free-feeding sawflies in the genus *Nematus* place eggs in pouches under the leaf

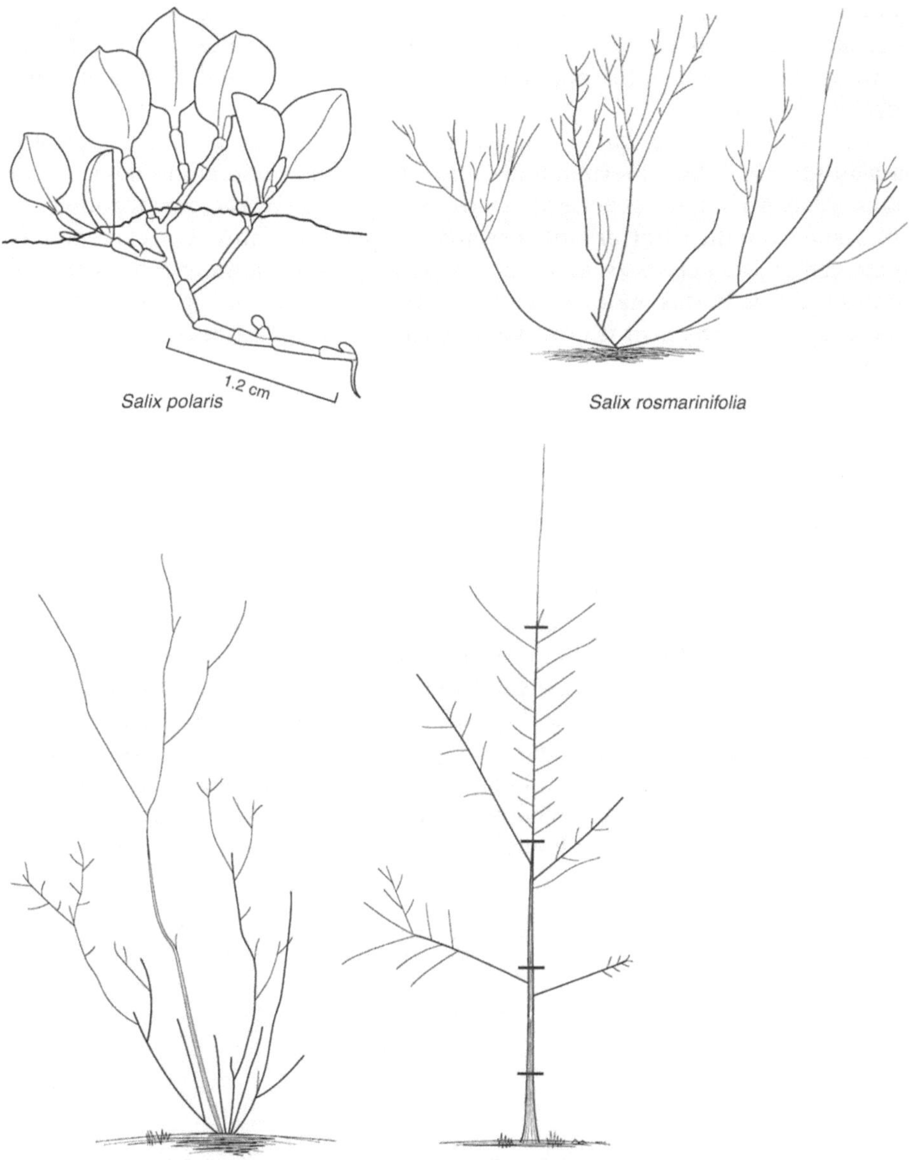

Fig. 18.1. A range of architectural types in the genus *Salix* (not to scale). *Salix polaris* from northern Russia creeps along the ground , covered mostly by moss, with this 9-year-old ramet reaching little more than 2 cm in height. *Salix rosmarinifolia* is a low shrub in Finland - the one illustrated is about 1 m tall, ramets are 5 years old or less, and some have been browsed by mountain hare. *Salix cinerea* growing on a lakeside site in Finland is about 2.15 m tall with ramets up to 5 years old. *Salix pentandra* in Finland grows to a large tree, but this specimen was only 5 years old, with growth per year marked, and reached 3.36 m tall. Figures were based on real specimens

epidermis. Of particular significance for our argument is the highly generalizable result that species in all these genera require young, rapidly growing plant modules for populations to persist (Price and Roininen 1993; Price et al. 1995).

The carnivores we have studied most in these trophic systems are parasitoids in the Pteromalidae, Eurytomidae, Ichneumonidae, and Braconidae. All but one species attack the sawfly larvae by piercing through the gall tissue from the exterior with a relatively long ovipositor. The exception is in the genus *Adelognathus*, in which the adult female chews into the gall and locates the host in this manner. If a host larva is located, an egg is placed in or on the larva and the host is eventually killed by the parasitoids.

18.3 Landscape Dynamics in Space

18.3.1 Disturbance

Landscapes differ profoundly in their rates and extent of disturbance, which influences the development of clear mineral soil, and in turn the extent of the regeneration niche for members of the Salicaceae. The three areas we have studied may well encompass the extremes from high to low disturbance (Fig. 18.2). In North Karelia in eastern Finland (62°N) the landscape is almost equal parts land and lake. Frequent changes in water level in the large Lake Saimaa, and disturbance by moving ice, create an extensive and changing lakeshore habitat for *Salix cinerea*. A glacial river such as the Tanana in central Alaska (64°N) carries enormous silt loads. Alluvial deposition is extensive and ero-

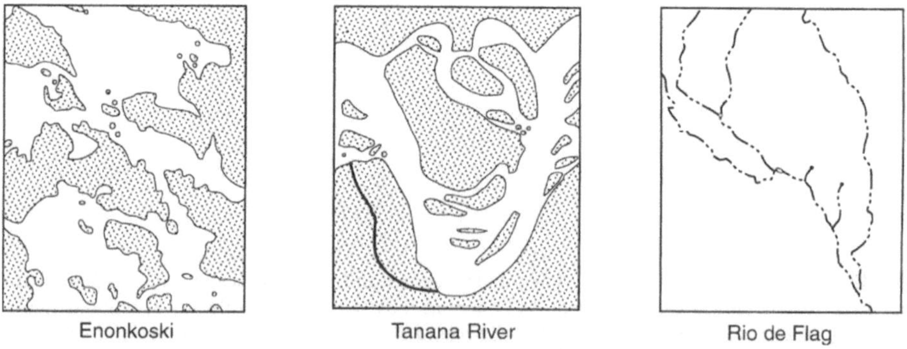

Fig. 18.2. Maps of the Enonkoski area in the Lakes Region of Finland (*left*), the Tanana River near Fairbanks, Alaska (*center*), and drainages of the Rio de Flag just north of Flagstaff, Arizona (*right*). Water is *white* and land is *stippled* in maps for the first two sites, and for the Arizona site land is *white* and *temporary streams* are indicated. Each map represents an area of about 2.5 × 3.2 km

sion is almost continuous. New and extensive mineral deposits are colonized by willows almost immediately, followed by poplar (Viereck et al. 1993). Plant succession is rapid with willows declining from dominance rapidly after the first 10–15 years after deposition of alluvium (Fig. 18.3).

These landscapes in Alaska and Finland contrast dramatically with those in many areas further south, such as the landscape on the Colorado Plateau near Flagstaff, Arizona (35°N; Fig. 18.2). Around Flagstaff there would be no permanent standing water without the intervention of humans. Rivers are intermittent, flowing perhaps for 2–8 weeks per year, and persistent springs are small and extremely rare. Disturbance on this landscape is minimal, mostly of human origin, and the regeneration niche for members of the Salicaceae is very limited.

18.3.2 Plant Responses

Dynamic landscapes, such as in glacial river floodplains or lakeshore environments, result in frequent creation of the regeneration niche for members of the Salicaceae. A patchwork quilt of regeneration episodes develops on the landscape; age structure is diverse and changing, making host-plant populations for herbivores heterogeneous. Stands of young plants are common and at a peak of vigorous growth.

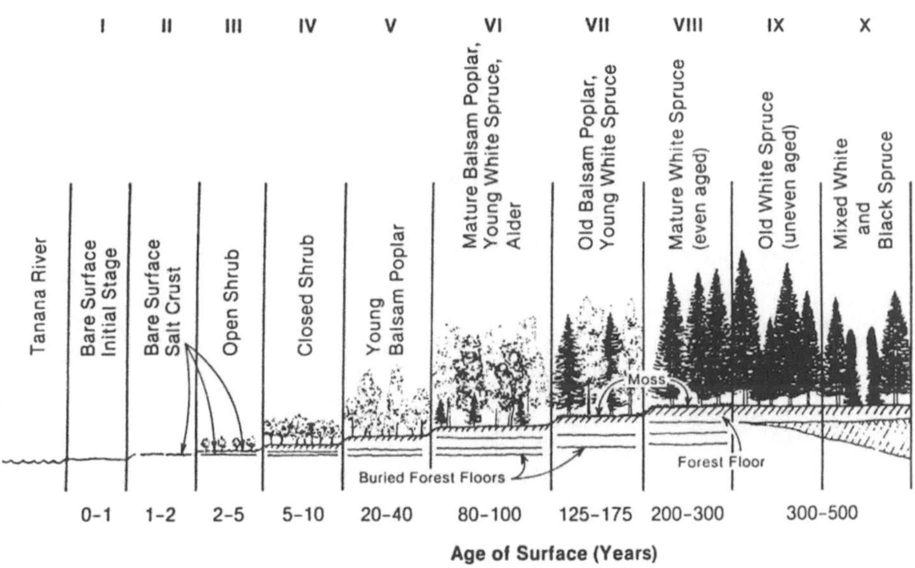

Fig. 18.3. Vegetational succession on the floodplain of the Tanana River, Alaska, with willows dominating stages *II–IV* followed by balsam poplar stage *V*, and alder stage *VI*. Willows are the first species to colonize new alluvial deposits. (Viereck 1989)

Stable landscapes with generally rare creation of the regeneration niche for Salicaceae tend toward a preponderance of long-established host-plant populations dominated by mature and overmature stands. Willows, especially, are restricted to narrow riparian belts, springs, and human devices such as ditches and pits. The frequency of new individuals or populations is very low with a scarcity on a landscape of the youngest, most vigorously growing age classes.

The inexorable aging of trees makes disturbance and regeneration of vital importance for insect herbivores which require rapidly growing plant modules as a resource. Among the willow shrubs, with more or less persistent development of basal shoots, a plant endures as a viable host for sawflies because of its own complex age structure (Fig. 18.4). Long, juvenile, and vegetative shoots develop at the base of the plant, they age over the years, become sexually reproductive and mature, and subsequently senesce and die. Old ramets are replaced by juveniles so that each individual shrub or genet develops an age-class structure of shoot modules from long, rapidly growing juvenile shoots to short, slow-growing but reproductive shoots. In shrubs, not only is there an age-class structure within each individual and patch, but there is age-class structure over a landscape with some young genets and probably many old genets. This contrasts with trees, which usually develop as even-aged stands, without development of basal shoots, such that age-class structure varies only over the extensive landscape. Exceptions are the clonal species of poplar such

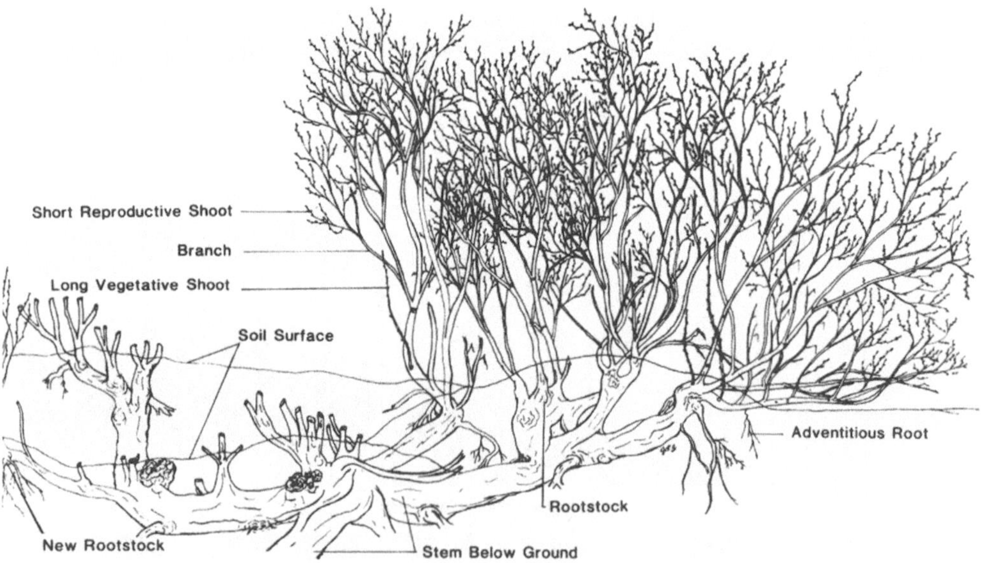

Fig. 18.4. Architecture of the shrub, *Salix lasiolepis*, showing the clonal habit and age structure of shoot modules within the clone available for herbivores such as the stem galler, *Euura lasiolepis*. Note the few long basal vegetative shoots and the many short reproductive shoots on older ramets. (Craig et al. 1988)

as quaking aspen, *Populus tremuloides*, which can develop a complex age structure within stands through vegetative reproduction. Without disturbance, however, these stands may also progress towards an even age structure and eventual senescence.

These fundamental differences between trees and shrubs are likely to result in much more patchy resources in space for sawflies utilizing trees than for those on shrubs. A cohort of trees will age in unison, pass through a susceptible stage while growing vigorously, and then become resistant to attack because of physiological aging. A population of shrubs, on the other hand, with genets of increasing age, will continually produce basal juvenile shoots available to attack by sawflies.

18.3.3 Herbivore and Carnivore Responses

We have followed sawfly population dynamics closely for two species, one on a shrub and another on a noncloning tree. In a less formal way we have also observed other sawfly species on shrubs and trees. We will treat the more detailed studies first and then move to our general observations.

The shrub *Salix lasiolepis*, depicted in Fig. 18.4, is utilized by the shoot-galling sawfly, *Euura lasiolepis*. In the vicinity of Flagstaff, we have recorded population size on 15 clones since 1983. Remarkably, populations per clone in 1983 effectively predicted populations 11 generations later in 1993, accounting for 99% of the variance (Price et al. 1995). Populations changed over only 2 orders of magnitude even during a period of 3 drier-than-usual years from 1987 to 1989 (Fig. 18.5). Populations are predictable and persistent at least over a decade or more. Vigorous ramets in wet sites support high sawfly populations while senescent ramets in dry sites support low populations. Because the local environments and willow responses are predictable, populations per clone persist at characteristic levels over a decade or more.

This predictability and persistence appears to be characteristic of many galling sawflies on willows. The bud-galling *Euura mucronata* on *Salix cinerea* in Joensuu, Finland, and environs still persists at densities similar to those we studied in 1986 (Price et al. 1987a,b). A related bud galler on *Salix scouleriana*, a shrub in the inner basin of the San Francisco Peaks, north of Flagstaff, is reliably present even on a small isolated population of host plants. An unnamed midrib-galling *Euura* species on *Salix exigua* has persisted on this shrubby willow along the Colorado River in northern Arizona since it was discovered over a decade ago. In general, when galling sawflies are found on willow shrubs, one can return year after year to find them. A few exceptions exist, such as *Pontania pustulata*, a leaf-galling sawfly on *Salix phylicifolia*, which is found exclusively on very young, vigorously growing willow plants.

The noncloning tree, *Salix pentandra*, is attacked by the shoot-galling sawfly, *Euura amerinae*. We have studied populations since 1986 and the

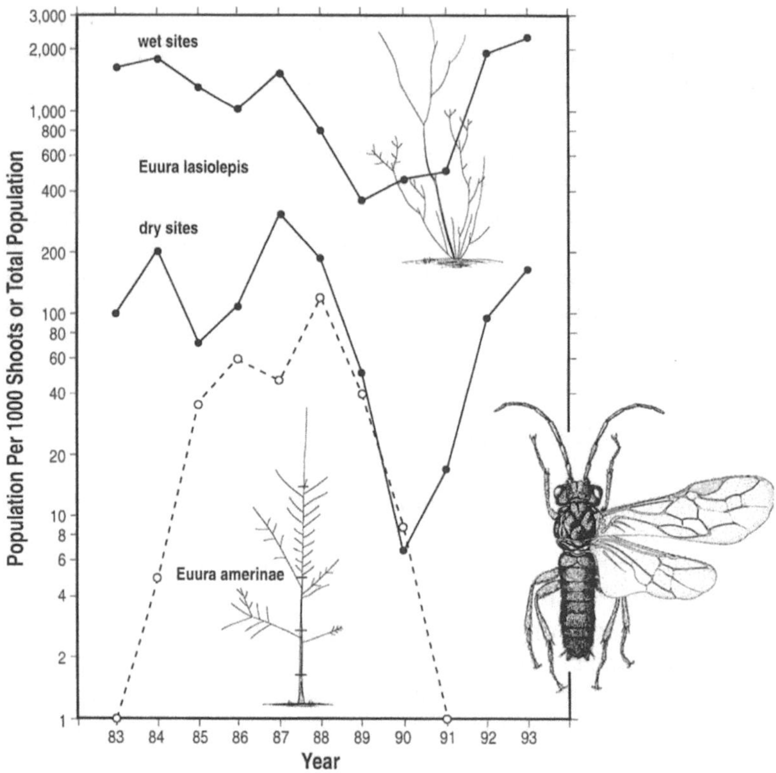

Fig. 18.5. Population change in *Euura lasiolepis* in wet and drier sites on the shrub, *Salix lasiolepis* in Arizona, USA (*solid lines*), and in *Euura amerinae* on the tree *Salix pentandra* in North Karelia, Finland (*dashed lines*). For *Euura lasiolepis* the population is expressed as the number of galls per 1000 shoots, and for *Euura amerinae* the total number of galls in the population is given. The *inset* is a male galling sawfly and the architecture of the respective willows is shown

persistent galls left a record of population size back to 1983. One gall was formed in 1983 by a colonizing sawfly when the young stand of trees was about 5 years old. The total population increased to 118 galls over the next 5 years and then declined rapidly to extinction in 1991 (Roininen et al. 1993). Persistence of the population in the stand we studied was only 8 years. The trees became highly resistant to sawfly attack by the time they were about 12 years old because of an unknown factor involved with ontogenetic or physiological aging. This pattern of interaction has been observed several times on these species, but we do not have enough observations on other species to enable generalizations. Evidence suggests, however, that other species, such as *Euura atra* on the trees *Salix alba* and *Salix fragilis*, persist only where pruning is heavy, such as along roadsides or after trees have been severely damaged (Price et al. 1997).

On the cloning tree, *Populus tremuloides*, the free-feeding sawfly, *Nematus iridescens*, attacks the youngest ramets in a host-plant population (Carr 1995). The sawfly is never found on the much more abundant shoots of older ramets, but oviposits on the largest leaves on the youngest ramets, where the larvae survive best.

This pattern of attack in which females show a strong preference for young and vigorous ramets as oviposition sites is characteristic of most of the species we have studied. This preference is usually linked to larval performance in the form of higher survival on younger ramets: a preference and performance linkage is observed commonly. Our studies in North America, Finland, and Japan covering several sawfly genera have revealed the same patterns, representing a generalization relevant to the Holarctic biogeographic realm (e.g., Price et al. 1987a,b, 1995; Craig et al. 1989; Carr 1995; Price and Ohgushi 1995). The bud-galling *Euura* species usually lumped into one taxon, *mucronata*, but probably representing a group of sibling species, all show the same preference-performance pattern from Arizona to Finland and Hokkaido (Price et al. 1995 and unpubl.).

The key to attaining a landscape view of sawfly dynamics clearly lies in the dynamics of module production in the form of rapidly developing juvenile shoots. Where disturbance is common, as in the Alaskan and Finnish sites, juvenile modules are common, providing plentiful resources for sawfly species, which are also common. At the more southerly end of the willow distribution in Arizona, conditions are drier and more stable. Many populations are old and senescent and several species are regarded as threatened because reproduction is sparse or absent (Price et al. 1996). As a consequence, rapidly growing modules are very rare over a landscape, and sawflies are generally rare. Higher populations may be found at springs, after human disturbance, or at high elevations where snow and rock slides break old ramets. Browsing by mammals, which prune back ramets, stimulates more juvenile regrowth and higher sawfly populations (Woods et al. 1996; Roininen et al. 1994, 1997).

Natural enemies of the sawflies we have studied have been generally unimportant in population dynamics. They appear to be very passive and opportunistic, without influencing greatly the strong preference-performance linkage and the bottom-up effects of resource supply (Price and Clancy 1986; Price et al. 1987b, 1995; Price 1988, 1989; Craig et al. 1990; Roininen et al. 1993; Price and Ohgushi 1995). Nevertheless, specific parasitoids, such as *Lathrostizus euurae*, depend on sawflies for their existence and are forced into increasingly patchier sites on sparser host populations. The upper trophic levels presumably become more threatened than the host plants and host herbivores in areas like northern Arizona, although we have conducted no specific studies on this topic.

In general, all landscapes are patchy in space for willows, sawflies, and associated carnivores. The interacting factors of disturbance, host-plant demography, and vegetational succession create a very dynamic landscape in northern latitudes, ideal for sawflies requiring vigorous modules. Even here, suitable resources for some sawflies, such as *Euura amerinae* on a nonclonal tree, are exceedingly patchy and sparse. In more southern latitudes, where willow populations are tending toward senescence, resources are frequently rare for sawflies both within and among host plant populations over a landscape.

18.4 Landscape Dynamics in Time

Certain temporal aspects of resource availability for sawflies involving plant aging and vegetational succession have been mentioned in this chapter. However, much larger scales of time, in terms of millennia, need consideration when discussing landscape dynamics.

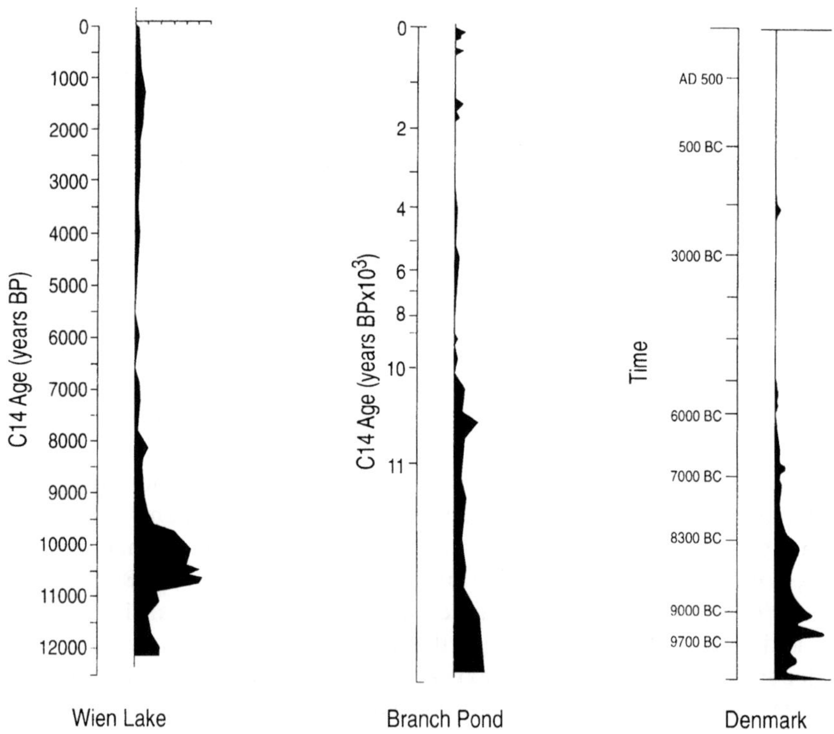

Fig. 18.6. Three examples of the pollen record for willows after the Pleistocene glaciations. Wien Lake is in central Alaska (Hu et al. 1993), Branch Pond is in northern Maine (Anderson et al. 1986), and the Danish record is a compilation by Iversen (1973) from eastern Denmark. Time is recorded as in the original publications and the *thickness of the pollen record* at any one time indicates the percentage of all pollen represented by *Salix* species. Therefore, the record affords only a crude view of the relative importance of willows in the vegetation after glaciation

What we see on a latitudinal gradient in space from the glaciated north in Alaska to the drier and less disturbed south in Arizona can also be reconstructed at any one locality in time using palynological techniques. A typical picture, whether in Europe or eastern or western North America, is abundant willow pollen present immediately after glaciation and persisting for several millennia, with a subsequent decline to trace amounts as other species become dominant (Fig. 18.6). The patterns illustrated are consistent with many other studies (e.g., Whitehead 1979; Heide 1984; Björck and Möller 1987; Matthews 1992).

In northern Arizona, glaciers developed on the San Francisco Peaks during the Pleistocene (Sharp 1942; Péwé and Updike 1976). A palynological record of the area has yet to be developed, but there is little doubt that prime habitat for willows was available and many species were abundant, following recession of the ice. Since post-Pleistocene times, there has almost certainly been an inevitable decline in the number of willow species and the abundance of species that remain, as conditions became warmer and drier and many species distributions moved northward. As time after glaciation increases to 11000 years into the interglacial, as we are now in southern latitudes like Arizona, willow species and populations are becoming increasingly stressed and limited. Populations are probably declining in number and geographic range. Regeneration is very poor or absent and populations tend toward senescence. The eventual result over the next millennium may well be the local extinction of species in Arizona and a northward retreat in geographic distribution.

The inevitable consequences for sawflies on willows is declining resources over the coming millennia, with marginal conditions for survival in the south, such as in Arizona. Rates of natural local extinction are likely to increase.

18.5 Predictions on the Richness of Sawflies on Host-Plant Species

18.5.1 Ecological Predictions

The ecological opportunities for sawflies appear to be greater in the north temperate and arctic than in the south, based on our analysis in this chapter. However, the actual patterns are inadequately documented. We make the following predictions, therefore, which we plan to test and we hope that this chapter may stimulate others to do so also.

1. Local sawfly species richness and density per species will decrease from north to south.
2. Patchiness of willows, herbivores, and specialized carnivore populations will increase from north to south.

3. Rarity and the threat of at least local extinction will increase from north to south. Any endemic species in the southern range of the Salicaceae are subject to species-level extinction.

Prediction 1 is not supported in the current literature, which shows peaks of sawfly in the 40°–50° N latitudinal range (Price and Roininen 1993). However, these data are derived from national distributions of species, undoubtedly underrepresented by studies in the north. No estimates are available on local richness of sawfly species on a latitudinal gradient. Such a study will provide a legitimate test of the prediction.

Predictions 2 and 3 are justified in this chapter but formal tests are needed.

18.5.2 Evolutionary Predictions

The ecological interactions discussed so far among sawflies and their host plants presumably exerted some influence on the evolutionary history of their relationships. For example, the number of sawfly species per host-plant species will depend on the evolutionary opportunities for colonization and subsequent speciation of lineages among hosts (Zwölfer 1982, 1987). Based on the foregoing discussion we make a set of predictions.

1. Sawflies will have colonized shrubs more frequently than trees because shrubs provide more persistent and reliable resources than trees.
2. Plant species in the Salicaceae in northern latitudes have been more likely to be colonized than those in southern latitudes. This is likely to result from greater species richness of the Salicaceae, as well as the more favorable landscape for willow growth and subsequent sawfly colonization and persistence. Local extinction of willows and sawflies has been more common in the south.
3. Plant species characteristic of early successional vegetation and those in riparian habitats have been more likely to be colonized by sawflies than plant species in other vegetation types.

Prediction 1 runs counter to much evidence that trees support more herbivore species than shrubs (Lawton 1983; Zwölfer 1987 and refs. therein). It will be interesting to see if the size effect in trees compensates for the particular problems that trees pose for sawflies. If trees do support more sawflies than shrubs, then is the increased richness reduced relative to other types of herbivores? And do small shrubs support fewer sawfly species than larger shrubs and/or trees?

Prediction 2 is consistent with Zwölfer's (1982, p. 293) concept of "plant taxa as radiation platforms" for insect-herbivore diversification. Centers of evolution of plant taxa are likely to produce more opportunities for radiation by insect herbivores (Zwölfer and Bush 1984; Zwölfer and Romstöck-Völkl 1991), and correlated biogeographic gradients of host plants and herbivores are to be expected (Zwölfer 1987).

Prediction 3 relates specifically to the plant vigor and sawfly herbivore relationship (e.g., Price 1991), such that species habitually utilizing habitats that permit vigorous growth will generally provide suitable sites for colonization through evolutionary time. Here, we become more specific with our prediction than is generally possible because of the special relationships among members of the Salicaceae and their sawfly herbivores.

Clearly, we have not fathomed the evolutionary aspects of the sawfly-Salicaceae relationships as well as Helmut Zwölfer has with the phytophagous insects on thistle hosts. Our next step is to test the predictions above, for which we have acquired the data very recently. As we proceed, the work of Helmut Zwölfer will continue to act as a stimulating guide for our investigations.

Acknowledgments. We sincerely appreciate the opportunity provided by the editors to contribute to this volume in honor of Professor Helmut Zwölfer. An earlier draft was reviewed by Drs. Timothy Craig, Stephen Jackson, and Christopher Sacchi, whose expertise and efforts we recognize. Funding for the research discussed herein has been provided most recently by the National Science Foundation (DEB-9318188), the Finnish Academy, and a National Science Foundation Graduate Fellowship.

References

Anderson RS, Davis RB, Miller NG, Stuckenrath R (1986) History of late- and post-glacial vegetation and disturbance around Upper South Branch Pond, northern Maine. Can J Bot 64:1977–1986

Björck S, Möller P (1987) Late Weichselian environmental history in southeastern Sweden during the deglaciation of the Scandinavian ice sheet. Quat Res 28:1–37

Carr TG (1995) Oviposition preference – larval performance relationships in three free-feeding sawflies. MS Thesis, Northern Arizona University, Flagstaff

Craig TP, Price PW, Clancy KM, Waring GM, Sacchi CF (1988) Forces preventing coevolution in a three-trophic-level system: willow, a gall-forming herbivore, and parasitoid. In: Spencer K (ed) Chemical mediation of coevolution. Academic Press, New York, pp 57–80

Craig TP, Itami JK, Price PW (1989) A strong relationship between oviposition preference and larval performance in a shoot-galling sawfly. Ecology 70:1691–1699

Craig TP, Itami JK, Price PW (1990) The window of vulnerability of a shoot-galling sawfly to attack by a parasitoid. Ecology 71:1471–1482

Grubb PJ (1977) The maintenance of species-richness in plant communities: the importance of the regeneration niche. Biol Rev 52:107–145

Heide K (1984) Holocene pollen stratigraphy from a lake and small hollow in north-central Wisconsin, USA. Palynology 8:3–20

Hu FS, Brubaker LB, Anderson PM (1993) A 12000-year record of vegetation change and soil development from Wien Lake, central Alaska. Can J Bot 71:1133–1142

Iversen J (1973) The development of Denmark's nature since the last glacial. Geological Survey of Denmark V Series No 7-C. Reitels, Copenhagen

Lawton JH (1983) Plant architecture and the diversity of phytophagous insects. Annu Rev Entomol 28:23–29

Matthews JA (1992) The ecology of recently deglaciated terrain: a geoecological approach to glacier forelands and primary succession. Cambridge University Press, Cambridge

Péwé TL, Updike RG (1976) San Francisco Peaks: a guidebook to the geology, 2nd edn. Northern Arizona Society of Science and Art, Flagstaff
Price PW (1988) Inversely density-dependent parasitism: the role of plant refuges for hosts. J Anim Ecol 57:89–96
Price PW (1989) Clonal development of coyote willow, *Salix exigua* (Salicaceae) and attack by the shoot-galling sawfly, *Euura exiguae* (Hymenoptera: Tenthredinidae). Environ Entomol 18:61–68
Price PW (1991) The plant vigor hypothesis and herbivore attack. Oikos 62:244–251
Price PW, Clancy KM (1986) Interactions among three trophic levels: gall size and parasitoid attack. Ecology 67:1593–1600
Price PW, Ohgushi T (1995) Preference and performance linkage in a *Phyllocolpa* sawfly on the willow, *Salix miyabeana*, on Hokkaido. Res Popul Ecol 37:23–28
Price PW, Roininen H (1993) The adaptive radiation in gall induction. In: Wagner MR, Raffa KF (eds) Sawfly life history adaptations to woody plants. Academic Press, Orlando, pp 229–257
Price PW, Roininen H, Tahvanainen J (1987a) Plant age and attack by the bud galler, *Euura mucronata*. Oecologia 73:334–337
Price PW, Roininen H, Tahvanainen J (1987b) Why does the bud-galling sawfly, *Euura mucronata*, attack long shoots? Oecologia 74:1–6
Price PW, Craig TP, Roininen H (1995) Working toward theory on galling sawfly population dynamics. In: Cappuccino N, Price PW (eds) Population dynamics: new approaches and synthesis. Academic Press, San Diego, pp 321–338
Price PW, Carr TG, Ormord AM (1996) Consequences of land management practices on willows and higher trophic levels. In: Maschinski J, Hammond HD, Holter L (eds) Southwestern rare and endangered plants: proceedings of the second conference. Flagstaff, Arizona, Sept 11–14, 1995. USDA For Serv Gen Tech Rep RM-GTR-283, pp 219–223
Price PW, Roininen H, Tahvanainen J (1997) Willow tree shoot module length and the attack and survival pattern of a shoot-galling sawfly, *Euura atra* L. (Hymenoptera: Tenthredinidae). Ann Entomol Fenn (in press)
Roininen H, Price PW, Tahvanainen J (1993) Colonization and extinction in a population of the shoot-galling sawfly, *Euura amerinae*. Oikos 68:448–454
Roininen H, Price PW, Tahvanainen J (1994) Does the willow bud galler, *Euura mucronata*, benefit from hare browsing on its host plant? In: Price PW, Mattson WJ, Baranchikov YN (eds) Ecology and evolution of gall-forming insects. USDA For Serv Northcentral For Exp Stn Gen Tech Rep NC-174:12–26
Roininen H, Price PW, Bryant JP (1997) Response of galling insects to natural browsing by mammals in Alaska. Oikos (in press)
Sacchi CF, Price PW (1992) The relative roles of abiotic and biotic factors in seedling demography of arroyo willow (*Salix lasiolepis*: Salicaceae). Am J Bot 79:395–405
Sharp RP (1942) Multiple Pleistocene glaciation on San Francisco Mountain, Arizona. J Geol 50:481–503
Southwood TRE (1977) Habitat, the templet for ecological strategies. J Anim Ecol 46:337–365
Viereck LA (1989) Flood-plain succession and vegetation classification in interior Alaska. In: Fergusen DE, Morgan P, Johnson FD (eds) Proceedings, land classifications based on vegetation: applications for resource management. USDA For Serv Gen Tech Rep PNW-106:197–203
Viereck LA, Dyrness CT, Foote MJ (1993) An overview of the vegetation and soils of the floodplain ecosystems of the Tanana River, interior Alaska. Can J For Res 23:889–898
Whitehead DR (1979) Late-glacial and postglacial vegetational history of the Berkshires, western Massachusetts. Quat Res 12:333–357
Woods JO, Carr TG, Price PW, Stevens LE, Cobb NS (1996) Growth of coyote willow and the attack and survival of a mid-rib galling sawfly, *Euura* sp. Oecologia 108:714–722.
Zwölfer H (1978) Mechanismen und Ergebnisse der Co-Evolution von Phytophagen und entomophagen Insekten und höheren Pflanzen. Sonderb Naturwiss Ver Hamb 2:7–50

Zwölfer H (1982) Patterns and driving forces in the evolution of plant insect systems. In: Visser JH, Minks AK (eds) Proc 5th Int Symp Insect-Plant Relationships. Pudoc, Wageningen, pp 287–296

Zwölfer H (1986) Insektenkomplexe an Disteln - ein Modell für die Selbstorganisation ökologischer Kleinsysteme. In: Dress A, Hendrichs H, Küppers G (eds) Selbstorganisation: die Entstehung von Ordnung in Natur und Gesellschaft. Piper, München, pp 181–217

Zwölfer H (1987) Species richness, species packing, and evolution in insect-plant systems. In: Schulze E-D, Zwölfer H (eds) Potentials and limitations of ecosystem analysis. Ecological studies, vol 61. Springer, Berlin Heidelberg New York, pp 301–319

Zwölfer H (1988) Evolutionary and ecological relationships of the insect fauna of thistles. Annu Rev Entomol 33:103–122

Zwölfer H, Arnold-Rinehart J (1993) The evolution of interactions and diversity in plant-insect systems: the *Urophora-Eurytoma* food web in galls on Palearctic Cardueae. In: Schulze E-D, Mooney HA (eds) Biodiversity and ecosystem function. Ecological studies, vol 99. Springer, Berlin Heidelberg New York, pp 211–233

Zwölfer H, Bush GL (1984) Sympatrische und parapatrische Artbildung. Z Zool Syst Evolutionsforsch 22:211–233

Zwölfer H, Romstöck-Völkl (1991) Biotypes and the evolution of niches in phytophagous insects on Cardueae hosts. In: Price PW, Lewinsohn TM, Fernandes GW, Benson WW (eds) Plant-animal interactions: evolutionary ecology in tropical and temperate regions. Wiley, New York, pp 487–507

Part F

Synopsis

19 Evolutionary Patterns and Driving Forces in Vertical Food Web Interactions

K. DETTNER, G. BAUER, and W. VÖLKL

19.1 Introduction

The evolution of life histories and the structure and function of food webs have been thoroughly reviewed in the last years (Price et al. 1980; Pimm 1982; Schoener 1989; Pimm et al. 1991; Roff 1992; Stearns 1992; Morin and Lawler 1995; Polis and Winemiller 1995). However, although there is considerable mutual interference of both areas, the influence of alternative life history strategies on food web structure and function has scarcely been addressed (Winemiller and Polis 1995). Therefore, we will first present some scenarios created by life-history variation of and interactions between resource and consumer, and we consider the symmetry of a mutual impact as far as it is determined by life histories; the exploitation strategy, i.e. predation or parasitism; the predictability of the resource; the relationship between generation times and the degree of synchronization of life cycles between resource and consumers. These scenarios provide a framework for the evolution of trophic relationships and we will show how particular patterns of life history will lead to particular patterns of interactions between trophic levels (e.g. defense mechanisms, mechanisms related to the rates of speciation, degree of resource utilization etc.). However, life history traits are rarely free to coevolve under purely demographic forces; they are usually more or less constrained. Thus, the evolution of trophic relationships will be also influenced by constraints, an important one being body size. Therefore we will show some examples about how size constraints operate in food web interactions.

As a general rule, members of higher trophic levels evolve strategies to increase consumption and, with the exception of non-interactive systems (see Sect. 19.2.1.1), members of lower levels evolve strategies to reduce feeding by their enemies. In these interactions, an important factor are chemical compounds. Therefore, a crucial point in our chapter is chemical ecology in trophic relationships, a research area which currently deserves increasing interest.

Finally, we try to draw some conclusions for biological control strategies arising from the analysis of vertical food web interactions.

Ecological Studies, Vol. 130
Dettner et al. (eds.) Vertical Food Web Interactions

19.2 The Influence of Life History Variation on the Evolution of Vertical Food Web Interactions

19.2.1 The Exploitation System

19.2.1.1 Interactive Versus Non-Interactive Systems

The type of food and mode of feeding defines the symmetry of the impact between resource and consumer. With respect to this, trophic relationships may be classified into two groups: interactive and non-interactive systems (sensu Caughley 1976). In non-interactive systems there may be a considerable impact from the resource on its consumer, but not vice versa. The consumer is not able to influence the resource availability for its subsequent generations, i.e. it is not able to influence the rate at which the resource is renewed. Examples include consumers of dead organic matter, but also consumers of living plant tissue when it is exclusively produced for animals. An example for the latter strategy are "illegal" fruit feeders, e.g. members of the dipteran genus *Rhagoletis*. Their larvae feed in pulpy fruits without affecting development of

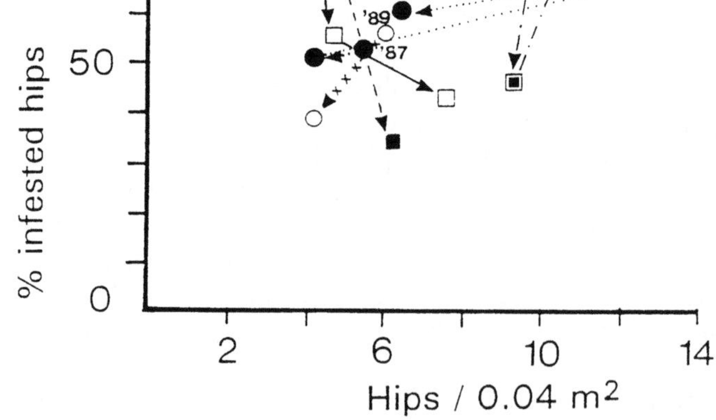

Fig. 19.1. Resource availability and resource utilization in a non-interactive system: *Rosa* and the rose hip-fly (*Rhagoletis alternata*). Since one hip is sufficient for one fly larva, the carrying capacity is 100% infestation. The *different symbols* refer to different sites in Upper Franconia. Subsequent generations (years) of one site are connected by *arrows*. The year of the first investigation is indicated for each site. (G. Bauer, unpubl.)

the seeds (Zwölfer 1982; Bauer 1986). Since there is no (or almost no) influence of the consumer on its resource, no defense mechanisms are evolved, and consumption rates of subsequent generations are frequently near the carrying capacity (Fig. 19.1). Such systems are usually donor controlled (in the sense of Pimm 1982). For mono- or oligophagous consumers of living resources, the mode of interaction should lead to sequential evolution unless there is a host shift as demonstrated by Bush and Smith (this Vol.).

All other interactions described in this volume must be considered interactive, i.e. there is some mutual, but often highly asymmetric impact. Defense mechanisms of the resource are frequently highly developed. Chemical defense is extensively treated elsewhere (see Sect. 19.3). Here, we will briefly deal with some general patterns which are frequently observed.

A simple defense strategy against natural enemies, which occur on almost all trophic levels, are morphological adaptations like spines. However, defense structures evolved in response to one group of consumers may be useless or even disadvantageous against others if they differ in their exploitation strategy. For example, the yellow star-thistle *Centaurea solstitialis* is effectively protected against grazing by vertebrate herbivores by its spines. However, these spiny bracts are used by its insect parasite, the tephritid fly *Urophora sirunaseva*, to recognize the host plant for oviposition (Zwölfer 1968).

Among animals, simple physical defense mechanisms are widely distributed (Gross 1993). In host-parasitoid or in predator-prey interactions, this defense is especially successful when the predator/parasitoid is smaller or of equal size to its prey/host. A typical example for such a system are aphid-parasitoid interactions (Stadler, this Vol.): physical defense behaviour like kicking with hind legs or raising and shaking the body is especially successful against smaller or similarly sized enemies like parasitoids or small predators (Gerling et al. 1990; Stadler, this Vol.; Völkl and Stadler 1996), while there is less success against larger enemies like big ladybird beetles (Brodsky and Barlow 1986; Stadler, this Vol.). In the latter case, aphids change their escape strategy from physical defense to flight response (Stadler, Chap. 14, this Vol.). Another behavioural defense strategy of aphids is colony formation, which may serve as some kind of protection against various predators or parasitoids (e.g. Turchin and Kareiva 1989).

Many insect species feed on special plant structures to escape natural enemies. Although there is a controversial discussion on the hypothesis that the endophytic way of life may be a protection against parasitoids (see references in Godfray 1994; Hawkins 1994; Hawkins and Sheehan 1994), there is no doubt that the host plant may provide structural refuges (Zwölfer and Arnold-Rinehart 1994). For example, flower-head dwelling insects may be protected by surrounding plant tissue against the attacks of predators or parasitoids (Zwölfer 1994). This could be demonstrated for various species of the tephritid fly genera *Urophora* and *Tephritis*, whose larvae develop in the flower heads of thistles (Michaelis 1984; Romstöck-Völkl 1990; Zwölfer and Arnold-Rinehart

1993; Zwölfer and Schlumprecht 1993; Zwölfer 1994). Similar structural refuges could be demonstrated for stem-dwelling insects (G. Freese 1995). The acquisition of a structural refuge and the resulting protection against natural enemies may be one case for a host shift which may lead to the evolution of biotypes (Feder 1995; Romstöck-Völkl, this Vol.).

Mutualistic interactions may also influence the pressure exerted by natural enemies. For example, the excretions and secretions of many insects and the exfloral nectaries of plants attract ants which collect the carbohydrate-rich exudations (Way 1954, 1963; Sudd 1987), producing associations which are either obligatory, like many ant-plant systems, or which vary both within and between seasons, resulting in a conditional mutualism that was shown for many homopteran species (Cushman and Addicott 1989, 1991; Cushman and Whitham 1989, 1991; Hölldobler and Wilson 1990; Bristow 1991). While collecting carbohydrates, the ants usually guard their source against phytophages, predators or parasitoids. Consumers that exploit systems with a conditional mutualism may be able to restrict feeding or oviposition to unattended resources, while species exploiting obligate mutualistic systems need to adapt to the new situation. A number of phytophages, predators and parastoids have evolved behavioural adaptions, camouflage or chemical mimicry to integrate themselves as "parasites of mutualism" (Letourneau 1990) into ant-guarded systems (Dettner and Liepert 1994; Völkl, this Vol.). These species also benefit from ant-attendance, and often adapt new ecological strategies. Many species concentrate their eggs within a few patches, rather than distributing them, to reduce mortality by natural enemies. Striking examples are aphid parasitoids (Völkl 1992, this Vol.). While aphidiid wasps that oviposit into unattended colonies usually lay only few eggs into a particular aphid colony and show high dispersal, species that exploit ant-attended resources are much less dispersive and concentrate their eggs in few colonies. The latter species benefit from ant-attendance by a reduced rate of hyperparasitism and predation. Similar patterns of resource utilization and differences between ant-attended and unattended resources were found for other homopteran-parasitoid systems (Way 1954; Novak 1994), for hyperparasitoids (Hübner and Völkl 1996), for aphidophagous ladybirds (Majerus 1994; Völkl 1995), for lycaenid butterflies (Fiedler 1995) and even for ant-plant systems (Letourneau 1990).

The rates of resource utilization are frequently low in interactive systems (e.g. Achziger, Peschken et al., Dixon, Völkl, this Vol.). The dynamics of these systems may be bottom-up, e.g. in thistle-*Urophora*-parasitoid systems (Zwölfer 1994; Peschken et al., this Vol.) or in shrub-hemipteran interactions (Achtziger, this Vol.); top-down (e.g. Hassell and May 1973; Dixon, this Vol.); or alternating (Stechmann, this Vol.). Accordingly, the type of evolution should be coevolution (e.g. specific sporozoan parasites of *Daphnia*; Ebert et al., this Vol.), diffuse coevolution (e.g. Stadler, Stechmann, Topp, this Vol.) or, if there is only a little impact of the consumer on its resource, sequential evolution (e.g. Achtziger, Romstöck-Völkl, this Vol.).

19.2.1.2 Predation Versus Parasitism

In contrast to most predators, parasites (in the sense of Price 1980) are frequently highly specific (e.g. Ebert et al., Bauer, Peschken et al., this Vol.). This close adaptation to the host may increase speciation rates (Carson 1975; Powell 1978; Price 1980). Among insects, approximately 70% are parasites of animals or phytoparasites (Price 1980). Important factors frequently related to high speciation rates in parasites are inbreeding or asexual reproduction, the selective pressure for coevolutionary modification (Price 1980), diversity of hosts, the size of the host target and the evolutionary time available (Zwölfer 1986). A further factor is demonstrated in Bush and Smith (this Vol.): new species of parasites may arise in a relative short period of time (*Rhagoletis pomonella* formed a new host race within approximately 100–150 years) due to colonization of a new host with a different phenology. This may quickly lead to sympatric evolution of new parasite species when the host serves as the site of courtship and mating. Mechanisms promoting host shifts, like host abundance, host quality or enemy free space, are discussed by Feder (1995) and Romstöck-Völkl (this Vol.).

This relationship between the mode of feeding, the degree of host specificity and the rate of speciation is also evident within hymenopterous parasitoids. Since they feed externally on their host and need no physiological adaptations for survival within a living host, as required for endoparasitoids, ectoparasitoids are in general much less host-specific than endoparasitoids (for reviews, see Godfray 1994; Hawkins 1994). A striking example for this pattern is found in the hyperparasitoid complex of aphids. In Europe, the hymenopterous genera *Dendrocerus* (6 spp.), *Asaphes* (2 spp.), *Pachyneuron* (3 spp.) and *Coruna* (1 spp.) altogether comprise only 12 species developing as ectohyperparasitoids. Most species have a very broad host range and attack a large number of aphidiid wasp species independent of the aphid host (Graham 1969; Stary 1977). By contrast, endohyperparasitic species of the genera *Alloxysta* and *Phaenoglyphis* are – with few exceptions – very host-specific, often attacking a single aphidiid or aphid genus. However, the species richness of aphid endohyperparasitoids is much higher, with more than 40 described species (6 *Phaenoglyphis* spp., approximately 35 *Alloxysta* spp.) and a number of cryptic yet undescribed species (Evenhuis and Barbotin 1987, and references therein; H.H. Evenhuis, pers. comm.).

19.2.1.3 Unpredictable Versus Predictable Resources

Short-lived resources are often highly unpredictable, exhibit high fecundity and frequently undergo tremendous density fluctuations. By contrast, long-lived resources often show the opposite pattern.

Annual and biennal plants species represent unpredictable short-lived resources, while perennial plants are long-lived and are more predictable and available at the same site during a longer period of time. These spatial and temporal differences influence the life-history patterns of many insect species feeding on them.

In temperate regions, holocyclic aphid species need oviposition sites on plants which will produce green shoots in the next season for a successful hibernation. Consequently, annuals and also most biennals do not support non host-alternating (monoecious) aphid species (Börner 1952; Dixon 1985), which are confined to perennial plants. Aphid species which are able to exploit annuals alternate between two host plant types: a perennial winter host (usually a bush or tree) where the sexual females deposit their eggs and fundatrices hatch in spring, and an annual summer host where only parthenogenetic virginoparous forms are produced.

There are also differences in dispersal capacity between insect species dwelling on annuals or on perennials. Phytophagous insects dwelling on annuals often need to relocate to new host plant stands in the next season and thus need a high dispersal capacity. Most of these species are typically r-selected. This is especially conspicuous in specialized species depending on their annual host, like phytophages exploiting chamomiles (A. Freese 1995), or in host-alternating aphid species which have to emigrate from their winter host. Insects feeding on perennial host plants are often site-specific and show a reduced tendency for dispersal. Boller et al. (this Vol.) stress the importance of perennial dicot plants for the number of arthropod taxa in vineyards. Many species exploiting predictable resources are K-selected. For example, chrysomelid beetles on willow or gall-formers on oak often remain within one plant stand for many generations, and xylobiontic cerambycid beetles (e.g. *Cerambyx cerdo*) feeding on dead wood often remain within particular forest sites for many years (Mikkola 1991).

19.2.1.4 Relationship Between the Generation Times of Resource and Consumer

Among specialized arthropod predators or parasitoids those which seem to have the greatest impact on their prey population density are those whose generation times are close to that of their prey (Hassell 1976). Such systems are marked by strong interactions. By contrast, the most ineffective are those with long generation times compared to their prey. This view is confirmed by Dixon (this Vol.), who compares a ladybird-aphid and ladybird-coccid system, which show striking differences. In the latter system, the generation time of predator and prey is comparable, whereas in the former the prey has a much shorter developmental time. Thus, coccidophagous species are able to track changes in the abundance of their prey closely and are efficient control agents, whereas the former are not. Similar patterns are found for syrphid flies: while univoltinous and bivoltinous species often only have a low impact on aphid populations, polyvoltinous species with low generation times like *Episyrphus balteatus* may considerably influence local and regional patterns of aphid density (Stechmann, this Vol.).

The host-parasite systems considered next exhibit the usual patterns with longer generation time of the host (most plant-herbivore systems, the Kudu-tick system: Petney and Horak, this Vol.; the *Daphnia*-sporozoan system: Ebert et al., this Vol.) or more or less equal generation times between host and parasite (some host-parasitoid systems; e.g. Mackauer et al., this Vol.). If the generation time is inversely correlated with the rate of evolution, the relationship between host and parasite should lead to completely different types of evolution of the system. An arms race due to (diffuse) coevolution is more likely to occur at similar generation times, whereas long lived hosts are unlikely to match the faster evolving attack possibilities of the parasites. In this case, competition within and between parasite species (Freeland 1986) or extremely high density fluctuations of the host (Ebert et al., this Vol.) may contribute to the long-term persistence of the host-parasite system.

An exceptional host-parasite system is the *Margaritifera*-Salmonid relationship (Bauer, this Vol.), where the parasites' long generation time exceeds that of its host by a factor of between 10 and 20. This must be considered a constraint for the evolution of the host-parasite relationship. However, the system is stable in space and time through a high degree of adaptation of the parasite to a pool of native host species. The longer generation time provides a considerable advantage for the parasite: the most susceptible hosts are young salmonids (Bauer 1987). Since both host and parasite populations spawn once a year, the availability of hosts for a parasite female increases as its generation time increases.

19.2.1.5 Synchronization Between Resource and Consumer

In many systems, the degree of specificity of the consumer determines its dependence on the resource's phenology. While specialized species need an exact synchronization with resource availability to assure successful reproduction, polyphagous species which exploit a broad range of resources are less dependent.

Synchronization with host plant phenology, i.e. the temporal availability of the resource within a season, is one important factor regulating the population dynamics of phytophagous insects. Striking examples are tephritid flies attacking flower heads of thistles. Members of the genera *Urophora* and *Tephritis*, which are often mono- or oligophagous, attack very early developmental stages of the flower head. Therefore, adults need to be present at the particular plant stand for oviposition at a time when the flower heads are at a suitable stage (Zwölfer 1987, 1988, 1994; Romstöck-Völkl and Wissel 1989). By contrast, polyphagous species with a broad host range often depend less on a particular resource which may be available for only a short time period. Thus, these species need less adaptations to spatial and temporal resource distribution. However, there are also examples of restricted host use by polyphagous herbivores due to a lack of synchrony under field conditions; certain plants are not used in the field, whereas in the laboratory the adults will oviposit and larvae will feed and develop normally on them (Barbosa 1988a).

The hibernation strategy of phytophagous insects may also depend on the temporal availability of its resources. Species which produce their next generation very early in the season often hibernate in the adult stage. Plants that are ready too early present a problem for the development of the next generation. These species often suffer from a high environmental risk due to a considerable mortality during dispersal, since suitable hibernation sites and plant stands are often very distant. In these cases, host plants need to be re-colonized each spring. Thus, these species need to produce high adult densities - as normally typical for r-selected species - even if they attack perennial host plants which are more K-selected. A typical example for such a species is *Tephritis conura* (Romstöck-Völkl 1990). By comparison, species attacking a resource occurring later in the season, like many *Urophora* species (Zwölfer 1987, 1988, 1994), often hibernate in the larval or pupal stage at the same location as their host plant. These species are able to complete their development in the next season and benefit from the reduced dispersal, since host plants are available within short distances.

19.2.2 Size Constraints

Warren and Lawton (1987) suggest that many food web patterns can be explained on the basis of body size since (1) the size of prey is usually smaller than the size of its predator and (2) there is a positive correlation between predator and prey size. These relationships are due to two opposing forces. On the one hand, the smaller the prey size relative to the predator, the easier for the latter to catch and handle it. This also holds for host-parasite systems where parasitoids, like predators, have to capture and subdue hosts. Thus koinobiosis is thought to have evolved from idiobiosis as an alternative strategy that allowed parasitoid wasps to exploit hosts at earlier stages of development (Gauld and Bolton 1987; Mackauer et al., this Vol.). On the other hand, there is a disadvantage if small prey is scattered, as demonstrated by Dixon (this Vol.). For example, ladybird species feeding on small aphids have to spend a long time searching for prey.

These size relationships do not hold for true host-parasite systems (excluding parasitoids). Here, the host is larger since it is not only the food but also the habitat for the parasite. Therefore the concept of island ecology (McArthur and Wilson 1967) might be applicable to some of these systems. If increased host size means increased diversity of habitats on the host (Price 1980, pp 30–32 and many references cited there) or increased colonization time due to host longevity (Dogiel 1961; Kennedy 1975), then there should be a positive relationship between host size and species richness of its parasite community. This is shown by Achziger (this Vol.) for hemiptera on blackthorn shrubs. However, such a relationship will not be expected for parasite communities which are strongly determined by stochastic processes, e.g. tick-communities (Petney and Horak, this Vol.). Also, if host size is not related to diversity and colonization time, it may not play a role for the parasite community (Zwölfer 1987), or there may even be a negative relationship between host size and number of parasite species (Price et al., this Vol.).

For parasites, the size of the host or the size of the used habitat on the host (e.g. fruits)

1. defines the amount of available resources if the feeding stage is not able to leave the consumed host and search for a new one. This has considerable consequences for the probability that competition may occur. The amount of resources is small for solitary parasitoids, for insects in small fruits, like *Carpocapsa* (Geier 1963) and *Rhagoletis* (Bauer 1986), or gall midges ovipositing in buds (Redfern and Cameron 1978) etc. For these consumers competition is likely to occur far below the carrying capacity, at infestation rates far below 100% if the eggs are randomly distributed among the hosts. Accordingly, subtle mechanisms to avoid competition are evolved, (frequently by host marking, Askew 1973; Bauer 1986) and competition is of the highly developed contest type to avoid overexploitation (Mackauer et al., Romstöck-Völkl, this Vol.). On the other hand, competition is less likely to occur on large hosts. Accordingly, in this group, competition is of the scramble type (Topp, this Vol.; Duffey 1968; Klomp 1968; Dempster 1971; Harcourt 1971).
2. may influence the spatial patterns of resource utilization. Many spatial refuges for endophytically dwelling insects result from the inability of their parasitoids to reach the host due to a too short ovipositor (Weis 1983; Romstöck-Völkl 1990; Godfray 1994).
3. may provoke morphological and physiological adaptations. For example, body structures of phytophagous insects which are used for oviposition (the rostrum length of weevils of the genus *Larinus* and the ovipositor of the tephritid genera *Urophora* and *Terellia*) are distinctly correlated with the dimensions of the exploited resource, flower heads of Cardueae (Zwölfer 1982, 1987, 1994). An example for physiological adaptations is shown by Mackauer et al. (this Vol.). If within a parasitoid species body size correlates positively with fitness, then a selection pressure for maximized size must be expected and accordingly different growth strategies have evolved among larvae of koino- and idiobiont parasitoids in order to gain as much energy as possible from the host.

Size scales allometrically with many other life history traits (Begon et al. 1986; Roff 1992; Stearns 1992) like developmental time. Achziger (this Vol.) suggests that this may be the reason for the negative size-density relationship which is frequently observed in animal communities (Blackburn and Lawton

1994). Small size means fast development, rapid response to favourable conditions and thus high densities.

Body size, on the one hand, constrains offspring size (Blueweis et al. 1978), while on the other hand, there is frequently a trade-off between offspring size and number (Roff 1992; Stearns 1992). In the latter case, offspring size may be optimized by selection, at least to some degree. Theoretical models on this selective compromise usually assume a positive correlation between offspring size and survival of offspring (Smith and Fretwell 1974; Brockelman 1975), as has been demonstrated for many groups, e.g. due to better competitive ability or desiccation resistance (review in Roff 1992). However, this assumption probably does not hold for aquatic parasites with passively dispersed propagules and Ebert et al. (this Vol.) suggest that propagules of aquatic sporozoans are produced as small as possible in order to increase their number and thus the infection intensity of the host population.

19.3 Chemical Ecology in Vertical Food Web Interactions

Members of lower trophic levels evolve strategies to reduce feeding by their enemies, whilst members of higher levels evolve strategies to increase consumption (see Sect. 19.2.1). In various publications, Price drew attention to the fact that members of alternate trophic levels may act in a mutualistic manner (for references, see Price 1986 and Price et al., this Vol.). It is of importance that plant defense or plant resistance is achieved in a twofold way. Intrinsic defense of the plant is physical (through trichomes, toughness) and/or chemical (secondary compounds; constitutive and inductive chemical defenses) and directly targets the higher trophic level of herbivores. In contrast, if a plant benefits from members of trophic levels above herbivores (e.g. predators or parasitoids), this represents an extrinsic defense of a plant.

Chemical ecology (sometimes also called ecological biochemistry) studies the structure, function and biosynthesis of interindividually acting natural compounds (semiochemicals) and investigates their ecological significance and evolutionary origin; a rapidly expanding area of study. This discipline focuses primarily on chemically mediated intra- and interspecific interactions but also considers the chemical strategies of an individual in order to cope with unfavourable abiotic conditions (e.g. the non-wettable coat of brood cells in ground-dwelling bees; Hefetz 1987).

Although an incredible variety of chemical interactions between individuals within a trophic level or between two trophic levels has been investigated, the studies on the role of semiochemicals within multitrophic systems and simple biocoenosis are still in their infancy. For example, overwhelming data exist on chemical ecology and natural product chemistry of plant-herbivore (Barbosa and Letourneau 1988; Bernays 1989–1994; Rosenthal and Berenbaum 1992; Bernays and Chapman 1994), predator-prey (Eisner 1970; Blum 1981; Witz

1990), and host-parasitoid (Godfray 1994) interactions which mainly focus on insects (Cardé and Bell 1995; Roitberg and Isman 1992) from terrestrial systems; e.g., the triterpene-rich latex of the Californian Cichorieae effectively protects them against attack by tephritid flies (Goeden, this Vol.). Several investigations on chemical mimicry and camouflage deal with simple or more complex intra- and interspecific interactions (Stowe 1988; Dettner and Liepert 1994; Howard and Akre 1995). Comparable studies from aquatic environments are rather scarce (marine systems) or rare (freshwater systems), which may be partly ascribed to physicochemical and chemical differences (molecular mass, volatility, polarity, diffusivity) between airborne and waterborne behaviour-modifying compounds (Wilson 1970) and the difficulties of detecting and analysing waterborne compounds. Although chemistry of marine secondary metabolites becomes more and more exciting (Pietra 1990; Chemical Reviews 1993) only few chemical ecological studies deal with mechanisms of chemical defense, plant-herbivore, predator-prey interactions or chemical induction of invertebrate larval settlement (Paul 1992) in marine environments. In this volume, Tomaschko presents actual data on a new type of marine chemical defense in pycnogonids. These primitive arthropods use their ecdysteroids both as hormones and as feeding deterrents against predatory crustaceans. This dual function of ecdysteroids resembles steroidal fish hormones which may be externalized in order to act simultaneously as pheromones and underscores the multifunctionality of such compounds.

However, in terrestrial systems, physicochemical differences between allelochemicals also create different situations. According to their distance effects on the behaviour of other organisms, these chemicals can either represent attractants (cause the receiving organism to make oriented movements towards the source of stimulus) or repellents (cause organisms to make oriented movements away from the source) which finally result in acceptance or rejection of the food. Higher molecular and non-volatile compounds may influence a receiving organism after contact with the releasing organism. Plant and animal epicuticular lipids and especially internal plant and animal secondary compounds represent such kinds of natural non-volatile compounds. According to their close-range effects on other organisms, (e.g. herbivorous insects: Loon 1996) these compounds may be classified as stimulants which elicit feeding (phagostimulants) or oviposition (oviposition stimulants) or in contrast they may inhibit both activities (deterrents).

This chapter presents some actual aspects of chemical ecology where intra- (e.g. pheromones) and interspecifically (allelochemicals) acting natural products such as allomones (favourable to emitter but not to the receiver), kairomones (favourable to the receiver but not to the emitter) or synomones (favourable to both emitter and receiver) may influence alternate trophic levels of food webs. First, the role of volatiles that directly affect the behaviour of organisms from various higher trophic levels is briefly considered (Sect. 19.3.1). In the next chapter (Sect. 19.3.2), specific non-volatile and usually toxic products from plant or animal origin are treated in more detail. Such non-volatile compounds may be transferred from one developmental stage to the next, intraspecifically from males to females or from females into the eggs, or even move through trophic levels, influencing other consumers. In this connection, the phenomenon of organisms adapted to sequester or bioaccumulate such chemicals is especially illustrated.

19.3.1 Volatiles From Plants or Herbivores Affecting Organisms on Higher Trophic Levels

There are various examples known where chemical cues from plants are important in attracting herbivores (see Bernays and Chapman 1994 for examples). At the same time plants may benefit from their secondary chemicals by attracting pollinators (Dettner and Liepert 1994), predators (including ants) or parasitoids which protect the plant from herbivores (Hölldobler and Wilson 1990; Rowell-Rahier and Pasteels 1992). In addition, predators and parasitoids develop searching strategies by using specific chemical cues from their herbivorous prey or hosts (Rowell-Rahier and Pasteels 1992). These kairomonal cues can be sex pheromones, aggregation pheromones, or defense secretions but may also include non-volatile components of honeydew (e.g. aphids, mealybugs, whiteflies, Budenberg 1990). Bark beetle aggregation pheromones for example may be attractive for bark beetle predators such as clerid or ostomid beetles or representatives of Rhaphidioptera (Vité and Francke 1985), whereas sex pheromones of *Matsucoccus* scales specifically attract anthocorid and hemerobiid predators (Mendel et al. 1995). Therefore the formulation or design of pheromone traps for pest species has to be modified in order to avoid simultaneous elimination of beneficial insects. In the same way lepidopteran or heteropteran pheromones, allomones (Aldrich 1995), or toxic haemolymph compounds of other, especially pharmacophagous, insects (e.g. pyrrolizidine alkaloids, cantharidin; see Dettner, this Vol.) may attract dipteran and/or hymenopteran parasitoids.

Apart from constitutive defenses of plants such as secondary non-volatile and volatile chemicals, there exist many data on how plants respond chemically to herbivory by induced defenses (Dicke 1995). Mechanical damage, defoliation, chewing, sucking or mining of herbivores may induce various kinds of chemical defense (Chessin and Zipf 1990; Harborne 1993). The volatile signal molecules methyl jasmonate, jasmonic acid and other compounds may induce production of digestability reducers or various secondary metabolites (e.g. various alkaloids, furanocoumarins, indole glucosinolates, phenolics, monoterpenes) which are harmful to herbivores.

Recently, it became more evident that herbivore's frass may provide stimuli for increased searching of parasitoids and predators. This phenomenon has been referred to as carnivores "listening" to "talking" plants. In general, it might be favourable for a searching parasitoid or predator to rely on herbivore-induced plant volatiles, because those are produced in larger amounts (up to micrograms per hour) than herbivore compounds (Dicke 1995). At the moment, two kinds of induced defense mechanisms are recognized (Dicke 1995): plants damaged by herbivores emit the same bouquet as that emitted by mechanically damaged plants, but in larger quantities and for a prolonged period; feeding by herbivores induces the emission of large amounts of de novo synthesized volatiles while mechanical wounding dose not (Hopke et al. 1994). In general, odour induction by jasmonic acid, according to the above types of induced defense mechanisms, seems a primitive character since it was reported from fens, mono- and dicotyledonous plants (Bolant et al. 1995). Remarkably, gaseous methyl jasmonate may also induce defense reactions in neighbouring plants which indicates a plant-to-plant signal transfer as an important plant defensive strategy (Bruin et al. 1995).

Up to 1997, studies on herbivore-induced carnivore attractants where carnivores always discriminated between herbivore-infested and mechanically damaged plants concerned more than 20 plant species and about 30 species of both herbivores and carnivores (Dicke 1995, 1997).

The famous experiments with the Lima bean plant *Phaseolus lunatus*, the spider mite *Tetranychus urticae* and the predatory mite *Phytoseiulus persimilis* showed that after damage by spider mites, the plant emits a 200-fold increase in amounts of volatiles such as (E)-β-ocimene, linalool, methylsalicylate or (E)-4,8-dimethyl-1,3,7-nonatriene (Dicke and Sabelis 1992). Probably due to linalool and other compounds, the spider mites *T. urticae* are attracted at low degrees of infestation, whereas high degrees of infestation repel them. Remarkably, the induced release of such volatiles and the observed attraction of parasitoids is systemic (an endogenous elicitor is systemically transported within a plant from damaged leaves to undamaged areas; Dicke 1995) and interindividual, as was shown when uninfested plants were placed in a wind tunnel downwind of infested plants (Oldham and Boland 1996). It is fascinating that production of plant volatiles often varies depending on different herbivore species or on different instars of the same attacking herbivore (Dicke 1995). The saliva of some herbivores with β-glucosidases and various other compounds while mechanical damage usually does not may specifically trigger the production of volatiles. Also, in other plant species the herbivore *T. urticae* may induce production of volatiles. Another example is that of *P. persimilis*; since in infested cucumber plants only young leaves are attractive to *P. persimilis*, it became clear that old leaves exclusively and additionally to the other volatiles produce 3-methylbutanal-O-methoxime which is probably responsible for masking the attractive volatiles (Oldham and Boland 1996).

In the meantime, similar effects with so-called infochemicals (Vet and Dicke 1992) are known from completely different multitrophic systems (Fig. 19.2). Corn seedlings (*Zea mays*) infested by beet armyworms (*Spodoptera exigua*) emit indole, homo- and sesquiterpenes during the daytime which attract the generalist braconid wasp *Cotesia marginiventris* and the specialized parasitoid *Microplitis croceipes*. Here, induction of volatile emissions may be triggered by saliva, salivary enzymes and regurgitated food. In addition, several other

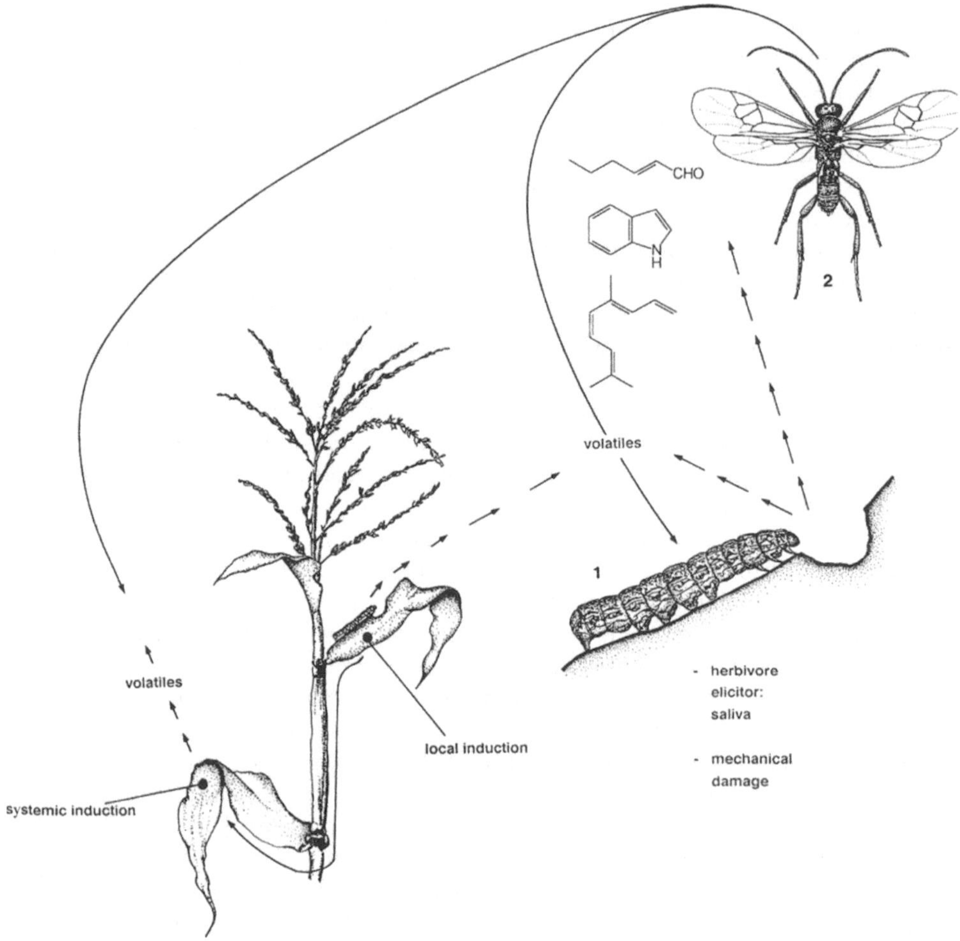

Fig. 19.2. Induced plant defense by volatiles [e.g. (E)-2-hexenal, indole and (E)-4,8-dimethyl-1,3,7-nonatriene] which affect organisms from higher trophic levels. *Zea mays* infested by *Spodoptera* larvae (*1*) may locally and systemically release volatiles (*broken arrows*) which results in attraction (*solid arrows*) of various parasitoids of caterpillars such as braconid wasps (*2*). (Turlings et al. 1993a)

lepidopteran species and grasshoppers may modify the production of volatiles (Turlings et al. 1993a). In sprout plants (*Brassica oleracea*) infested by cabbage white caterpillars or treated with the caterpillar's regurgitated food, a 200–8000-fold increase in green leaf volatiles such as (E)-2-hexenal or (E)-2-hexenyl acetate could be recorded (Mattiacci et al. 1994). It is important to note also that normal and especially wet faeces of caterpillars from different stages and different *Pieris* species may elicit a modified host searching behaviour in *Cotesia rubecula* parasitoids (Agelopoulos et al. 1995).

Further modifications of such kinds of induced defenses occur in *Gossypium* plants infested by *Helicoverpa zea* caterpillars. In this system old damaged plants in particular emit a blend of 22 terpenes and aldehydes (McCall et al. 1994). In addition, the braconid *Cotesia rubecula* is strongly attracted to cabbage previously infested by either caterpillars of *Pieris rapae* (its host) or of *Plutella xylostella* (non-host). Even mechanical damage, *Plutella* frass/regurgitated food or cabbage feeding by snails may elicit moderate attraction of the parasitoid (Agelopoulous et al. 1994a,b). Finally, recent data support the suggestion that release of volatile infochemicals does not only occur after attack by chewing herbivores. Sucking aphids of the genera *Diuraphis* and *Brevicoryne* may induce odour production in wheat and cabbage which attracts the aphidiid parasitoid *Diaeretiella rapae* (Reed et al. 1995). Even concealed feeding behaviour as in *Liriomyza* leaf miners from bean plants can evoke emission of 15 compounds including (Z)-hexen-1-ol and 4-hydroxy-4-methyl-2-pentanone which serve for host location of the larval ectoparasitoid *Diglyphus isaea* (Finidori-Logli et al. 1996).

Remarkably, emission of induced volatiles by herbivore-injured plants may vary qualitatively and quantitatively during the photoperiod and during the time that the plants suffer from herbivore attack (Loughrin et al. 1994). A further report might support the hypothesis that the susceptibility of host and parasitoid differs to plant volatiles: the antenna of parasitoid *Microplitis demolitor* reacts much more sensitively to several plant volatiles and especially 7-carbon compounds than that of its host, the herbivore *Pseudoplusia includens* (Ramachandran and Norris 1991). Most available data indicate that the induced plant volatiles are not innately attractive to female parasitoids, but rather females first require experience of the caterpillar-corn complex (Turlings et al. 1993b).

19.3.2 Chemical Defenses of Insects

Many insects from all trophic levels use chemical defenses (Witz 1990) for protection from predators, parasitoids and insect pathogenic microorganisms by de novo synthesis of defensive chemicals or by sequestering of toxic compounds acquired from larval or adult food. Due to the progress in trace analysis of organic compounds, hundreds of low molecular defensive chemicals, haemolymph toxins and exogenously derived compounds have been identified from most insect orders (Eisner 1970; Blum 1981; Evans and Schmidt 1990; Pasteels et al. 1983; Entognatha: Dettner et al. 1996; Orthoptera: Whitman 1990; Isoptera: Prestwich 1984; Thysanoptera: Ananthakrishnan and Gopichandran 1993; Hemiptera: Aldrich 1988; Coleoptera: Dettner 1987; Lepidoptera: Bowers 1993, Deml and Dettner 1997, Rothschild 1985; Hymenoptera: Hermann 1984, Codella and Raffa 1993, Wheeler and Duffield

1988). Chemical defenses seem to be absent in Odonata and in parasitic groups such as Mallophaga, Anoplura or Siphonaptera.

The observed chemical patterns are highly variable depending on taxa but it seems premature to argue that secondary metabolism of insects would be more influenced by their ecology than by phylogeny (Berenbaum and Seigler 1992).

If insects with defense glands are fed with various foreign compounds, there may be a considerable change in gland constituents if the ingested compounds are biosynthetically accepted (Huth et al. 1993). Even in omnivorous and carnivorous species the qualitative and quantitative intraspecific constancy of defensive secretions seems to be due to a constant ingestion of definite precursors that means a narrowly defined nutritional biology. Therefore exocrine glands may function to some extent at the same time as excretory mechanisms in order to eliminate and metabolize toxic constituents from the food (Huth et al. 1993). On contact with predators, de novo synthesized and sequestered haemolymph toxins are usually released via haemolymph depletion (reflex bleeding). Therefore insects containing systemically acting poisons often show warning colorations (Eisner 1970), emit warning odours (Dettner and Liepert 1994) and may exhibit a gregarious behaviour (Evans and Schmidt 1990). In general, the efficiency of chemical defense, i.e. individual titres of defensive chemicals, usually varies among a population and ensures survival of an insect after encountering predators or parasitoids.

19.3.2.1 Strategies of Insects Ingesting and Sequestering Toxicants

What kind of evolutionary strategies are observed when insects feed on toxic plants or animals? Studies on herbivores (Brower 1984; Blum 1992; Wink 1992) indicate that polyphagous generalists detect the allelochemical present and either avoid feeding or eliminate ingested toxicants by faeces. Prior to elimination there may also be degradation of toxicants. Otherwise these non-adapted species may have longer developmental times, smaller sizes, and reduced fecundity. Since the compounds do not penetrate the gut, a limited temporary defensive capacity is achieved by regurgitation of toxic gut material as observed in acridids (Eisner 1970; Blum 1992) or caterpillars (Bowers 1993; Brower 1984). Even anal discharges in adults and meconium in endopterygote pupae may represent efficient defense strategies (Blum 1992). If toxic compounds penetrate the gut membrane, they are converted to more hydrophilic metabolites that can be readily excreted by mixed-function oxidases and in particular cytochrome P-450 (Blum 1992).

Mono- and oligophagous specialists which are adapted to certain toxic chemicals are able to process these compounds. There are two different mechanisms: (1) insects selectively detoxify ingested compounds and subsequently eliminate them by faeces. This is illustrated by the chrysomelid beetle *Trirhabda geminata* and the insecticidal chromene encecalin of its food plant

(Kunze et al. 1996). In contrast, the polyphagous generalist *Spodoptera littoralis* metabolizes encecalin into toxic compounds, which represents a suicide metabolism similar to that of pyrrolizidine alkaloids in mammalian liver cells. (2) The insects sequester and accumulate toxic compounds for their own purposes. These toxic compounds include non-protein amino acids, various alkaloids, terpenoids, carminic acid and lichen anthraquinones, coumarins, butenolides, glucosinolates, or cyanogenic glycosides.

Sequestering of compounds is found in the following insect orders (arranged in decreasing abundance of herbivory): Lepidoptera, Orthoptera, Coleoptera, Hemiptera, Homoptera, Hymenoptera, and few examples in Diptera and Neuroptera (Bowers 1990; Rowell-Rahier and Pasteels 1992; Brown and Trigo 1995).

During digestion of plant material, these specialist herbivores must first be able to selectively absorb toxicants actively or passively through the gut membrane and sequester allelochemicals which then are used for their own defense (acquired defense), whereas other compounds are rapidly eliminated by defecation and excretion. Distinct differences exist between representatives of various insect orders with respect to gut permeabilities for certain compounds such as cardenolides (Rowell-Rahier and Pasteels 1992). The titre of sequestered toxicants is a function of concentrations of compounds in plant or animal sources, feeding rates and tolerance by animals. These factors modify the amounts of toxicant that reaches a particular site of action, where tissue damage and specific toxic effects actually occur.

After sequestration, the compound may additionally be metabolized and degraded. Processes like absorption, sequestration, pharmacodynamics and generally mobility of toxicants in an animal are complicated and depend mainly on physicochemical and biophysical conditions such as water solubility, the partition coefficient, volatility, size, shape and molecular mass of the toxicant (Duffey 1980). Moreover, toxicants in haemolymphs or tissue must be detoxified in a biochemical way and/or sequestered within specialized cuticular cavities as in cardenolide-sequestering lygaeid bugs (Blum 1992). Apart from increased haemolymph titres of allelochemicals, these compounds may be channelled to the integument or into wing scales in order to defend against avian predators. In general, chewing herbivores presumably consume higher amounts of toxic compounds from their food than phloem-sucking species, therefore adaptations to toxic chemicals should be better developed in chewing rather than sucking herbivores (Mullin 1986). For example, chewing herbivores such as lepidopterans and coleopterans are often less susceptible to pesticides than phloem-sucking aphids (Mullin 1986).

When specialists show further adaptations to acquired allelochemicals from plants and animals we may reach the extreme case of pharmacophagous insects where especially toxic secondary compounds are well integrated in the physiology, biochemistry, or development of these insects in order to serve a diversity of important functions (Boppré 1984, 1986; see Dettner, this Vol.). Independent from foraging for nutrients, these insects selectively try to find

Table 19.1. Characteristics and modes of adaptation in four types of pharmacophagous insects sequestering the toxicants clerodendrin (Nishida and Fukami 1990), cucurbitacines (Metcalf and Metcalf 1992), pyrrolizidine alkaloids (Boppré 1986, 1995) and cantharidin (Dettner, this Vol.). If various degrees of adaptation exist, the extreme case is indicated

	Clerodendrin D	Cucurbitacines	Pyrrolizidine alkaloids (PA)	Cantharidin
Source of toxin	*Clerodendron*-plants (Verbenaceae)	Cucurbitaceae plants	Asteraceae- and Boraginaceae plants	Meloid and oedemerid beetles
Attractive agent	Long-range attraction unknown	Volatiles of Cucurbitaceae blossoms	PAs or degradation products	Cantharidin or degradation products
Pharmacophagous taxa	Adults of sawfly *Athalia*	Diabroticine chrysomelid beetles	Lepidoptera, Coleoptera Orthoptera, Diptera	Coleoptera, Diptera Hymenoptera Heteroptera
Attracted sex	Adult males and females	Adult males and females	Males and/or females	Males and/or females
Phagostimulancy	Adult females	Adult females/males	Males and/or females	Males and/or females
Selective uptake and sequestration	+	+	+	+ +
Metabolism of toxin	?	+	+	+
Increased survival (predators/parasitoids)	+	+	+	+

Transfer of toxin:				
through developmental stages (a)[a]	?	+	+	+
from females into eggs (b)	?	+	+	+
from males into females (c)	?	?	+	+
interspecific (d)	–	–	+	+
Translocation into exocrine glands	–	–	+	+
Importance of toxin as sex attractant or in courtship behaviour	?	?	+	+
Increased mating chance and sexual selection	? Females become attractive	?	+	+
Increase in toxin titre by cannibalism	?	?	+	+
Morphogenetic effects of toxin	–	–	+	–

[a] (a)–(d) denote various kinds of toxin transfer between individuals.

allelochemicals. Such specific chemicals represent attractants and phagostimulants. They are taken up, sequestered and thus render those insects unpalatable. These compounds are utilized for specific purposes other than primary metabolism or (merely) host recognition and as a whole may increase their fitness (Boppré 1986, 1995; Dettner, this Vol.). As noted by Bernays and Chapman (1994) pharmacophagy may represent one extreme of a spectrum which grades into those situations where the specific chemical is also a normal component of the food.

Apart from female euglossine bees, which selectively collect fragrances from flowers (which do not contain nutrients) in order to attract conspecific males (Boppré 1984), the status of pharmacophagous insects sequestering polar toxicants either from plants as clerodendrin (adult *Athalia* sawflies; Nishida et al. 1989; Nishida and Fukami 1990), cucurbitacines (diabroticine chrysomelid beetles; Metcalf and Metcalf 1992), pyrrolizidine alkaloids (Lepidoptera and representatives of other insect orders; Boppré 1986, 1995) or from animals as cantharidin (Coleoptera and representatives of other insect orders; Dettner, this Vol.) has been investigated in more detail. As illustrated by Table 19.1, the degree of adaptation to the toxin increases from the clerodendrin/cucurbitacine to the alkaloid/cantharidin system. In the first system, long-range attraction seems to be achieved by other volatiles, whereas pyrrolizidine alkaloids or cantharidin alone or by their unknown degradation products directly attract. Moreover, in the second system, various unrelated insects from different orders could independently adapt to different degrees to the toxicant, whereas only representatives of a genus or a tribe could adapt to the toxic clerodendrins or cucurbitacines. Whereas sequestration, metabolism, and transfer of the toxicants seem characteristic for all the systems mentioned above, the following peculiarities are restricted to single taxa within alkaloid- and/or cantharidin-containing insects: transfer of toxicants from males into females in conjunction with a translocation of toxins into exocrine glands, which may illustrate the importance of these compounds as sex attractant or in courtship behaviour. An increase in fitness by higher titres of toxins is therefore not only achieved by better survival against predators and parasitoids in both sexes of pharmacophagous insects and their offspring, but through an increased mating chance of the males which is due to modes of chemically based sexual selection by females. An increase in individual fitness may be even achieved by uptake of toxic compounds through egg and pupal cannibalism (Bogner and Eisner 1991, 1992). Finally, the high integration of toxins into the physiology of pharmacophagous insects is reflected by morphogenetic effects of pyrrolizidine alkaloids in determining the size of the pheromone-disseminating organs (coremata) in adult males of certain arctiid Lepidoptera.

The significance of the above observations and generalizations often apply only to narrowly defined experimental situations in the laboratory, when definite amounts of a specific toxicant in the food of a herbivore are considered. As mentioned by Topp (this Vol.), if developmental success of the herbivorous leaf beetle *Chrysomela vigintipunctata* on different food plant species with deviating chemistry (e.g. phenolglycosides, flavonoids) must be assessed in the field during different seasons, the interpretation of the results is usually very difficult, since temporarily variable mixtures of different plant constituents simultaneously influence feeding, development and fecundity of the herbivore investigated.

19.3.2.2 De Novo Synthesis Versus Exogenously Derived Toxicants

Within insects it appears that de novo synthesis of chemical defensive secretions is more widespread than sequestration. Since sequestration is usually associated with selective uptake of secondary plant or animal compounds, the way to become unpalatable by sequestion is well marked within insect orders dominated by the phytophagous life-style (e.g. Lepidoptera, Orthoptera). Moreover, especially in adult Lepidoptera, where the body surface is covered with scales, the evolution of exocrine defensive glands is improbable. It is not possible to deplete such a gland and deposit droplets onto targets without contaminating the scaled body surface of the producer.

Even within predominantly phytophagous taxa such as chrysomelid beetles or Heteroptera, de novo synthesis of defensive secretions obviously seems to represent an ancestral (plesiomorphic) and widespread type, whilst sequestration is always restricted to certain specialized taxa (Aldrich 1988; Pasteels et al. 1988). In general, biosynthesis of toxicants may represent a preadaptation for storage of similar or deviating compounds from food (Rowell-Rahier and Pasteels 1992). This is illustrated by many Chrysomelinae larvae which produce de novo cyclopentanoid monoterpenes in their segmental defensive glands (Lorenz et al. 1993), whereas few Salicaceae-feeding species additionally or exclusively derive their glandular salicylaldehyde from the salicin of their host plants (Pasteels et al. 1984). In addition, many adult chrysomelid beetles synthesize cardenolides in their defensive glands, whereas few *Oreina* species feeding on Apiaceae, Asteraceae or Senecioneae rely on autogenous chemical defenses and/or enrich their defensive secretion with pyrrolizidine alkaloids from their food plants (Pasteels et al. 1995).

Until now it was only been possible to give some general ideas on the evolution of unpalatability/availability of defensive chemicals and to assess the costs of chemical defense if de novo synthesis is compared with sequestration (Bowers 1992; Rowell-Rahier and Pasteels 1992).

Haemolymph toxins offer one crucial advantage compared with the production and storage of more or less volatile chemicals within exocrine glands. Larval exocrine glands with their epicuticular lining must be replaced during moultings and holometabolous taxa are characterized by a pupal stage devoid of exocrine structures whilst adults may possess non-homologous glands. Haemolymph defensive compounds, once upon a time sequestered during a larval stage, may be transferred into the forthcoming stages where the titre may ever be augmented. Adults with other feeding strategies (e.g. nectar feeding) which have no access to allelochemicals may sequester compounds that had been stored during their larval stages. Haemolymph toxins offer the possibility of being accumulated within male accessory glands in order to be transferred into females. Females mate frequently in order to obtain considerable amounts of allelochemicals which can be allocated to different egg batches. In order to improve the protective value of eggs containing internal toxicants they are often intensively coloured and deposited not singly but as clusters. That alkaloid-containing eggs of certain coccinellid beetles are avoided by various predators (Dixon, this Vol.) underscores the importance of haemolymph toxins for non-movable developmental stages. Moreover, it is interesting that females avoid feeding on their own eggs, whereas eggs of conspecific females are often eaten (Dixon, this Vol.). The chemical basis of this ability is unknown but it may lead to recognition of individuals by their chemical identity (Smith and Breed 1995).

19.3.2.3 Transfer of Toxicants Through Developmental Stages of an Individual

Another important aspect of sequestration is the titre of toxicants in the different developmental stages of a holometabolous insect. Whereas hemimetabolous insects may sequester toxicants from larval to adult stages, for most holometabolus species the pupal stage may represent an evolutionary barrier (Bowers 1992). Due to massive reorganization of internal structures during the pupal stage, holometaboles run the risk of increased autotoxicity by sequestered compounds. Moreover, the excretory Malpighian tubules of larvae and adult holometaboles may have very different transport characteristics for toxicants ingested during the larval stage. Therefore last stage larvae of many sequestering holometabolous species may eliminate toxicants partially or completely via pupal meconium or incorporation into cocoon silk. Thus, adults are usually devoid of these compounds (e.g. certain Nymphalidae and Sphingidae sequestering iridoid glycosides: Bowers 1992; pyralid larvae sequestering pyrrolizidine alkaloids: Wink et al. 1991). Indeed, most pupae are cryptic and pupation takes place singly. Chemical defense of this stage does not seem to be necessary, especially if one considers the fact that pupal stages are generally killed when they are attacked by visually hunting vertebrate predators (Bowers 1993). On the other hand, it may be adaptively important to save sequestered larval toxicants into adults stages, because there is a drastic nutritional change from larvae to adults. Moreover, due to their scaled surfaces, adult Lepidoptera are usually devoid of exocrine defensive glands and therefore may depend particularly on haemolymph toxins acquired during the larval stage.

A transfer of exogenously derived larval toxicants into adults may therefore be highly advantageous and may reflect the degree of adaptation of a sequestering species to cope with these compounds. Accordingly, many sequestering holometabolous species exist, where pupal and adult stages are chemically defended by the same toxicant. Examples are comparatively high titres of *Veratrum* alkaloids in sawflies (Schaffner et al. 1994), butenolides in Yponomeutidae (Fung et al. 1988), iridoid glycosides in certain nymphalids (L'Empereur and Stermitz 1990) or sawfly larvae (Bowers et al. 1993) and cardenolides in the monarch (Brower 1984).

Since pharmacophagous insects are adapted to handle sequestered toxins, a transfer of such compounds without loss from larval into adult stage should be observed frequently in those specialists. In pyrrolizidine alkaloid-sequestering arctiid moths, no or only small amounts of the sequestered alkaloid N-oxides are lost during ecdysis (von Nickisch-Rosenegk et al. 1990; Trigo et al. 1993). Moreover, toxic cucurbitacins from plants are transferred from specialized galerucine beetle larvae to the adult stages (Metcalf and Metcalf 1992). Finally, most deuterated cantharidin which was transferred from female anthicid beetles into their eggs was transported through all larval stages and appeared in female and male adult beetles (K. Dettner and S. Thießen, unpubl.).

These data indicate that the capacity to carry over significant amounts of sequestered compounds from larvae via pupae into adults may characterize and separate extremely adapted insects and in particular pharmacophagous species from moderately adapted holometaboles. This may be corroberated by those highly adapted insects with de novo synthesis of haemolymph toxicants. Although both catabolism and anabolism of toxicants may occur during the pupal stage, there is no significant decrease in chemical defense from pupal to adult stage if titres of cyanogenic glucosides of various Lepidoptera (Davis and Nahrstedt 1985), cantharidin in oedemerid beetles (Holz et al. 1994) or pederin in *Paederus* rove beetles (Kellner and Dettner 1995) are compared. It would be highly desirable to receive more data on titres of de novo synthesized and exogenously derived toxicants during the whole life cycle.

19.3.2.4 Transfer of Toxicants Through Higher Trophic Levels

If secondary compounds are sequestered within an individual or are continously stored through developmental stages, the question may be raised as to whether these compounds may flow through higher trophic levels and whether members of these higher levels may be negatively or positively influenced by these compounds.

A flow of allelochemicals from plants, through herbivores into predators and/or parasitoids, has been observed several times. Plant-derived quinolizidine and pyrrolizidine alkaloids (Fig. 19.3) in honeydew from aphids were taken by ants (see Wink 1992), by ladybirds after feeding on alkaloid-containing aphids (Witte et al. 1990) or by braconid and chalcidoid parasitoids after consuming seed-feeding *Bruchidius* beetles (Szentesi and Wink 1991) or arctiid *Nyctemera* moths (Benn et al. 1979). Various publications report on the

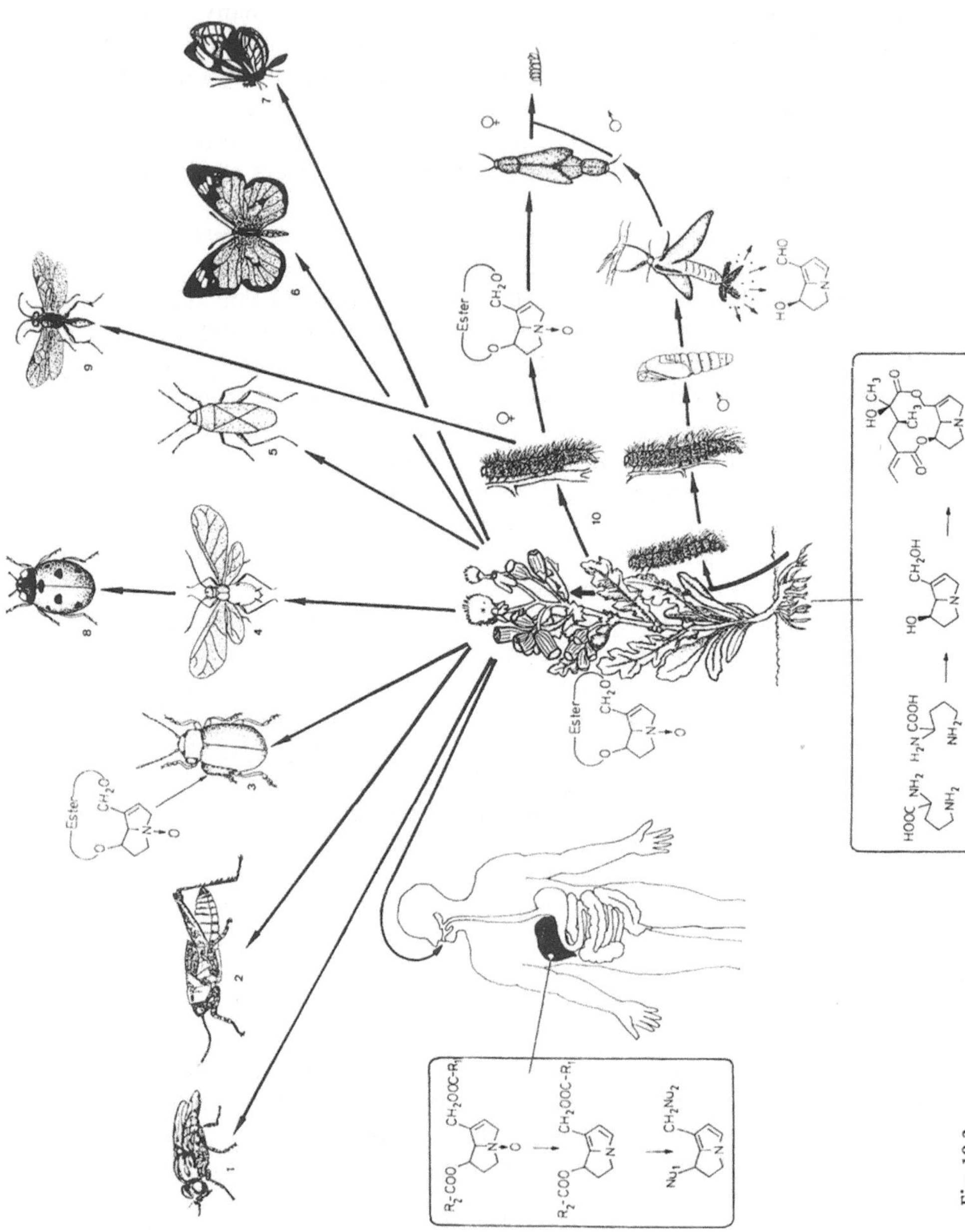

Fig. 19.3

flow of pyrrolizidine alkaloids through multitrophic levels (Fig. 19.3). In the roots of Asteraceae and other plants, these toxins are synthesized from the amino acid ornithine and are subsequently distributed within the plant as N-oxides. Several pharmacophagous insects such as chloropid flies, pyrgomorphid grasshoppers, danaid and ithomiid butterflies or specialized arctiid moths are attracted to pyrrolizidine alkaloid sources and selectively sequester the toxins (Boppré 1995). In arctiid moths the pyrrolizidine alkaloids are transferred from caterpillars to adults, from males to females and from females to eggs. An alkaloid-derived male pheromone may also serve as an attractant for females. Other herbivore species may sequester pyrrolizidine alkaloids in their haemolymph (Heteroptera) or translocate food alkaloids into defensive glands (*Oreina* leaf beetles). Finally, an unintentional transfer of pyrrolizidine alkaloids into mammals and humans occurs via contaminated honey (from the honeydew of aphids), or via milk or teas (containing alkaloidal plants). In the liver of mammals the alkaloids are metabolized into heptotoxic and carcinogenic compounds (suicide metabolism; Fig. 19.3).

A flow of *Aphis nerii* cardenolides into predators like ladybirds or lacewings or into tachinids has been reported by Rothschild et al. (1973). In the same way, toxic products of animal origin may flow from prey into predator and even into parasitoids. Carminic acid from cochineal bugs is sequestered by carnivorous larvae of pyralids, coccinellids and flies (Eisner et al. 1994). Cantharidin from meloid and oedemerid beetles was found to be enriched by various canthariphiles such as anthicid, pyrochroid and staphylinid beetles or ceratopogonid and anthomyid flies, respectively (Dettner, this Vol.). Probably this terpene anhydrid is also ingested by canthariphiles from other insect orders.

Fig. 19.3. Interspecific flow (*arrows*) of pyrrolizidine alkaloids through trophic levels of herbivores, carnivores and parasitoids. Biosynthesis of alkaloids from amino acid ornithine in the roots of an Asteracease (*below*) and their intraplant distribution as N-oxides. Pharmacophagous insects which are attracted to pyrrolizidine alkaloid sources and selectively sequester the toxicants: chloropid flies (*1*), pyrgomorphid grasshoppers (*2*), danaid (*6*) and ithomiid (*7*) butterflies, specialized arctiid moths (*10*). Herbivorous and carnivorous insects sequestering pyrrolizidine alkaloids: Homoptera (*4*), Heteroptera (*5*), coccinellid beetles (*8*), braconid Hymenoptera (*9*). Translocation of food alkaloids into defensive glands: *Oreina* leaf beetles (*3*). In certain arctiid moths (*10*) the transfer of pyrrolizidine alkaloids from caterpillars into adults, the production of an alkaloid-derived male pheromone and the transfer of alkaloids from males into females and into eggs is illustrated. Unintentional transfer of pyrrolizidine alkaloids into humans via contaminated honey, milk or teas transform alkaloids in the human liver into heptotoxic and carcinogenic compounds which react with nucleophiles (*Nu*; *left*). (Boppré 1986, 1995; Dussourd et al. 1988; Witte et al. 1990; Trigo et al. 1993; Brown and Trigo 1995; Pasteels et al. 1995)

19.3.2.4.1 Effects of Sequestered Compounds Against Predators

Several reports exist on reduced growth rates and increased mortalities of predators whose herbivorous prey was fed with resistant plants (Shepard and Dahlman 1988). In a few cases these reports may be ascribed to toxic food plant constituents as in coccinellid beetles showing atrophies of elytra and wings, higher larval and adult mortalities, lengthened postembryonic development and lower female fecundity, if the beetles feed exclusively on *Aphis nerii* from the toxic asclepiadacean plant *Cionura* (Pasteels 1978).

Sequestered secondary compounds are directed against both vertebrate (especially birds, lizards and small mammals) and invertebrate predators but specialized and immune target organisms may always exist (Brower 1984; Rowell-Rahier and Pasteels 1992). Among invertebrates, ants, wasps, beetles and spiders have been used as targets. For example, Wink (1992) reported high mortalities of *Coccinella* and *Episyrphus* larvae after feeding on aphids previously fed with quinolizidine alkaloid-containing plants. Moreover, he found that *Uresephita* larvae containing these alkaloids exhibited deterrent activities on ants and paper wasps. In addition, *Acyrthosiphon* iridoids may deter ladybird beetles (Nishida and Fukami 1989) and plant-derived cardenolides from *Aphis nerii* can represent psychoactive drugs for spiders which result in severely disrupted webs (Malcom 1989). Also, in pharmacophagous insects, the sequestered toxins may effectively protect against all kind of predators. In a spectacular way, ithomiine butterflies containing plant pyrrolizidine alkaloids are cut out from the webs by giant tropical *Nephila* orb spiders (Brown 1984). Increased titres of pyrrolizidine alkaloids in eggs of pharmacophagous arctiid Lepidoptera rendered these eggs relatively unacceptable to coccinellid beetles of the genus *Coleomegilla* (Dussourd et al. 1988). The high cucurbitacin titres of Luperini leaf beetles are sufficient to deter arthropod and bird predators (Metcalf and Metcalf 1992). In the same way, cantharidin containing developmental stages of cantharidin-producing and canthariphilous insects are protected against predatory arthropods (Dettner, this Vol.). Another rare example for the sequestration and utilization of defensive chemicals from animal sources concerns various insects which take up carminic acid from cochineal insects in order to deter ants and other predators (Eisner et al. 1994).

Finally, it must be mentioned that sequestration of exogenous toxins and translocation into exocrine defensive glands may be of high protective value. For example, sequestration of host plant mono- and sesquiterpenes (*Pinus*, *Eucalyptus*) in pouches of the foregut by sawfly larvae has deterrent effects on ants and spiders (Eisner 1970). That exocrine glands at the same time may represent some kind of excretory organ in order to eliminate foreign compounds is highlighted by the following data. The deterrency of the defensive secretion of the grasshopper *Romalea guttata* against ants was increased if the hoppers were previously fed with toxic plants in contrast to their normal polyphagous diet (Blum 1992). Other grasshoppers or the larvae of swallowtail butterflies fortify their defensive secretions with cardenolides or aristolochic acids from their food plants (see Blum 1992). Salicylaldehyde of some chrysomelid larvae which is derived from the ingested plant allelochemical salicin is a potent feeding deterrent against predators (Rowell-Rahier and Pasteels 1992), and the metasternal defensive glands of *Oncopeltus* bugs may contain both de novo synthesized defensive alkaloids and cardenolides from their milkweed host plant (see Blum 1992).

19.3.2.4.2 Effects of Sequestered Compounds Against Parasitoids

As in predators, oviposition, survival and adult longevity of braconid or tachinid parasitoids were often reported to be negatively influenced by those herbivorous hosts feeding on resistant and/or toxic plants (Shephard and Dahlman 1988).

Reduced growth, retarded development, smaller size, shortened adult longevity and even adult antennal and ovipositor deformities were reported for various parasitoids that were less likely to parasitize, when these species parasitized caterpillars which previously sequestered allelochemicals such as nicotin, tomatine, pyrrolizidine alkaloids or cardenolides (Barbosa and Saudners 1985; Duffey et al. 1986; Barbosa 1988b). Moreover, these influences of allelochemicals on higher trophic levels may be modified by the action of other compounds or nutrients of the plant (Duffey et al. 1986), which might be highly relevant for plant breeding programs or for constructing resistant transgenic plants. It is not clear whether these effects are due to direct toxic action of the allelochemicals against different developmental stages of parasitoids or whether the reduced quality of the host is responsible for these observations.

Data from Barbosa (1988b) might eventually support the first hypothesis that plant allelochemicals sequestered by host insects may poison parasitoids. Since allelochemical effects depend on whether the herbivorous insect is a specialist or a generalist, the degree of host specificity of parasitoids seems to determine conclusively the effects of plant allelochemicals on the parasitoid's biology. If more specialist (*Manduca*) and more generalist herbivores (*Spodoptera*) are fed with nicotine, and if interactions of monophagous and variably polyphagous parasitoids with these herbivores are studied, the least detrimental effects on parasitoids occurred in monophagous species (*Cotesia congregata*), whereas most polyphagous parasitoids such as the ichneumonid *Hyposoter annulipes* suffered the most severe effects (Barbosa 1988b). In another specialist parasitoid, *Campoletis*, the seequiterpenoic cotton allelochemical gossypol in the diet of its noctuid host had no detrimental effects on it; on the contrary, medium titres of the allelochemical promoted increased body size in *Campoletis* adults (Williams et al. 1988).

Parasitoids may be protected from host allelochemicals by the presence of detoxifying enzymes. For example, a tachinid *Zenillia* species and *Apanteles zygenarum*, a typical parasitoid of cyanide-containing Zygaenidae, contain rhodanese for cyanide detoxification (Davis and Nahrstedt 1985). Additionally, alkaloids such as nicotine from the host insect can be transferred either into larval faeces or may be subsequently incorporated into the pupal cocoon of the parasitoid (Barbosa and Saunders 1985; Duffey et al. 1986). In the eulophid egg parasitoid *Oomyzus galerucivorus* the deterrent anthraquinones and anthrones from the eggs of its chrysomelid beetle host *Galeruca tanaceti* are eliminated prior to pupation by the faeces of last stage parasitoid larvae (Meiners et al. 1996).

In general, plant allelochemicals ingested by herbivores seem to have a significant negative impact on non-specialized parasitoids, whereas specialized parasitoids are adapted and can handle the toxic metabolites without harm.

19.3.2.5 Driving Forces of Toxin Transfer Through Trophic Levels

A first interpretation of allelochemical flow through trophic levels may be helpful if a mechanistic approach is considered. If toxic compounds can be easily handled (e.g. two kinds of derivatives with lipophilic/hydrophilic properties) and detoxified (toxic/nontoxic) as in pharmacophagous species (e.g. pyrrolizidine alkaloids as free alkaloids or N-oxides; free and bound cantharidin), there is a strong tendency that these compounds may be both sequestered and transferred through multitrophic levels. However, another prerequisite for multitrophic transfer of toxins seems to be of importance: the primary toxicity of these selected allelochemicals against eukaryotic targets should be universal in order to be used for defensive purposes against various unrelated target organisms. As in those allelochemicals which bind to eukaryotic protein phosphatases, a multitrophic transfer is also possible when chemically unrelated toxic compounds may block this central enzyme (Fig. 19.4). A transfer of different biologically active protein phosphatase inhibitors may then evolve in completely unrelated organisms in very different habitats. Potent inhibitors of protein phosphatase type PP2A represent the chemically related toxins okadaic acid, dinophysistoxin and acanthifolicin, which are de novo synthesized by marine dinoflagellates of the genera *Prorocentrum* and *Dinophysis* (see Dettner, this Vol.). It is remarkable that these toxins and tumour promoters may flow from dinoflagellates into either marine sponges

→

Fig. 19.4. Inhibition of protein phosphatase 2A by different toxins such as okadaic acid (*above*; marine systems) and cantharidin (*below*; terrestrial systems) and interspecific flow (*arrows*) of these toxins through trophic levels up to humans. Okadaic acid-producing organisms: dinoflagellates *Dinophysis* (*1*), *Prorocentrum* (*2*), sequestering organisms: sponge *Halichondria* (*3*), mussel *Mytilus* (*4*). Cantharidin-producing organisms: oedemerid (*5*) and meloid (*6*) beetles, sequestering organisms: pharmacophagous anthicid beetles (*9*) and ceratopopogonid flies (*10*), cantharidin-containing frogs (*7*) and geese (*8*). (Dettner, this Vol.; Eisner et al. 1990; Yasumoto and Murata 1993; MacKintosh and MacKintosh 1994)

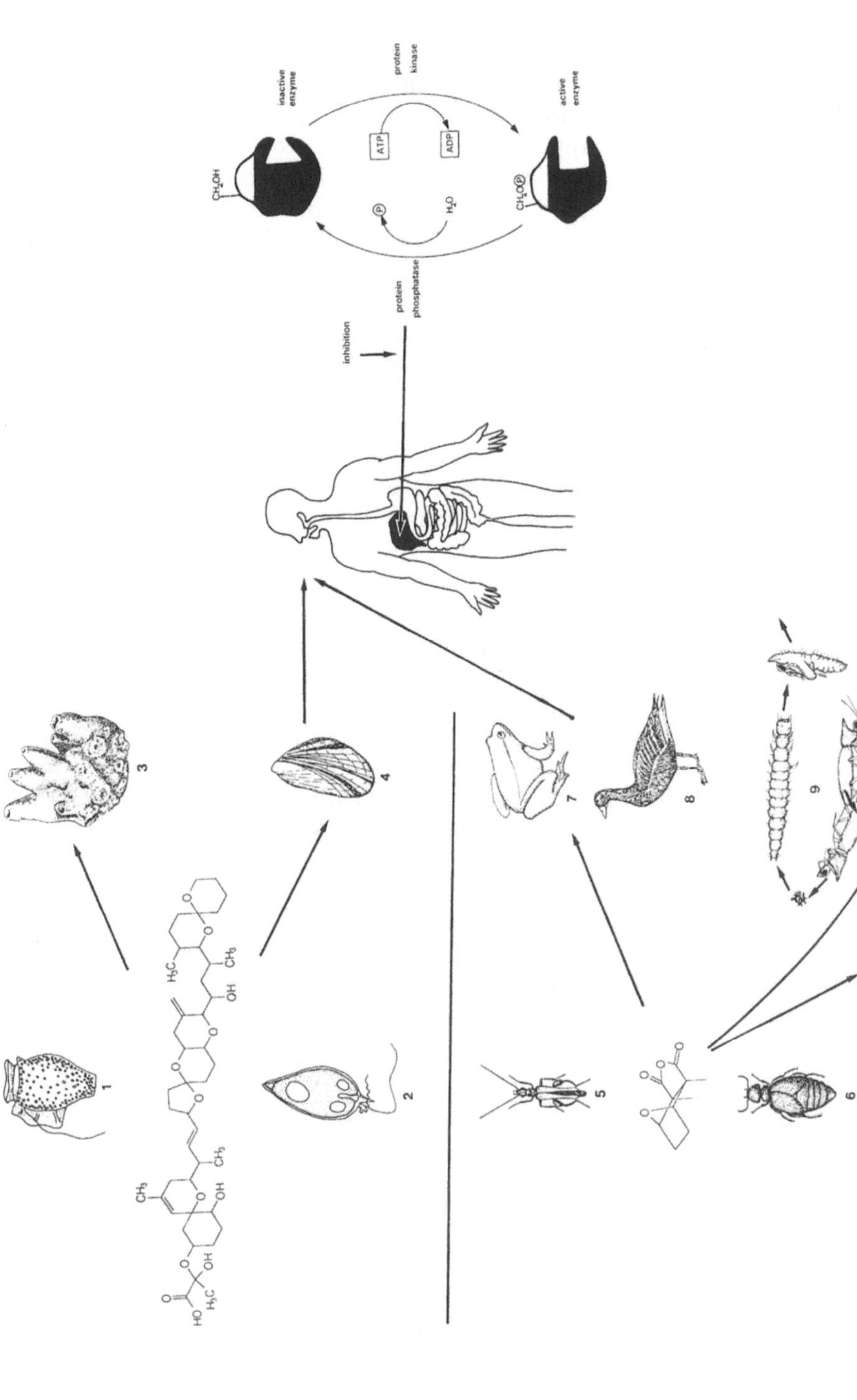
inactive enzyme
protein kinase
ATP
ADP
active enzyme
protein phosphatase
inhibition
1
2
3
4
5
6
7
8
9
10

Fig. 19.4

(e.g. *Halichondria*) or marine bivalve molluscs (mussels, scallops, clams) which are both excellent filter feeders. It is not surprising that these toxins could be first isolated from sponges (okadaic acid: *Halichondria okadaii*) and digestive glands of mussels (dinophysistoxin) of these filter-feeding dinoflagellate consumers. Apparently, these bound toxins cause no visible harm to the physiological functions of the consuming organisms which seem therefore to be well adapted. As in other dinoflagellate poisons the hepatopancreas and siphon of molluscs can bind and partially destroy or excrete these toxins (Schantz 1971). The molluscs which are normally culinary delicacies may accumulate these flagellate toxins in their hepatopancreas and become highly toxic after ingestion by humans. In Japan or northwestern Europe, this diarrhetic shellfish poisoning with gastrointestinal disorders is a serious problem to both public health and the shellfish industry (Yasumoto and Murata 1993). Just as in cantharidin-containing human food, shellfish poisons are not destroyed by heating, as carried out in ordinary food-processing techniques. In the same way as okadaic acid and its derivatives in marine systems, the toxic terpene anhydride cantharidin from meloid and oedemerid beetles may flow through trophic levels in terrestrial systems (Dettner, this Vol.) and finally may even cause death in humans if cantharindin-containing food (e.g. frogs, geese) is ingested (Eisner et al. 1990). Like okadaic acid, cantharidin is hydrophobic and cell-permeable and can inhibit the important cytosolic protein phosphatases in the intact eurakyotic cell. Moreover, cantharidin and okadaic acid apparently bind to the same site of the extremely conservative protein phosphatases (MacKintosh and MacKintosh 1994; Barford 1996) which may illustrate that evolutionary conservation of an important enzyme and its conservation of toxin sensitivity may be a driving force for the invention of various unrelated peptides and secondary metabolites and the creation of adapted organisms which can handle and may use such compounds. Therefore it is not surprising that unrelated organisms independently invented chemically diverse toxins (Shenolikar 1994) such as calyculin (marine sponge), tautomycin (*Streptomyces*), microcystin (cyanobacteria) or nodularin (*Nodularia*) which all inhibit the universally present target enzymes, phosphatases.

19.4 Consequences for Biological Control

A successful biological control agent should be able to reduce the density of the target below an economic threshold. This usually requires a high degree of resource utilization.

Differences in generation times may have a great impact on the success of biological control. Aphidophagous ladybirds, for example, usually have much longer generation times than their aphid prey. Therefore, most aphidophagous ladybirds fail to control aphid populations despite their high voracity (Dixon, this Vol.). By contrast, coccidophagous ladybirds are usually much better synchronized with their prey and have similar generation times. This results in a great impact on coccid population density, and indeed there are some spectacular examples in biological control among coccidophagous ladybirds, e.g. the Vedalia beetle, *Rodolia cardinalis* (Dixon 1996, this Vol.).

Multitrophic interactions may also prevent the economic success of single partners; ants often prevent a successful parasitization of aphid colonies by aphidiid parasitoids (Völkl, this Vol.). The economic success of *Aphidius colemani*, which was introduced and established on some South Pacific islands for the control of the banana aphid, *Pentalonia nigronervosa*, may be limited by the high degree of ant attendance in banana aphid colonies (Stechmann et al. 1996). Also, aphid parasitoids which are tolerated by ants differ in their resource utilization between ant-attended and unattended colonies, often laying considerably fewer eggs into unattended colonies (Völkl, this Vol.). This factor may contribute to the low impact of *Lysiphlebus fabarum* in the control of the black bean aphid, *Aphis fabae*, on bean and sugar beet despite the high control potential of *L. fabarum*.

The patterns of resource utilization depend on the egg-laying and foraging strategies of the control organism. Many parasitoids, for example, lay only a few eggs into available host patches and exploit their available resource only to a low degree. The reasons for this behaviour may be manifold: egg limitation or time limitation of the foraging female may set an upper limit for the number of ovipositions (for reviews, Godfray 1994; Hawkins 1994). Other species, like many aphid parasitoids, lay only a few eggs within one colony despite a high number of mature eggs (Mackauer and Völkl 1993). In these species, the average number of ovipositions per colony is determined by their species-specific behaviour which follows a strategy of "spreading the offspring mortality risk" among many host colonies. In natural habitats, these species usually have no important impact on the population dynamics of their hosts and thus often fail as biological control agents (Mackauer and Völkl 1993). This situation may be, however, different under greenhouse conditions with a limited dispersal. Some aphid parasitoids, for example *Aphidius colemani*, *Aphidius matricariae* and *Ephedrus cerasicola*, are successful biological control agents in greenhouses despite their low impact in the field.

Different habitat requirements between control agents and target organisms reduce the success of biocontrol attempts. *Urophora cardui*, which was introduced in Canada for the control of *Cirsium arvense*, is sensitive to dry conditions. Adult flies obviously prefer plant stands for oviposition which are similarly structured to their natural habitats along river flood plains (Zwölfer 1994), and additionally galls show reduced survival at dry sites (Peschken et al., this Vol.). Accordingly, *U. cardui* exerts highest pressure on Canada thistle at moist sites, while its impact at dry sites is negligible (Peschken et al., this Vol.). One explanation for the failure of *Tephritis dilacerata* as a biological control agent against sow thistle, *Sonchus arvense*, in North America may be found in its life history: like all *Tephritis* spp., *T. dilacerata* hibernates in the adult stage.

The lack of suitable hibernation sites in the vicinity of the resource may have prevented a successful recolonization of plant stands in the next season by a high number of adult flies. Thus, *T. dilacerata* may have failed to build up the necessary high population densities to establish permanent populations.

Chemoecological aspects discussed here may be of importance for biological control and plant protection. When herbivore-induced volatiles from crop plants have been analysed, it should be possible, in principle, to manipulate responses of natural enemies. It seems possible to select for natural enemy strains which may differ in their ability to use infochemicals. Moreover, the natural enemy's foraging behaviour could be improved through learning of plant volatiles (Vet and Dicke 1992). Another possibility could be to manipulate sources of infochemicals in the crop. However, the disseminating of infochemicals was not very successful (see Vet and Dicke 1992). In contrast a selection of plant cultivars that enhance foraging efficiency of natural enemies could be rather promising.

At the moment, the extent to which agricultural diversification (polyculture; as compared with monocultures) affects the efficiency of natural enemy species for example through masking effects is not known. According to Vet and Dicke (1992), the searching behaviour of natural enemies seems less likely to be negatively influenced by associated plants as compared to that of herbivores. Also, specific volatiles of animal origin may be of importance for integrated pest management by using pheromone traps. Kairomonal effects of sex or aggregation pheromones from pest insects on natural predators or parasitoids could be considered. This may be significant if pheromone traps which attract and kill insects serve to detect and to monitor pest populations or if as many of the pest population as possible have to be removed. In this case the design of the traps must be modified in order to prevent simultanous elimination of natural predators or parasitoids. If pheromone dispensers have to disrupt the pest's mate finding or aggregation behaviour, kairomonal effects on natural antagonists should be not detrimental but rather beneficial.

Like volatiles, polar natural compounds may be of relevance for biological control and plant protection. Among pharmacophagous insects there are unfortunately only a few pest species. However, there are promising results with the management of diabroticine chrysomelid beetles and with African *Zonocerus* grasshoppers by applying the "catch and kill" technique. Since cucurbitacins are effective arrestants and phagostimulants are of high molecular weight, formulated cucurbitacin baits for integrated pest management of cucumber beetles and corn rootworms must be mixed with both volatile kairomonal attractans (e.g. 4-methoxycinnamaldehyde) and powerful insecticides or glues in order to eliminate diabroticid beetles (Metcalf and Metcalf 1992). Comparative research projects deal with the African grasshopper *Zonocerus variegatus* which became a pest species in conjunction with the increasing distribution of neophyte, pyrrolizidine alkaloid-containing *Chromolaena* plants. Since synthetic pyrrolizidine alkaloids represent both long-range attractants and phagostimulants for the grasshoppers, there are plans for monitoring and controlling *Zonocerus* grasshoppers by combining alkaloid-containing traps with powerful insecticides (Boppré 1995).

Worldwide, many research projects deal with the construction of transgenic plants which produce and enrich natural insecticidal plant or animal compounds (Brandt 1995). As a rule, effects of such compounds on trophic levels above herbivores have never been considered (as they were not in synthetic pesticides). Apart from risk assessment, it seems important to increase research on the positive or negative role of such compounds on specialized and polyphagous parasitoids or predators (Wink 1988).

Finally, the creation of resistant cultured plants by plant breeding or by genetic engineering usually focuses on the primary target herbivore but does not consider the effects of plants on the third trophic level. Based on the model experiments of Duffey et al. (1986) it seems important to also consider the negative or positive effects of plant toxins in admixture with other components on parasitoids. In principle, the simultaneous titre of certain nutritients may modify the toxic action of secondary compounds on parasitoid targets.

Acknowledgement. This chapter is dedicated to our honoured colleague, and teacher Prof. Dr. H. Zwölfer.

References

Agelopoulos NG, Keller MA (1994a) Plant-natural-enemy association in the tritrophic system, *Cotesia rubecula–Pieris rapae*–Brassiceae (Cruciferae). II. Sources of infochemicals. J Chem Ecol 20:1725–1734

Agelopoulos NG, Keller MA (1994b) Plant-natural enemy association in the tritrophic system, *Cotesia rubecula–Pieris rapae*–Brassiceae (Cruciferae). II. Preference of *C. rubecela* for landing and searching. J Chem Ecol 20:1735–1748

Agelopoulos NG, Dicke M, Posthumus MA (1995) Role of volatile infochemicals emitted by feces of larvae in host-searching behavior of parasitoid *Cotesia rubecela* (Hymenoptera: Braconidae): a behavioral and chemical study. J Chem Ecol 21:1789–1811

Aldrich JR (1988) Chemical ecology of the Heteroptera. Annu Rev Entomol 33:211–238

Aldrich JR (1995) Chemical communication in the true bugs and parasitoid exploitation. In: Cardé PT, Bell WJ (eds) Chemical ecology of insects 2. Chapman & Hall, New York, pp 318–363

Ananthakrishnan TN, Gopichandran R (1993) Chemical ecology in thrips-host plant interactions. International Science Publ, New York

Askew RR (1973) Parasitic insects. Heineman Educational Books, London

Barbosa P (1988a) Some thoughts on the evolution of host range. Ecology 69:912–915

Barbosa P (1988b) Natural enemies and herbivore-plant interactions: influence of plant allelochemicals and host specificity. In: Barbosa P, Letourneau DK (eds) Novel aspects of insect-plant interactions. Wiley, New York, pp 201–229

Barbosa P, Letourneau DK (1988) Novel aspects of insect-plant interactions. Wiley, New York

Barbosa P, Saunders JA (1985) Plant allelochemicals: linkages between herbivores and their natural enemies. In: Cooper-Driver GA, Swain T, Conn EE (eds) Chemically mediated interactions between plants and other organisms. Recent advances in phytochemistry, vol 19. Plenum Press, New York, pp 107–137

Barford D (1996) Molecular mechanisms of the protein serine/threonine phosphatases. TIBS 21:407–412

Bauer G (1986) Life history strategy of *Rhagoletis alternata* (Diptera: Trypetidae), a fruit fly operating in a "non-interactive" system. J Anim Ecol 55:785–794
Bauer G (1987) The parasitic stage of the freshwater pearl mussel. II. Susceptibility of brown trout. Arch Hydrobiol 76:393–402
Begon M, Harper JL, Townsend CR (1986) Ecology. Blackwell, Oxford
Benn M, DeGrave J, Gnanasunderam C, Hutchins R (1979) Host-plant pyrrolizidine alkaloids in *Nyctemera annulata* Boisduval: their persistence through the life-cycle and transfer to a parasite. Experientia 35:731–732
Berenbaum M, Seigler D (1992) Biochemicals: engineering problems for natural selection. In: Roitberg BD, Isman MB (eds) Insect chemical ecology. Chapman & Hall, New York, pp 89–121
Bernays E (1989–1994) Insect-plant interactions, 5 vols. CRC Press, Boca Raton
Bernays EA, Chapman RF (1994) Host-plant selection by phytophagous insects. Chapman & Hall, New York
Blackburn TM, Lawton JH (1994) Population abundance and body size in animal assemblages. Philos Trans R Soc B 343:33–39
Blueweiss L, Fox H, Kudzma V, Nakashima RP, Sams S (1978) Relationships between body size and some life history parameters. Oecologia 37:257–272
Blum MS (1981) Chemical defenses of arthropods. Academic Press, New York
Blum MS (1992) Ingested allelochemicals: in insect wonderland: a menu of remarkable functions. Am Entomol 38:222–234
Bogner F, Eisner T (1991) Chemical basis of egg cannibalism in a caterpillar (*Utetheisa ornatrix*). J Chem Ecol 17:2063–2075
Bogner F, Eisner T (1992) Chemical basis of pupal cannibalism in a caterpillar (*Utethteisa ornatrix*). Experientia 48:97–102
Boland W, Hopke J, Donath J, Nüske J, Bublitz F (1995) Jasmonic acid and coronatin induced odor production in plants. Angew Chem Int Ed Engl 34:1600–1602
Boppré M (1984) Redefining "pharmacophagy". J Chem Ecol 10:1151–1154
Boppré M (1986) Insects pharmacophagously utilizing defensive plant chemicals (pyrrolizidine alkaloids). Naturwissenschaften 73:17–26
Boppré M (1995) Pharmakophagie: Drogen, Sex und Schmetterlinge. Biol Unserer Zeit 25:8–17
Börner CB (1952) Aphidae Europae Centralis. Mitt Thür Bot Ges 4:1–184
Bowers MD (1990) Recycling plant natural products for insect defense. In: Evans DL, Schmidt JO (eds) Insect defenses. SUNY, New York, pp 353–386
Bowers MD (1992) The evolution of unpalatability and the cost of chemical defense in insects. In: Roitberg BD, Isman MB (eds) Insect chemical ecology. Chapman & Hall, New York, pp 216–244
Bowers MD (1993) Aposematic caterpillars: life-styles of the warningly colored and unpalatable. In: Stamp NE, Casey TM (eds) Caterpillars – ecological and evolutionary constraints on foraging. Chapman & Hall, New York, pp 331–371
Bowers MD, Boockvar K, Collinge SK (1993) Iridoid glycosides of *Chelone glabra* (Scrophulariaceae) and their sequestration by larvae of a sawfly, *Tenthredo grandis* (Tenthredinidae). J Chem Ecol 19:815–823
Brandt P (1995) Transgene Pflanzen. Birkhäuser, Basel
Bristow C (1991) Why are so few aphids ant-attended? In: Huxley CR, Cutler DF (eds) Ant-plant interactions. Oxford University Press, Oxford, pp 104–119
Brockelman WY (1975) Competition, the fitness of offspring and optimal clutch size. Am Nat 109:677–699
Brodsky LM, Barlow CA (1986) Escape responses of the pea aphid, *Acyrthosiphon pisum* (Harris) (Homoptera: Aphididae): influence of predator type and temperature. Can J Zool 64:937–939
Brower LP (1984) Chemical defence in butterflies. In: Vane-Wright RI, Ackery PR (eds) The biology of butterflies. Academic Press, London, pp 109–134
Brown KS (1984) Adult-obtained pyrrolizidine alkaloids defend ithomiine butterflies against spider predators. Nature 309:707–709

Brown KS, Trigo JR (1995) The ecological activities of alkaloids. In: Cordell GA (ed) The alkaloids, vol 47. Academic Press, San Diego, pp 227–354
Bruin J, Sabelis MW, Dicke M (1995) Do plants tap SOS signals from their infested neighbours? Tree 10:167–170
Budenberg WJ (1990) Honeydew as a contact kairomone for aphid parasitoids. Entomol Exp Appl 55:139–148
Cardé RT, Bell WJ (1995) Chemical ecology of insects 2. Chapman & Hall, New York
Carson HL (1975) The genetics of speciation at the diploid level. Am Nat 109:83–92
Caughley (1976) Plant-herbivore-systems. In: May RM (ed) Theoretical ecology. Blackwell, Oxford, pp 94–113
Chessin M, Zipf AE (1990) Alarm systems in higher plants. Bot Rev 56:193–235
Codella SG, Raffa KF (1993) Defensive strategies of folivorous sawflies. In: Wagner MR, Raffa KF (eds) Sawfly life history: adaptations to woody plants. Academic Press, San Diego, pp 261–294
Cushman JH, Addicott JF (1989) Inter- and intraspecific competition for mutualists: ants as a limited and limiting resource for aphids. Oecologia 79:315–321
Cushman JH, Addicott JF (1991) Conditional interactions in ant-herbivore mutualisms. In: Huxley CR, Cutler DF (eds) Ant-plant interactions. Oxford University Press, Oxford, pp 92–103
Cushman JH, Whitham TG (1989) Conditional mutualism in a membracid-ant association: temporal, age-specific and density-dependent effects. Ecology 70:1040–1047
Cushman JH, Whitham TG (1991) Competition mediating the dynamics of a mutualism: protective services of ants as a limiting resource for membracids. Am Nat 138:851–865
Davis RH, Nahrstedt A (1985) Cyanogenesis in insects. In: Kerkut GA, Gilbert LJ (eds) Comprehensive Insect physiology, biochemistry and pharmacology, vol 11. Pergamon Press, Oxford, pp 635–654
Deml R, Dettner K (1997) Chemical defence of Saturniidae and Lymantriidae (Lepidoptera). Entomol Gen (in press)
Dempster JP (1971) The population ecology of the Cinnabar Moth *Tyria jacobea*. Oecologia 7:26–47
Dettner K (1987) Chemosystematics and evolution of beetle chemical defenses. Annu Rev Entomol 32:17–48
Dettner K, Liepert C (1994) Chemical mimicry and camouflage. Annu Rev Entomol 39:125–150
Dettner K, Scheuerlein A, Fabian P, Schulz S, Francke W (1996) Chemical defense of giant springtail *Tetrodontophora bielanensis* (WAGA) (Insecta: Collembola). J Chem Ecol 22:1051–1074
Dicke M (1995) Why do plants "talk"? Chemoecology 5/6:159–165
Dicke M (1997) Evolution of induced indirect defense of plants. In: Harvell CD, Tollrian R (eds) Evolution of induced defenses. Princetown University Press (in press)
Dicke M, Sabelis MW (1992) Costs and benefits of chemical information conveyance: proximate and ultimate factors. In: Roitberg BD, Isman MB (eds) Insect chemical ecology – an evolutionary approach. Chapman & Hall, New York, pp 122–155
Dixon AFG (1985) Aphid ecology. Blackie, Glasgow
Dogiel VA (1961) Ecology of the parasites of freshwater fishes. In: Dogiel VA, Petrushevski GK, Polansky YI (eds) Parasitology of fishes. Oliver & Boyd, Edinburgh, pp 1–47
Duffey E (1968) Ecological studies on the large copper butterfly *Lycaena dispar*. J Appl Ecol 5:69–96
Duffey SS (1980) Sequestration of plant natural products by insects. Annu Rev Entomol 25:447–477
Duffey SS, Bloem KA, Campbell BC (1986) Consequences of sequestration of plant natural products in plant-insect-parasitoid interactions. In: Boethel DJ, Eikenbary RD (eds) Interactions of plant resistance and parasitoids and predators of insects. Elis Horwood, New York, pp 31–60

Dussourd DE, Ubik K, Harvis C, Resch J, Meinwald J, Eisner T (1988) Biparental defensive endowment of eggs with acquired plant alkaloid in the moth *Utetheisa ornatrix*. Proc Natl Acad Sci USA 85:5992–5996
Eisner T (1970) Chemical defense against predation in arthropods. In: Sontheimer E, Simeone JB (eds) Chemical ecology. Academic Press, New York
Eisner T, Conner J, Carrel JE, McCormick JP, Slagle AJ, Gans C, O'Reilly JC (1990) Systemic retention of ingested cantharidin by frogs. Chemoecology 1:57–62
Eisner T, Ziegler R, McCormick L, Eisner M, Hoebeke ER, Meinwald J (1994) Defensive use of an acquired substance (carminic acid) by predaceous insect larvae. Experientia 50:610–615
Evans DL, Schmidt JO (1990) Insect defenses. State University Press, New York
Evenhuis HH, Barbotin F (1987) Types des espèces d'Alloxystidae (Hymenoptera: Cynipoidea) de la collection Carpentier, décrits par J.J. Kieffer, avec synonymes nouveaux et une nomen novum. Bull Ann Soc R Belge Entomol 123:211–224
Feder JL (1995) The effects of parasitoids on sympatric host races of *Rhagoletis pomonella* (Diptera: Tephritidae). Ecology 76:801–813
Fiedler K (1995) Lebenszyklen tropischer Bläulinge – von Interaktionen mit Ameisen geprägt. Rundgespr Kom Ökol 10:199–214
Finidori-Logli V, Bagnères AG, Clément JL (1996) Role of plant volatiles in the search for a host by parasitoid *Diglyphus isaea* (Hymenotera: Eulophidae). J Chem Ecol 22:541–558
Freeland B (1986) Arms races and covenants: the evolution of parasite communities. In: Kikkawa J, Anderson DJ (eds) Community ecology: pattern and process. Blackwell, Oxford, pp 289–303
Freese A (1995) Die Phytophagenfauna ausgewählter europäischer Anthemideen: Eine vergleichende Analyse zu Gildenstruktur und Ressourcennutzung unter besonderer Berücksichtigung der Wirtspflanzenevolution. Agrarökologie 16:1–153
Freese G (1995) Structural refuges in two stem-boring weevils on *Rumex crispus*. Ecol Entomol 20:351–358
Fung SY, Herrebout WM, Verpoorte R, Fischer FC (1988) Butenolides in small ermine moths, *Yponomeuta* spp. (Lepidoptera: Yponomeutidae), and spindle-tree, *Euonymus europaeus* (Celastraceae). J Chem Ecol 14:1099–1111
Gauld ID, Bolton B (1987) The Hymenoptera. Oxford University Press, Oxford
Geier P (1963) The life history of codling moth, *Cydia pomonella* in the Australian capital territory. Aust J Zool 11:323–367
Gerling DH, Roitberg BD, Mackauer M (1990) Instar-specific defense of the pea aphid, *Acyrthosiphon pisum*: influence on oviposition success of the hymenopterous parasite *Aphelinus asychis*. J Insect Behav 3:501–514
Godfray HCJ (1994) Parasitoids. Behavioral and evolutionary ecology. Princeton University Press, Princeton
Graham MWR de V (1969) The Pteromalidae of north western Europe (Hymenoptera, Chalcidoidea). Bull Br Mus (Nat Hist) Entomol Suppl 16, 897 pp
Gross P (1993) Insect behavioural and morphological defenses against parasitoids. Annu Rev Entomol 38:251–273
Harborne JB (1993) Advances in chemical ecology. Nat Prod Rep 10:327–348
Harcourt DC (1971) Population dynamics of *Leptinotarsa decemlineata*. Can Entomol 103:1049–1061
Hassell MP (1976) Arthropod predator-prey-systems. In: May RM (ed) Theoretical ecology. Blackwell, Oxford, pp 71–93
Hassell MP, May R (1973) Stability in insect host-parasite models. J Anim Ecol 42:693–726
Hawkins BA (1994) Pattern and process in host-parasitoid interactions. Cambridge University Press, Cambridge
Hawkins BA, Sheehan WH (1994) Parasitoid community ecology. Oxford University Press, Oxford
Hefetz A (1987) The role of Dufour's gland secretion in bees. Physiol Entomol 12:243–253
Hermann HR (1984) Defensive mechanisms in social insects. Praeger, New York

Hölldobler B, Wilson EO (1990) The ants. Springer, Berlin Heidelberg New York

Holz C, Streil G, Dettner K, Dötemeyer J, Boland W (1994) Intersexual transfer of a toxic terpenoid during copulation and its paternal allocation to developmental stages: Quantification of cantharidin in cantharidin-producing Oedemerids (Coleoptera: Oedemeridae) and canthariphilous Pyrochroids (Coleoptera: Pyrochroidae). Z Naturforsch 49c:856–864

Hopke J, Donath J, Blechert J, Boland W (1994) Herbivore-induced volatiles: the emission of acyclic homoterpenes from leaves of *Phaseolus lunatus* and *Zea mays* can be triggered by a β-glucosidase and jasmonic acid. FEBS Lett 352:146–150

Howard RW, Akre RD (1995) Propaganda, crypsis, and slave-making. In: Cardé RT, Bell WJ (eds) Chemical ecology of insects 2. Chapman & Hall, New York, pp 364–424

Hübner G, Völkl W (1996) Behavioral strategies of aphid hyperparasitoids to escape aggression by honeydew-collecting ants. J Insect Behav 9:143–157

Huth A, Dettner K, Frößl C, Boland W (1993) Feeding of xenobiotic w-phenylalkanoic acids remarkably changes the chemistry and toxicity of the defensive secretion of *Oxytelus sculpturatus* Grav. (Coleoptera: Staphylinidae: Oxytelinae). Insect Biochem Mol Biol 23:927–935

Kellner RLL, Dettner K (1995) Allocation of pederin during lifetime of *Paederus* rove beetles (Coleoptera: Staphylinidae): evidence for a polymorphism of hemolymph toxin. J Chem Ecol 21:1719–1733

Kennedy CR (1975) Ecological animal parasitology. Blackwells Oxford

Klomp H (1968) A seventeen-year study of the abundance of the pine looper. In: Southwood TRE (ed) Insect abundance. Blackwell, Oxford

Kunze A, Aregullin M, Rodriguez E, Proksch P (1996) Fate of the chromene encecalin in the interaction of *Encelia farinosa* and its specialized herbivore *Trirhabda geminata*. J Chem Ecol 22:491–498

L'Empereur KM, Stermitz FR (1990) Iridoid glycoside metabolism and sequestration by *Poladryas minuta* (Lepidoptera: Nymphalidae) feeding on *Penstemon virgatus* (Scrophulariaceae). J Chem Ecol 16:1495–1506

Letourneau DK (1990) Code of ant-plant mutualism broken by parasite. Science 248:215–217

Loon JJA van (1996) Chemosensory basis of feeding and oviposition behaviour in herbivorous insects: a glance at the periphery. Entomol Exper Appl 80:7–13

Lorenz M, Boland W, Dettner K (1993) Biosynthesis of iridodials in the defensive glands of beetle larvae (Chrysomelinae). Angew Cham Int Ed 32:912–914

Loughrin JH, Manukion A, Heath RR, Turlings TCJ, Tumlinson JH (1994) Diurnal cycle of emission of induced volatile terpenoids by herbivore-injured cotton plants. Proc Natl Acad Sci 91:11836–11840

MacArthur RH, Wilson EO (1967) The theory of island biogeography. Princeton University Press, Princeton

Mackauer M, Völkl W (1993) Regulation of aphid populations by aphidiid wasps: does aphidiid foraging behaviour or hyperparasitism limit impact? Oecologia 94:339–350

MacKintosh C, MacKintosh RW (1994) Inhibitors of proteins kinases and phosphates. TIBS 19:444–448

Majerus MEN (1994) Ladybirds. Harper Collins, London

Malcolm SB (1989) Disruption of web structure and predatory behavior of a spider by plant-derived chemical defenses of an aposematic aphid. J Chem Ecol 15:1699–1716

Mattiacci L, Dicke M, Posthumus MA (1994) Induction of parasitoid attracting synomone in brussel sprout plants by feeding of *Pieris brassicae* larvae: role of mechanical damage and herbivore elicitor. J Chem Ecol 20:2229–2247

McCall PJ, Turlings TCJ, Loughrin J, Proveaux AT, Tumlinson JH (1994) Herbivore-induced volatile emissions from cotton (*Gossypium hirsutum* L.) seedlings. J Chem Ecol 20:3039–3050

Meiners T, Köpf A, Stein C, Hilker M (1997) Chemical signals mediating interactions between *Galeruca tanaceti* L. (Coleoptera, Chrysomelidae) and its egg parasitoid *Oomyzus galerucivorus* (Hedqvits) (Hymenoptera, Eulophidae). J Insect Behav (in press)

Mendel Z, Zegelmann L, Hassner A, Assael F, Harel M, Tam S, Dunkelblum E (1995) Outdoor attractancy of males of *Matsucoccus josephi* (Homoptera: Matsucoccidae) and *Elatophilus hebraicus* (Hemiptera: Anthocoridae) to the synthetic female sex pheromone of *M. josephi.* J Chem Ecol 21:331–341
Metcalf RO, Metcalf ER (1992) Plant kairomones in insect ecology and control. Chapman & Hall, New York, 168 pp
Michaelis H (1984) Struktur- und Funktionsuntersuchungen zum Nahrungsnetz in den Blütenköpfen von *Cirsium arvense.* PhD Thesis, University of Bayreuth, Bayreuth
Mikkola K (1991) The conservation of insects and their habitats in northern and eastern Europe. In: Collins NM, Thomas JA (eds) The conservation of insects and their habitats. Academic Press, London, pp 109–119
Morin PJ, Lawler SP (1995) Food web architecture and population dynamics: theory and empirical evidence. Annu Rev Ecol Syst 26:505–529
Mullin CA (1986) Adaptive divergence of chewing and sucking arthropods to plant allelochemicals. In: Brattsten LB, Ahmad S (eds) Molecular aspects of insect-plant associations. Plenum Press, New York, pp 175–209
Nishida R, Fukami H (1989) Host plant iridoid-based chemical defense of an aphid, *Acyrthosiphon nipponicus*, against ladybird beetles. J Chem Ecol 15:1837–1845
Nishida R, Fukami H (1990) Sequestration of distasteful compounds by some pharmacophagous insects. J Chem Ecol 16:151–164
Nishida R, Fukami H, Miyata T, Takeda M (1989) Clerodendrins: feeding stimulants for the adult turnip sawfly, *Athalia ruficornis*, from *Clerodendron trichotomum* (Verbenaceae). Agric Biol Chem 53:1641–1645
Novak H (1994) The influence of an attendance on larval parasitism in hawthorn psyllids (Homoptera: Psyllidae). Oecologia 99:72–78
Oldham NJ, Boland W (1996) Chemical ecology: multifunctional compounds and multitrophic interactions. Naturwissenschaften 83:248–254
Pasteels JM (1978) Apterous and brachypterous coccinellids at the end of the food chain, *Cionura erecta* (Asclepiadaceae)-*Aphis nerii*. Entomol Exp Appl 24:379–384
Pasteels JM, Grégoire JC, Rowell-Rahier M (1983) The chemical ecology of defense in arthropods. Annu Rev Entomol 28:263–289
Pasteels JM, Rowell-Rahier M, Braekman JC, Daloze D (1984) Chemical defence in leaf beetles and their larvae: the ecological, evolutionary and taxonomic significance. Biochem Syst Ecol 12:395–406
Pasteels JM, Rowell-Rahier M, Raupp MJ (1988) Plant-derived defense in chrysomelid beetles. In: Barbosa P, Letourneau DK (eds) Novel aspects of insect-plant interactions. Wiley, New York, pp 235–272
Pasteels JM, Dobler S, Rowell-Rahier M, Ehmke A, Hartmann T (1995) Distribution of autogenous and host-derived chemical defenses in *Oreina* leaf beetles (Coleoptera: Chrysomelidae). J Chem Ecol 21:1163–1179
Paul VJ (1992) Ecological roles of marine natural products. Comstock, Ithaca, 245 pp
Pietra F (1990) A secret world – natural products of marine life. Birkhäuser, Basel, 279 pp
Pimm SL (1982) Food webs. Chapman & Hall, London
Pimm SL, Lawton JH, Cohen JE (1991) Food web patterns and their consequences. Nature 350:669–674
Polis GA, Winemiller KO (eds) (1996) Food webs. Integration of patterns and dynamics. Chapman & Hall, London
Powell JR (1978) The founder-flush speciation theory. Evolution 32:464–474
Prestwich GD (1984) Defensive mechanisms of termites. Annu Rev Entomol 29:201–232
Price PW (1980) Evolutionary ecology of parasites. Plenum Press, New York
Price PW (1986) Ecological aspects of host plant resistance and biological control: interactions among three trophic levels. In: Boethel DJ, Eikenbary RD (eds) Interactions of plant resistance and parasitoids and predators of insects. Elis Horwood, New York, pp 11–29

Price PW, Bouton CE, Gross P, McPheron BA, Thompson JN, Weis AE (1980) Interactions among three trophic levels: influence of plants on interactions between insect herbivores and natural enemies. Annu Rev Ecol Syst 11:41–64

Ramachandran R, Norris DM (1991) Volatiles mediating plant-herbivore-natural-enemy interactions: electroantennogram responses of soybean looper, *Pseudoplusia includens*, and a parasitoid *Microplitis demolitor*, to green leaf volatiles. J Chem Ecol 17:1665–1690

Redfern M, Cameron RAD (1978) Population dynamics of the yew gall midge *Taxomya taxi*. Ecol Entomol 3:251–262

Reed HC, Tan SH, Haapanen K, Killmon M, Reed DK, Elliott NC (1995) Olfactory responses of the parasitoid *Diaeretiella rapae* (Hymenoptera: Aphidiidae) to odor of plants, aphids, and plant-aphid complexes. J Chem Ecol 21:407–418

Roff DA (1992) The evolution of life histories – theory and analysis. Chapman & Hall, London

Roitberg BD, Isman MB (1992) Insect chemical ecology – an evolutionary approach. Chapman & Hall, New York

Romstöck-Völkl M (1990) Host refuges and spatial patterns of parasitism in an endophytic host-parasitoid system. Ecol Entomol 15:321–331

Romstöck-Völkl M, Wissel C (1989) Spatial and seasonal patterns in the egg distribution of *Tephritis conura* Loew (Diptera: Tephritidae). Oikos 55:165–174

Rosenthal GA, Berenbaum MR (1992) Herbivores – their interactions with secondary plant metabolites, vols I–II. Academic Press, San Diego

Rothschild M (1985) British aposematic Lepidoptera. In: Heath J, Emmet AM (eds) The moths and butterflies of Great Britain and Ireland. Harley, Essex, pp 9–62

Rothschild M, Euw J, Reichstein T (1973) Cardiac glycosides in a scale insect (*Aspidiotus*), a ladybird (*Coccinella*) and a lacewing (*Chrysopa*). J Entomol 48:89–90

Rowell-Rahier M, Pasteels JM (1992) Third trophic level influences of plant allelochemicals. In: Rosenthal GA, Berenbaum MR (eds) Herbivores – their interactions with secondary plant metabolites, vol II. Academic Press, San Diego, pp 243–277

Schaffner U, Boevé L, Gfeller H, Schlunegger UP (1994) Sequestration of *Veratrum* alkaloids by specialist *Rhadinoceraea nodicornis* Konow (Hymenoptera, Tenthredinidae) and its ecoethological implications. J Chem Ecol 20:3233–3250

Schantz EJ (1971) The dinoflagellate poisons. In: Kadis S, Ciegler A, Ajl SJ (eds) Microbial Toxins, vol VII. Academic Press, New York, pp 3–26

Schoener TW (1989) Food webs from the small to the large. Ecology 70(6):1559–1589

Shenolikar S (1994) Protein serin/threonine phosphatases – new avenues for cell regulation. Annu Rev Cell Biol 10:55–86

Shepard M, Dahlman DL (1988) Plant-induced stresses as factors in natural enemy efficacy. In: Heinrichs EA (ed) Plant-stress-insect interactions. Wiley, New York, pp 363–379

Smith BH, Breed MD (1995) The chemical basis of nestmate recognition and mate discrimination in social insects. In: Cardé RT, Bell WJ (eds) Chemical ecology of insects 2. Chapman & Hall, New York, pp 287–317

Smith CC, Fretwell SD (1974) The optimal balance between size and number of offspring. Am Nat 108:499–506

Stary P (1977) *Dendrocerus*-hyperparasites of aphids in Czechoslovakia (Hymenoptera: Ceraphronoidea). Acta Entomol Bohemoslov 74:1–9

Stearns SC (1992) The evolution of life histories. Oxford University Press, Oxford

Stechmann DH, Völkl W, Stary P (1996) Ants as a critical factor in the biological control of the banana aphid *Pentalonia nigronervosa* in Oceania. J Appl Entomol 120:119–123

Stowe MK (1988) Chemical mimicry. In: Spencer KC (ed) Chemical mediation of coevolution. Academic Press, San Diego, pp 513–580

Sudd JH (1987) Ant aphid mutualism. In: Minks AK, Harrewijn P (eds) Aphids. Their biology, natural enemies and control, vol 2A. Elsevier, Amsterdam, pp 355–365

Szentesi A, Wink M (1991) Fate of quinolizidine alkaloids through three trophic levels: *Laburnum anagyroides* (Leguminosae) and associated organisms. J Chem Ecol 17:1557–1573

Trigo JR, Witte L, Brown KS, Hartmann T, Barata LES (1993) Pyrrolizidine alkaloids in the arctiid moth *Hyalurga syma*. J Chem Ecol 19:669–679

Turchin P, Kareiva P (1989) Aggregation in *Aphis varians*: an effective strategy for reducing predation risk. Ecology 70:1008–1016

Turlings TCJ, McCall PJ, Alborn HT, Tumlinson JH (1993a) An elicitor in caterpillar oral secretions that induces corn seedlings to emit chemical signals attractive to parasitic wasps. J Chem Ecol 19:411–425

Turlings TCJ, Wäckers FL, Vet LEM, Lewis WJ, Tumlinson JH (1993b) Learning of host-finding cues by hymenopterous parasitoids. In: Papaj DR, Lewis AC (eds) Insect learning: ecology and evolutionary perspectives. Chapman & Hall, New York, pp 51–78

Vet LEM, Dicke M (1992) Ecology of infochemicals used by natural enemies in a tritrophic context. Annu Rev Entomol 37:141–172

Vité JP, Francke W (1985) Waldschutz gegen Borkenkäfer: vom Fangbaum zur Falle. Chem Unserer Zeit 19:11–21

Völkl W (1992) Aphids or their parasitoids: who actually benefits from ant-attendance? J Anim Ecol 61:273–281

Völkl W (1995) The exploitation of ant-attended resources by the coccinellid *Platynaspis luteorubra*: patterns and benefits. J Insect Behav 8:653–670

Völkl W, Stadler B (1996) Colony orientation and successful defence behaviour in the conifer aphid, *Schizolachnus pineti*. Entomol Exp Appl 78:197–200

Von Nickisch-Rosenegk E, Schneider D, Wink M (1990) Time-course of pyrrolizidine alkaloid processing in the alkaloid exploiting arctiid moth, *Creatonotos transiens*. Z Naturforsch 45c:881–894

Warren PH, Lawton JH (1987) Invertebrate predator-prey body size relationships: an explanation for upper triangular food webs and patterns in food web structure? Oecologia 74:231–235

Way MJ (1954) Studies on the association of the ant, *Oecophylla longinoda* (Latr.) (Formicidae) with the scale insect *Saissetia zanzibarensis* Williams (Coccidae). Bull Entomol Res 45:113–154

Way MJ (1963) Mutualism between ants and honeydew-producing Homoptera. Annu Rev Entomol 8:307–344

Weis AE (1983) Pattern of parasitism by *Torymus capite* on its host distributed in small patches. J Anim Ecol 52:867–878

Wheeler JW, Duffield RM (1988) Pheromones of Hymenoptera and Isoptera. In: Morgan ED, Mandava NB (eds) CRC Handbook of natural pesticides, vol IV. Pheromones, part B. CRC Press, Boca Raton, pp 59–206

Whitman DW (1990) Grasshopper chemical communication. In: Chapman RF, Joern A (eds) Biology of grasshoppers. Wiley, New York, pp 357–391

Williams HJ, Elzen GW, Vinson SB (1988) Parasitoid-host-plant interactions emphasizing cotton (*Gossypium*). In: Barbosa P, Letourneau DK (eds) Novel aspects of insect-plant interactions. Wiley, New York, pp 171–200

Wilson EO (1970) Chemical communication within animal species. In: Sondheimer E, Simeone JB (eds) Chemical ecology. Academic Press, New York, pp 133–155

Winemiller KO, Polis GA (1996) Food webs: what do they tell us about the world? In: Polis GA, Winemiller KO (eds) Food webs. Integration of patterns and dynamics. Chapman & Hall, London, pp 1–22

Wink M (1988) Plant breeding: importance of plant secondary metabolites for protection against pathogens and herbivores. Theor Appl Genet 75:225–233

Wink M (1992) The role of quinolizidine alkaloids in plant-insect interactions. In: Bernays E (ed) Insect-plant interactions, vol IV. CRC Press, Boca Raton, pp 131–166

Wink M, Montllor CB, Bernays EA, Witte L (1991) *Uresiphita reversalis* (Lepidoptera: Pyralidae): carrier-mediated uptake and sequestration of quinolizidine alkaloids obtained from the host plant *Teline monspessulana*. Z Naturforsch 46c:1080–1088

Witte L, Ehmke A, Hartmann T (1990) Interspecific flow of pyrrolizidine alkaloids. Naturwissenschaften 77:540–543

Witz BW (1990) Antipredator mechanisms in arthropods: a twenty year literature survey. Fla Entomol 73:71–99

Yasumoto T, Murata M (1993) Marine toxins. Chem Rev 93:1897–1944

Zwölfer H (1968) Untersuchungen zur biologischen Bekämpfung von *Centaurea solstitialis* L. Strukturmerkmale der Wirtspflanze als Auslöser des Eiablageverhaltens von *Urophora sirunaseva* (Hg.) (Dipt., Trypetidae). Z Angew Entomol 61:119–130

Zwölfer H (1982) Life systems and strategies of resource exploitation in Tephritids. In: Cavalloro R (ed) Fruit flies of economic importance. Proc CEC/IOBC Int Symp Balkema, Rotterdam, pp 16–30

Zwölfer H (1986) Insektenkomplexe an Disteln - ein Modell für die Selbstorganisation ökologischer Kleinsystems. In: Dress A, Hendrichs H, Küppers G (eds) Selbstorganisation. Die Entstehung von Ordnung in Natur und Gesellschaft. Piper, München, pp 181–217

Zwölfer H (1987) Species richness, species packing, and evolution in insect-plant systems. In: Schulze E-D, Zwölfer H (eds) Potentials and limitations of ecosystem analysis. Ecological studies, vol 61. Springer, Berlin Heidelberg New York, pp 301–319

Zwölfer H (1988) Evolutionary and ecological relationships among the insect fauna of thistles. Annu Rev Entomol 33:103–122

Zwölfer H (1994) Structure and biomass transfer in food webs: stability, fluctuations, and network control. In: Schulze ED (ed) Flux control in ecological systems. Academic Press, San Diego, pp 365–419

Zwölfer H, Arnold-Rinehart J (1993) The evolution of interactions and diversity in plant-insect systems: the *Urophora-Eurytoma* food web in galls on Palearctic Cardueae. Ecological studies, vol 99. Springer, Berlin Heidelberg New York, pp 211–233

Zwölfer H, Arnold-Rinehart J (1994) Parasitoids as a driving force in the evolution of the gall size of *Urophora* in Cardueae hosts. In: Williams MAJ (ed) Plant galls. Clarendon Press, Oxford, pp 245–257

Zwölfer H, Schlumprecht H (1993) Resource utilization, populations structure and population dynamics of *Urophora cardui* L. (Dipt.: Tephritidae), a gall former in stems of *Cirsium arvense*. In: den Boer PJ, Mols PJM, Szyszko J (eds) Dynamics of populations. Proc Meet of Population problems. Agricultural University Warsaw, Warsaw, pp 55–58

Subject Index

Species Index

Ecological Studies

Volumes published since 1992

Volume 89
Plantago: A Multidisciplinary Study (1992)
P.J.C. Kuiper and M. Bos (Eds.)

Volume 90
Biogeochemistry of a Subalpine Ecosystem: Loch Vale Watershed (1992)
J. Baron (Ed.)

Volume 91
Atmospheric Deposition and Forest Nutrient Cycling (1992)
D.W. Johnson and S.E. Lindberg (Eds.)

Volume 92
Landscape Boundaries: Consequences for Biotic Diversity and Ecological Flows (1992)
A.J. Hansen and F. di Castri (Eds.)

Volume 93
Fire in South African Mountain Fynbos: Ecosystem, Community, and Species Response at Swartboskloof (1992)
B.W. van Wilgen et al. (Eds.)

Volume 94
The Ecology of Aquatic Hyphomycetes (1992)
F. Bärlocher (Ed.)

Volume 95
Palms in Forest Ecosystems of Amazonia (1992)
F. Kahn and J.-J. DeGranville

Volume 96
Ecology and Decline of Red Spruce in the Eastern United States (1992)
C. Eagar and M.B. Adams (Eds.)

Volume 97
The Response of Western Forests to Air Pollution (1992)
R.K. Olson, D. Binkley, and M. Böhm (Eds.)

Volume 98
Plankton Regulation Dynamics (1993)
N. Walz (Ed.)

Volume 99
Biodiversity and Ecosystem Function (1993)
E.-D. Schulze and H.A. Mooney (Eds.)

Volume 100
Ecophysiology of Photosynthesis (1994)
E.-D. Schulze and M.M. Caldwell (Eds.)

Volume 101
Effects of Land Use Change on Atmospheric CO_2 Concentrations: South and South East Asia as a Case Study (1993)
V.H. Dale (Ed.)

Volume 102
Coral Reef Ecology (1993)
Y.I. Sorokin

Volume 103
Rocky Shores: Exploitation in Chile and South Africa (1993)
W.R. Siegfried (Ed.)

Volume 104
Long-Term Experiments With Acid Rain in Norwegian Forest Ecosystems (1993)
G. Abrahamsen et al. (Eds.)

Volume 105
Microbial Ecology of Lake Plußsee (1993)
J. Overbeck and R.J. Chrost (Eds.)

Volume 106
Minimum Animal Populations (1994)
H. Remmert (Ed.)

Volume 107
The Role of Fire in Mediterranean-Type Ecosystems (1994)
J.M. Moreno and W.C. Oechel

Volume 108
Ecology and Biogeography of Mediterranean Ecosystems in Chile, California and Australia (1994)
M.T.K. Arroyo, P.H. Zedler, and M.D. Fox (Eds.)

Volume 109
Mediterranean-Type Ecosystems. The Function of Biodiversity (1995)
G.W. Davis and D.M. Richardson (Eds.)

Volume 110
Tropical Montane Cloud Forests (1995)
L.S. Hamilton, J.O. Juvik, and F.N. Scatena (Eds.)

Ecological Studies

Volumes published since 1992

Volume 111
Peatland Forestry. Ecology and Principles (1995)
E. Paavilainen and J. Päivänen

Volume 112
Tropical Forests: Management and Ecology (1995)
A.E. Lugo and C. Lowe (Eds.)

Volume 113
Arctic and Alpine Biodiversity. Patterns, Causes and Ecosystem Consequences (1995)
F.S. Chapin III and C. Körner (Eds.)

Volume 114
Crassulacean Acid Metabolism. Biochemistry, Ecophysiology and Evolution (1996)
K. Winter and J.A.C. Smith (Eds.)

Volume 115
Islands. Biological Diversity and Ecosystem Function (1995)
P.M. Vitousek, L.L. Loope, and H. Adsersen (Eds.)

Volume 116
High Latitude Rainforests and Associated Ecosystems of the West Coast of the Americas: Climate, Hydrology, Ecology and Conservation (1995)
R.G. Lawford, P. Alaback, and E.R. Fuentes (Eds.)

Volume 117
Anticipated Effects of a Changing Global Environment on Mediterranean-Type Ecosystems (1995)
J. Moreno and W.C. Oechel (Eds.)

Volume 118
Impact of Air Pollutants on Southern Pine Forests (1995)
S. Fox and R.A. Mickler (Eds.)

Volume 119
Freshwater Ecosystems of Alaska (1997)
A.M. Milner and M.W. Oswood (Eds.)

Volume 120
Landscape Function and Disturbance in Arctic Tundra (1996)
J.F. Reynolds and J.D. Tenhunen (Eds.)

Volume 121
Biodiversity and Savanna Ecosystem Processes. A Global Perspective (1996)
O.T. Solbrig, E. Medina, and J.F. Silva (Eds.)

Volume 122
Biodiversity and Ecosystem Processes in Tropical Forests (1996)
G.H. Orians, R. Dirzo, and J.H. Cushman (Eds.)

Volume 123
Marine Benthic Vegetation. Recent Changes and the Effects of Eutrophication (1996)
W. Schramm and P.H. Nienhuis (Eds.)

Volume 124
Global Change and Arctic Terrestrial Ecosystems (1997)
W.C. Oechel (Ed.)

Volume 125
Ecology and Conservation of Great Plains Vertebrates (1997)
F.L. Knopf and F.B. Samson (Eds.)

Volume 126
The Central Amazon Floodplain: Ecology of a Pulsing System (1997)
W.J. Junk (Ed.)

Volume 127
Forest Decline and Ozone: A Comparison of Controlled Chamber and Field Experiments (1997)
H. Sandermann, A.R. Wellburn, and R.L. Heath (Eds.)

Volume 128
Productivity and Sustainability of Southern Forest Ecosystems (1997)
S. Fox and R.A. Mickler (Eds.)

Volume 129
Pelagic Nutrient Cycles: Herbivores as Sources and Sinks (1997)
T. Andersen

Volume 130
Vertical Food Web Interactions: Evolutionary Patterns and Driving Forces (1997)
K. Dettner, G. Bauer, and W. Völkl (Eds.)

MIX
Papier aus verantwortungsvollen Quellen
Paper from responsible sources
FSC® C105338

If you have any concerns about our products,
you can contact us on
ProductSafety@springernature.com

In case Publisher is established outside the EU,
the EU authorized representative is:
Springer Nature Customer Service Center GmbH
Europaplatz 3, 69115 Heidelberg, Germany

Printed by Libri Plureos GmbH
in Hamburg, Germany